Roles of Transporters and Receptors in Drug Delivery to the Brain in Health and Disease

Roles of Transporters and Receptors in Drug Delivery to the Brain in Health and Disease

Guest Editors

Gert Fricker
Elena Puris

Basel • Beijing • Wuhan • Barcelona • Belgrade • Novi Sad • Cluj • Manchester

Guest Editors

Gert Fricker	Elena Puris
Institute of Pharmacy and	A.I. Virtanen Institute for
Molecular Biotechnology	Molecular Sciences
Heidelberg University	University of Eastern Finland
Heidelberg	Kuopio
Germany	Finland

Editorial Office
MDPI AG
Grosspeteranlage 5
4052 Basel, Switzerland

This is a reprint of the Special Issue, published open access by the journal *Pharmaceutics* (ISSN 1999-4923), freely accessible at: www.mdpi.com/journal/pharmaceutics/special_issues/transporters_receptors.

For citation purposes, cite each article independently as indicated on the article page online and using the guide below:

Lastname, A.A.; Lastname, B.B. Article Title. *Journal Name* **Year**, *Volume Number*, Page Range.

ISBN 978-3-7258-2506-6 (Hbk)
ISBN 978-3-7258-2505-9 (PDF)
https://doi.org/10.3390/books978-3-7258-2505-9

Contents

About the Editors vii

Preface ix

Wandong Zhang, Qing Yan Liu, Arsalan S. Haqqani, Ziying Liu, Caroline Sodja and Sonia Leclerc et al.
Differential Expression of *ABC* Transporter Genes in Brain Vessels vs. Peripheral Tissues and Vessels from Human, Mouse and Rat
Reprinted from: *Pharmaceutics* **2023**, *15*, 1563, https://doi.org/10.3390/pharmaceutics15051563 . **1**

Emőke Sóskuti, Nóra Szilvásy, Csilla Temesszentandrási-Ambrus, Zoltán Urbán, Olivér Csíkvári and Zoltán Szabó et al.
Applicability of MDR1 Overexpressing Abcb1KO-MDCKII Cell Lines for Investigating In Vitro Species Differences and Brain Penetration Prediction
Reprinted from: *Pharmaceutics* **2024**, *16*, 736, https://doi.org/10.3390/pharmaceutics16060736 . **26**

Jae Pyun, HuiJing Koay, Pranav Runwal, Celeste Mawal, Ashley I. Bush and Yijun Pan et al.
Cu(ATSM) Increases P-Glycoprotein Expression and Function at the Blood-Brain Barrier in C57BL6/J Mice
Reprinted from: *Pharmaceutics* **2023**, *15*, 2084, https://doi.org/10.3390/pharmaceutics15082084 . **47**

Yijun Pan, Yoshiteru Kagawa, Jiaqi Sun, Bradley J. Turner, Cheng Huang and Anup D. Shah et al.
Altered Blood–Brain Barrier Dynamics in the C9orf72 Hexanucleotide Repeat Expansion Mouse Model of Amyotrophic Lateral Sclerosis
Reprinted from: *Pharmaceutics* **2022**, *14*, 2803, https://doi.org/10.3390/pharmaceutics14122803 . **68**

Sejal Sharma, Yong Zhang, Khondker Ayesha Akter, Saeideh Nozohouri, Sabrina Rahman Archie and Dhavalkumar Patel et al.
Permeability of Metformin across an In Vitro Blood–Brain Barrier Model during Normoxia and Oxygen-Glucose Deprivation Conditions: Role of Organic Cation Transporters (Octs)
Reprinted from: *Pharmaceutics* **2023**, *15*, 1357, https://doi.org/10.3390/pharmaceutics15051357 . **85**

Jianwei Jiang, Lijun Luo, Ziqian Zhang, Xiao Liu, Naihong Chen and Yan Li et al.
The Active Glucuronide Metabolite of the Brain Protectant IMM-H004 with Poor Blood–Brain Barrier Permeability Demonstrates a High Partition in the Rat Brain via Multiple Mechanisms
Reprinted from: *Pharmaceutics* **2024**, *16*, 330, https://doi.org/10.3390/pharmaceutics16030330 . **105**

Md Masud Parvez, Armin Sadighi, Yeseul Ahn, Steve F. Keller and Julius O. Enoru
Uptake Transporters at the Blood–Brain Barrier and Their Role in Brain Drug Disposition
Reprinted from: *Pharmaceutics* **2023**, *15*, 2473, https://doi.org/10.3390/pharmaceutics15102473 . **126**

Mercedes Fernandez, Manuela Nigro, Alessia Travagli, Silvia Pasquini, Fabrizio Vincenzi and Katia Varani et al.
Strategies for Drug Delivery into the Brain: A Review on Adenosine Receptors Modulation for Central Nervous System Diseases Therapy
Reprinted from: *Pharmaceutics* **2023**, *15*, 2441, https://doi.org/10.3390/pharmaceutics15102441 . **159**

Alexander D. Mazura and Claus U. Pietrzik
Endocrine Regulation of Microvascular Receptor—Mediated Transcytosis and Its Therapeutic Opportunities: Insights by PCSK9—Mediated Regulation
Reprinted from: *Pharmaceutics* **2023**, *15*, 1268, https://doi.org/10.3390/pharmaceutics15041268 . **178**

Masaki Ueno, Yoichi Chiba, Ryuta Murakami, Yumi Miyai, Koichi Matsumoto and Keiji Wakamatsu et al.
Distribution of Monocarboxylate Transporters in Brain and Choroid Plexus Epithelium
Reprinted from: *Pharmaceutics* **2023**, *15*, 2062, https://doi.org/10.3390/pharmaceutics15082062 . **193**

About the Editors

Gert Fricker

Prof. Gert Fricker, PhD is a Full Professor at the Institute of Pharmacy and Molecular Biotechnology in the Faculty of Engineering Sciences, Ruprecht-Karls University Heidelberg, Germany. After studying at the University of Freiburg, Germany, and finishing his PhD in 1986, he worked as a Post-Doc in the Department of Clinical Pharmacology, University Hospital Zurich, Switzerland. In 1988, he became A Research Scientist at Sandoz Pharma AG, Basel, Switzerland. As a member of the Drug Delivery Systems Department, he studied mechanisms of drug permeation across barrier tissues. In the year 1995, he joined Heidelberg University. His main research interests include the permeation of drugs across barrier tissues with particular emphasis on the intestines and blood brain barrier, as well as the development of drug-targeting systems by colloidal drug carriers.

Elena Puris

Elena Puris, PhD is an Academy Research Fellow at the A.I. Virtanen Institute for Molecular Sciences, University of Eastern Finland (Kuopio, Finland). She received her PhD in Drug Research from the University of Eastern Finland in 2019. After graduation, she worked on her projects as a Postdoctoral Researcher and later as a Principal Investigator in the group of Prof. Fricker at the University of Heidelberg. Her research projects focused on investigating the roles of membrane transporters in drug delivery and pathogenesis of diseases, including Alzheimer's disease, and these have been supported by the Alexander von Humboldt Foundation, Alzheimer Forschung Initiative e.V., Taconic Biosciences, Inc., and the Orion Research Foundation. Since September 2024, with the support of the Research Council of Finland, she established her group at the University of Eastern Finland, aiming to explore the roles of membrane transporters and their impact on drug delivery and biochemical processes in health and pathological conditions, the development of novel drug delivery strategies, and the identification of the molecular mechanisms underlying the pathogenesis of CNS diseases.

Preface

Dear Colleagues,

Drug delivery to the brain, restricted by the blood-brain barrier (BBB), remains one of the major obstacles in central nervous system (CNS) drug development. Solute carriers and ATP-binding cassette family (ABC) transporters expressed at the BBB play crucial roles in drug delivery to the brain. Knowledge about brain transporters and receptors expressed has opened up new opportunities for improving drug delivery to the brain and understanding the pathogenesis of CNS diseases. This Special Issue, entitled "Roles of Transporters and Receptors in Drug Delivery to the Brain in Health and Disease", aims to provide information about the latest developments in the field of transporter-mediated drug delivery to the brain, the roles of transporters and receptors in CNS diseases, as well as advances in the methodologies for the evaluation of carrier-mediated drug delivery to the brain.

We feel honored to include a paper by our colleague Dr. Danica Stanimirovic, who passed away suddenly on January 17, 2024. Danica has long been a key leader in the BBB research area, and her publications have had a profound and lasting impact on the field. She was a deeply trusted scientific partner and friend to many of us. In her study by Zhang et al., species differences in the expression of ABC transporter genes in brain microvessels and peripheral tissues were investigated, providing important information for translating animal data to humans. Sóskuti et al. studied the applicability of using Abcb1 knock-out MDCKII cells overexpressing ABCB1 for studying species differences in BBB penetration of drug candidates. Pyun et al. revealed that the copper complex can increase ABCB1 expression and function at the BBB in vivo, with the potential application to facilitate amyloid beta clearance across the BBB in Alzheimer's disease. Importantly, Pan et al. characterized BBB transporter expression and function in a transgenic mouse model of amyotrophic lateral sclerosis. Sharma et al. studied the role of organic cation transporter 1 and the effect of oxygen-glucose deprivation in the delivery of metformin across the BBB. The study by Jiang et al. shed light on the mechanism of the BBB permeation of the active glucuronide metabolite of the brain protectant IMM-H004 against ischemic stroke. Parvez et al. summarized the current knowledge of uptake drug transporters in the brain and species differences in transporter-mediated drug disposition. Fernandez et al. presented insights into the modulation of the BBB's permeability via targeting adenosine receptor signaling in CNS diseases. Mazura and Pietrzik discussed the therapeutic potential of targeting receptor-mediated transcytosis at the BBB via endocrine regulation. Ueno et al. summarized recent knowledge of the distribution and significance of monocarboxylate transporters in the brain choroid plexus epithelial cells.

We thank all authors for their valuable contributions.

Gert Fricker and Elena Puris
Guest Editors

 *pharmaceutics*

Article

Differential Expression of *ABC* Transporter Genes in Brain Vessels vs. Peripheral Tissues and Vessels from Human, Mouse and Rat

Wandong Zhang [1],*, Qing Yan Liu [1], Arsalan S. Haqqani [1], Ziying Liu [2], Caroline Sodja [1], Sonia Leclerc [1], Ewa Baumann [1], Christie E. Delaney [1], Eric Brunette [1] and Danica B. Stanimirovic [1],*

[1] Human Health Therapeutics Research Centre, National Research Council of Canada, Ottawa, ON K1A 0R6, Canada
[2] Scientific Data Mining/Digital Technology Research Centre, National Research Council of Canada, Ottawa, ON K1A 0R6, Canada
* Correspondence: wandong.zhang@nrc-cnrc.gc.ca or wzhan2@uottawa.ca (W.Z.); danica.stanimirovic@nrc-cnrc.gc.ca (D.B.S.); Tel.: +1-613-993-5988 (W.Z.); +1-613-993-3730 (D.B.S.)

Abstract: Background: ATP-binding cassette (ABC) transporters comprise a superfamily of genes encoding membrane proteins with nucleotide-binding domains (NBD). These transporters, including drug efflux across the blood–brain barrier (BBB), carry a variety of substrates through plasma membranes against substrate gradients, fueled by hydrolyzing ATP. The expression patterns/enrichment of *ABC* transporter genes in brain microvessels compared to peripheral vessels and tissues are largely uncharacterized. Methods: In this study, the expression patterns of *ABC* transporter genes in brain microvessels, peripheral tissues (lung, liver and spleen) and lung vessels were investigated using RNA-seq and Wes[TM] analyses in three species: human, mouse and rat. Results: The study demonstrated that *ABC* drug efflux transporter genes (including *ABCB1*, *ABCG2*, *ABCC4* and *ABCC5*) were highly expressed in isolated brain microvessels in all three species studied; the expression of *ABCB1*, *ABCG2*, *ABCC1*, *ABCC4* and *ABCC5* was generally higher in rodent brain microvessels compared to those of humans. In contrast, *ABCC2* and *ABCC3* expression was low in brain microvessels, but high in rodent liver and lung vessels. Overall, most *ABC* transporters (with the exception of drug efflux transporters) were enriched in peripheral tissues compared to brain microvessels in humans, while in rodent species, additional *ABC* transporters were found to be enriched in brain microvessels. Conclusions: This study furthers the understanding of species similarities and differences in the expression patterns of *ABC* transporter genes; this is important for translational studies in drug development. In particular, CNS drug delivery and toxicity may vary among species depending on their unique profiles of *ABC* transporter expression in brain microvessels and BBB.

Keywords: ABC transporters; brain microvessels; peripheral tissues and vessels; gene expression patterns; RNA-seq; Wes[TM] analysis; across species

Citation: Zhang, W.; Liu, Q.Y.; Haqqani, A.S.; Liu, Z.; Sodja, C.; Leclerc, S.; Baumann, E.; Delaney, C.E.; Brunette, E.; Stanimirovic, D.B. Differential Expression of *ABC* Transporter Genes in Brain Vessels vs. Peripheral Tissues and Vessels from Human, Mouse and Rat. *Pharmaceutics* **2023**, *15*, 1563. https://doi.org/10.3390/pharmaceutics15051563

Academic Editors: Gert Fricker, Elena Puris and Kenneth K. W. To

Received: 23 March 2023
Revised: 13 May 2023
Accepted: 18 May 2023
Published: 22 May 2023

1. Introduction

ATP-binding cassette (ABC) transporters comprise a superfamily of transmembrane proteins that are conserved in evolution and found in many species from prokaryotes to humans [1–3]. Functional transporter proteins typically contain one or two ATP-binding domains (also known as nucleotide-binding domain (NBD)) and transmembrane (TM) domains. Each TM domain contains six membrane-spanning alpha-helices and determines specificity for substrates [1–3]. Eukaryotic ABC transporters are either full transporters containing two NBDs and two TM domains or half transporters with one NBD and one TM domain. The half transporters may be functional as homodimers, heterodimers or multimeric complexes [1,4]. ABC transporters hydrolyze ATP to provide energy to shuttle a variety of substrates through plasma membranes against substrate concentration gradients [3,5]. The superfamily of 51 ABC transporters in humans is divided into seven subfamilies based on similarity in gene structure (half vs. full transporters), order of the domains and sequence homology in the NBD and TM domain (Accessed on 18 May 2023; https://www.genenames.org/cgi-bin/genefamilies/set/417) [1–3].

ABC transporters are widely expressed in different tissues and carry out important physiological functions through the transport of metabolites, ions, lipids/cholesterol/steroids, drugs/xenobiotics, antibiotics, toxins and peptides across biological membranes (known functions of ABC transporters are listed in Supplementary Table S1). Notably, ABCB1/MDR-1 (multidrug resistance 1) P-glycoprotein (ABCB1/Pgp), ABCG2/breast cancer resistant protein (ABCG2/BCRP) and ABCC subfamily members/multidrug resistant proteins (ABCC1-5/MRP1-5) play critical roles in drug efflux transport and in forming barrier functions in the human body, including the blood–brain barrier (BBB), blood-cerebrospinal fluid barrier (BCSFB), blood-retinal barrier, blood-testis barrier and placenta barrier [3,6–10]. These barriers protect organs and tissues from blood-borne toxic substances and metabolic waste. However, they can also impede the delivery of therapeutics into 'protected' organs and cells; for example, the BBB-expressed ABC transporters actively extrude substrate drugs, making them inaccessible to the targets in the central nervous system (CNS) [3,11–13]. The expression and function of ABC transporters are regulated by a spectrum of endogenous and exogenous factors, including nuclear receptors, xenobiotics/drugs and a variety of inflammatory molecules [12]. Long-term administrations of certain xenobiotics/drugs, such as chemotherapeutic and epilepsy drugs, can up-regulate the expression of ABC drug transporters, leading to multidrug-resistant (MDR) phenotypes, particularly in cancer chemotherapy and at the BBB [3,12]. The mutations in *ABC* transporter genes resulting in disease phenotypes have been described previously [2,3] (Table S1).

ABC transporter transgenic and gene knockout (KO) animals, including *Abcb1a/1b* KO, *Abcg2/Bcrp* KO and *Abcc1/Mrp1* KO mice [14–17], have been widely used in studies of drug development and the mechanisms of drug transport. However, it has been observed that therapeutics developed and tested in animal models often display different efficacy and toxicological profiles in clinical studies in humans [18] due, at least partially, to differences in the expression levels of genes encoding ABC transporters among different species. These species differences undermine the translational predictability of animal study data to humans [19]. The current study employs the next-generation sequencing (NGS) approach to compare the levels of *ABC* transporter gene expression in different tissues across species, particularly focusing on *ABC* drug transporter genes in brain vessels compared to peripheral vessels and tissues. This study furthers the understanding of species similarities and differences in the expression patterns of *ABC* transporter genes, which is particularly important for translational considerations from rodents to humans in drug development.

2. Materials and Methods

2.1. Human and Rodent Tissues

The use of human tissues in this study was approved by the Research Ethics Board of the National Research Council of Canada. Human brain and lung tissue samples from three individuals were used in the study as described previously [20]. Human post-mortem brain frontal cortex tissues were obtained from the Human Brain and Spinal Fluid Resource Center, VAMC, Los Angeles, CA, USA), which is sponsored by NINDS/NIMN, National Multiple Sclerosis Society, VA Greater Los Angeles Healthcare System, and Veterans Health Services and Research Administration, Department of Veteran Affairs. All patients had signed informed consent. Brain samples from three patients, two female (aged 73 and 76 years) and one male (aged 76 years), ethnicity unknown, all deceased from non-brain-related pathologies (cardiomyopathy, coronary artery and obstructive pulmonary disease), were used in this study. Human lung tissues were from the normal adjacent lung tissues of three NSCLC (non-small cell lung cancer) patients undergoing surgical resections. The surgical samples were deposited in the Lung Tumor Bank managed by the CDHA-Capital District Health Authority in Halifax, Nova Scotia, Canada. All patients signed informed consent as per CDHA-RS/2013-271, which allowed their tissues to be archived in the CDHA Lung Tumor Bank for molecular studies. No additional information about the patients could be disclosed. The use of animals (mice and rats) in this study was approved by the Animal Care Committee of the Human Health Therapeutics Research Centre at the National Research Council of Canada. Three mice (C57BL/6 J strain, males, 9 months old) and three Wistar rats (males, 7 to 8 months old) were used in the study. Rodent species (mouse and rat) are the most commonly used in studies of CNS diseases and the BBB. They are also typical toxicology species used for regulatory approvals. These two rodent species were selected for comparative analyses with humans in this study.

2.2. Isolation of Microvessels and Capillaries from Tissues

Microvessels (20–300 μm) were isolated from naïve mouse and rat brains and lungs as well as from human brain tissues, as described previously [20,21]. Three animals of each rodent species were used in the study. All human tissues were frozen and stored at −80 °C until vessel isolation. Rodent vessels were isolated from fresh tissues. Human and rodent tissue homogenization and vessel separations were performed on ice using phosphate-buffered saline (PBS) (Wisent, St-Bruno, QC, Canada) containing a protease inhibitor cocktail (Sigma Aldrich, St. Louis, MO, USA), with instruments pre-chilled on ice. Respective tissues were chopped using a razor blade placed in a 5-mL Wheaton Dounce homogenization tube (Fisher Scientific, Hampton, NH, USA), and 5 mL of PBS buffer was added per tube. Tissues were homogenized with 10 strokes of the pestle connected to the Eberbach Con-Torque Homogenizer (Fisher Scientific, Hampton, NH, USA). Homogenized tissues were transferred using gentle suction and filtered through a stack of series of pluriStrainers in a 50-mL conical tube with a connector ring and strainers (pluriSelect, San Diego, CA, USA) in descending order, as follows: connector ring, 300 μm, 100 μm and 20 μm strainers. We would like to point out that the tissue storage (fresh vs. frozen) and processing may have affected the quality of isolated RNA.

2.3. RNA Isolation

RNA was extracted from microvessels, lung, liver and spleen tissue homogenates from humans, rats and mice using the RNeasy Plus Mini kit (Qiagen Inc., Toronto, ON, Canada) and NucleoSpin RNA plus kit (Macherey-Nagel GmbH & Co. KG, Dueren, Germany), respectively, according to the manufacturer's instructions as described previously [20]. Genomic DNA contamination was removed by the Turbo DNA-Free Kit (Life Technologies, Burlington, ON, Canada). RNA quality was assessed using Agilent Bioanalyzer 2100 (Agilent, Santa Clara, CA, USA).

2.4. RNA-Seq

RNA-Seq Libraries were generated using the TruSeq strand RNA kit (Illumina, San Diego, CA, USA), as described previously [20]. The RNA-Seq libraries were quantified by Qbit and qPCR according to the Illumina Sequencing Library qPCR Quantification Guide and the quality of the libraries was evaluated on Agilent Bioanalyzer 2100 using the Agilent DNA-100 chip (Santa Clara, CA, USA). RNA-Seq library sequencing was performed using Illumina Next-Seq500. FASTQ file format was processed by trimming the adaptor sequences, filtering low-quality reads (Phred Score $\leq$ 20) and eliminating short reads (length $\leq$ 20 bps) using the software package FASTX-toolkit [Accessed on 18 May 2023; http://hannonlab.cshl.edu/fastx_toolkit/]. STAR (Spliced Transcripts Alignment to a Reference) (v2.5.3a) [22] was used for the alignment of the reads to the reference genome and to generate gene-level read counts. RSEM (RNA-Seq by Expectation-Maximization) (version 1.3.3) [23] was used for alignment to generate transcripts per million (TPM) counts. A mouse reference genome (version GRCm38.p6, M24), a human reference genome (version GRCh38.p13, Genecode 33) and corresponding annotations were used as references for

RNA-seq data alignment processes. DESeq2 [24] was used for data normalization and differentially expressed gene identification for each pair-wise comparison.

2.5. Public Datasets and Analysis

RNA-seq and microarray data in the public domains were obtained to compare/benchmark the data generated from this study for quality and comparability purposes, as described [20]. For RNA-seq data, raw data corresponding to normal lung and brain samples were obtained from the Sequence Read Archive [25] from the Genomics Data Commons [26]. GTEx (Genotype-Tissue Expression) data were processed using GDC (genomic data commons) reference files using the GDC mRNA analysis pipeline (STAR two-pass) [22]. These data were combined with 12 samples analyzed at NRC and processed using DESeq2 [24].

2.6. Automated Western Blot Analysis (Wes^{TM})

Human and rodent brain vessel pellets were lysed in Cellytic MT buffer (Sigma-Aldrich, Oakville, ON, Canada) with $1\times$ complete protease inhibitor (Roche Canada, Mississauga, ON, Canada) on ice. The lysates were incubated on ice for 30 min, vortexed, then centrifuged at $21,000\times g$ for 10 min in a Sorvall Legend Micro 21R centrifuge. Protein concentrations were determined using the Quantipro BCA (bicinchoninic acid) Assay Kit (Sigma-Aldrich, Oakville, ON, Canada). Wes^{TM} was run using the 12 to 230 kDa separation module (ProteinSimple) and the mouse or rabbit detection module (ProteinSimple Inc., San Jose, CA, USA). Samples (protein at 0.8 mg/mL) were prepared by combining Master Mix with samples at a 1:4 ratio. Samples and Biotinylated Ladder were heated in an Accublock digital dry bath at 95 °C for 5 min. Samples were cooled to room temperature, vortexed and then centrifuged in a Mandel mini microfuge. Biotinylated ladder, samples, primary and secondary antibodies, and luminol were loaded on the plate and Wes^{TM} was run using the standard protocol. The primary antibodies were rabbit anti-P-glycoprotein (Pgp) (Abcam, Toronto, ON, Canada; Cat# Ab170904, lot GR299351-2, dilution 1:50), mouse anti-MRP1 (Novus Biologicals/Bio-Techne Canada, Toronto, ON; Cat# NB400-156, dilution 1:100), rabbit anti-ABCG2 (Biobyrt, Cambridge, UK; Cat# ORB155559, dilution 1:100) and anti-actin-HRP (Sigma-Aldrich, Oakville, ON, Canada; Cat# A3854, dilution 1:100). HRP (horseradish peroxidase)-conjugated anti-mouse or anti-rabbit secondary antibodies from Wes^{TM} detection modules (ProteinSimple Inc., San Jose, CA, USA) were used for detection. Streptavidin-HRP was used to detect the ladder proteins. The Wes^{TM} densitometry software compass measured the chemiluminescence signal during the Wes run and plotted it as chemiluminescence vs. molecular weight. Data for each sample were first normalized to β-actin in the same lane. The level of the protein in human brain microvessels (BMV) was set as one-fold. The fold change of mouse and rat proteins was calculated relative to that of human proteins (means $\pm$ SD).

2.7. Statistical Analysis

The data were analyzed and compared by one-way ANOVA among multiple groups; this was followed by Tukey's multiple comparisons tests or unpaired two-tailed Student's t-tests between the two groups. $p < 0.05$ was considered significant.

3. Results

3.1. RNA-Seq Datasets: Quality, Comparability and Validation

RNA samples were isolated from brain and peripheral vessels and tissues. Due to limited availability, only cerebral microvessels and lungs were analyzed from humans, whereas cerebral and lung vessels, liver and spleen were analyzed from mice and rats. The quality of the cerebral vessel preparation was confirmed by the expression of brain vessel marker proteins, including TfR (transferrin receptor), IGF1R (insulin-like growth factor receptor-1), CD31/PECAM-1 (Platelet endothelial cell adhesion molecule-1), GFAP (glial fibrillar acid protein) and Collagen IV using immunofluorescence analyses, as described previously [20]. RNA-seq libraries were then generated and sequenced.

To confirm the comparability among datasets, it was essential to evaluate the quality of the RNA-seq datasets generated in this study with similar RNA-seq datasets in the public domain. The analysis showed that RNA-seq datasets generated from total human brains and lungs were highly correlated to RNA-seq datasets in public domains concerning the same tissues, with a correlation co-efficient of 0.96 [20]. The comparability of different datasets generated in our own studies was also analyzed, giving a correlation co-efficient between 0.94 and 0.97. These analyses confirmed the quality and comparability of the datasets generated in the study, as demonstrated in our previous report [20], mitigating the impact of low sample numbers.

3.2. The Expression of ABC Transporter Genes in Brain Vessels across Species

The first objective of this study was to compare the expression of *ABC* transporter genes in isolated brain vessels from different species to understand potential species differences in BBB function in xenobiotic and lipid transport. The brain microvessels analyzed in this study were highly enriched in brain endothelial cells (BEC), but also contained pericytes and astrocyte end-feet, as described previously [20]. It is important to note that a very high expression level of a gene in 'contaminating' cells coupled with low or absent expression in a dominant cell type tissue could result in misinterpretation of data obtained from mixed cell tissue. While this is a confounding factor in this study, it has been shown previously that transporters expressed in both BEC and other cells of the neurovascular unit contribute to drug influx and efflux across the BBB [27]. Therefore, conclusions concerning the impact of *ABC* transporter gene expressions in microvessel preparations on the BBB function could be drawn. To compare the levels of gene expression across species, the normalized read counts of RNA-seq data were converted to transcripts per million (TPM) counts [20]. The levels of *ABC* transporter gene expression were then compared across humans, mice, and rats based on TPM data analysis, and the results are presented in Table 1. The data on the major *ABC* drug efflux transporter genes were extracted from Table 1, analyzed and presented in Figure 1. ABCB1/MDR1/P-glycoprotein (Pgp) is a drug efflux transporter responsible for multidrug resistance in cancer chemotherapy and drug efflux at the BBB. There is one *ABCB1* gene in the human genome but there are two isoforms, *Abcb1a* and *Abcb1b*, in that of rodents. The TPM values of *Abcb1a* and *Abcb1b* RNA-seq data were combined and then compared to human *ABCB1* (Table 1; Figure 1). It was noted that the level of *Abcb1b* expression was lower compared to that of *Abcb1a* in mouse and rat cerebral vessels, suggesting that *Abcb1a* is the main isoform of a drug efflux transporter in cerebral vessels in rodents (Table 1). It was observed that *Abcb1* expression levels in the cerebral vessels of mice (295.34 ± 26.58) were three-fold and eightfold higher than those in rats (93.78 ± 47.90) and humans (36.62 ± 54) (Figure 1; Table 1) (one-way ANOVA, ** $p < 0.01$), respectively. The expression levels of *ABCB1* in human brain vessels were significantly lower compared to those in either rat or mouse brain vessels.

Table 1. The levels of *ABC* transporter gene expression in cerebral vessels of humans, mice and rats (RNA-seq analysis by transcripts per million counts (TPM)) [§].

ABC Transporter Genes	Human (Mean ± SD)	Mouse (Mean ± SD)	Rat (Mean ± SD)
ABCA1	0.69 ± 0.66	19.45 ± 1.72 [h**; r***]	5.90 ± 1.47 [h***]
ABCA2	208.09 ± 70.27 [r*]	107.57 ± 15.42	71.63 ± 25.12
ABCA3	10.85 ± 6.45	35.98 ± 2.22 [h**]	26.49 ± 7.73 [h*]
ABCA4	0.19 ± 0.15	8.22 ± 1.19 [h*]	7.85 ± 4.64 [h*]
ABCA5	4.51 ± 1.34	9.77 ± 1.16 [r***; h**]	2.62 ± 0.72

Table 1. *Cont.*

ABC Transporter Genes	Human (Mean ± SD)	Mouse (Mean ± SD)	Rat (Mean ± SD)
ABCA6	0.88 ± 0.08 [m*; r****]	0.67 ± 0.10 [r***]	0.03 ± 0.01
ABCA7	13.04 ± 3.68	19.29 ± 1.78	9.86 ± 4.96
ABCA8	4.84 ± 1.54	13.82 ± 3.03 [h*]	13.52 ± 2.97 [h*]
ABCA9	0.81 ± 0.91	7.12 ± 1.23 [r***; h***]	0.78 ± 0.26
ABCA10	1.93 ± 1.26	0.00 ± 0.00	0.00 ± 0.00
ABCA12	0.01 ± 0.00	0.03 ± 0.01	0.05 ± 0.02
ABCA13	0.02 ± 0.01	0.03 ± 0.01	0.05 ± 0.02
Abca14	Not present	0.00 ± 0.01	0.00 ± 0.00
Abca15	Not present	0.00 ± 0.00	0.00 ± 0.00
Abca16	Not present	0.00 ± 0.00	0.04 ± 0.04
Abca17	Not present	0.20 ± 0.13	0.04 ± 0.02
Abcb1a	Homologous to ABCB1	290.67 ± 26.70 [r##]	93.50 ± 47.78
Abcb1b	Homologous to ABCB1	4.67 ± 0.45 [r###]	0.28 ± 0.12
ABCB1 (Abcb1a + b1b)	36.62 ± 54.00	295.34 ± 26.58 [h***; r**]	93.78 ± 47.90
ABCB2/TAP1	19.84 ± 19.30	14.98 ± 1.72	18.94 ± 5.33
ABCB3/TAP2	6.99 ± 8.53	47.64 ± 4.68 [r*; h*]	12.64 ± 20.52
ABCB4	0.84 ± 0.33	1.71 ± 0.04	3.21 ± 1.00 [h**]
ABCB5	0.08 ± 0.03	0.00 ± 0.00	0.00 ± 0.00
ABCB6	19.99 ± 7.05	10.93 ± 1.61	17.39 ± 2.98
ABCB7	1.72 ± 0.41	9.95 ± 1.09 [h**; r*]	5.79 ± 2.54 [h*]
ABCB8	7.08 ± 3.07	33.10 ± 9.46 [h**]	40.43 ± 5.68 [h**]
ABCB9	5.26 ± 2.20	21.47 ± 4.52 [h**]	17.09 ± 4.28 [h*]
ABCB10	1.88 ± 1.37	16.11 ± 1.24 [h**]	14.00 ± 2.80 [h**]
ABCB11	0.31 ± 0.25	0.14 ± 0.21	0.03 ± 0.01
ABCC1	0.90 ± 0.80	7.35 ± 0.19 [h**]	6.52 ± 4.65 [h*]
ABCC2	0.18 ± 0.11 [m*]	0.03 ± 0.01	0.35 ± 0.05 [m***; h**]
ABCC3	0.41 ± 0.37	1.21 ± 0.59 [h*]	0.98 ± 0.34
ABCC4	1.00 ± 1.02	77.02 ± 25.83 [h**; r*]	26.29 ± 6.88
ABCC5	10.00 ± 5.01	146.28 ± 22.30 [h****; r****]	17.69 ± 7.92
ABCC6	1.04 ± 0.74	15.72 ± 4.27 [h**; r*]	4.69 ± 3.00
ABCC7	0.13 ± 0.07	0.14 ± 0.02	0.04 ± 0.04
ABCC8	43.31 ± 37.99	14.67 ± 2.43	4.95 ± 1.34
ABCC9	2.40 ± 1.95	26.48 ± 10.26 [h**; r*]	8.14 ± 2.74
ABCC10	10.30 ± 4.29 [r*]	5.89 ± 0.27	3.48 ± 1.13
ABCC11	0.23 ± 0.10	0.00 ± 0.00	0.00 ± 0.00
ABCC12	0.24 ± 0.20	1.99 ± 0.39 [h***; r***]	0.11 ± 0.02
ABCC13	0.01 ± 0.01	0.00 ± 0.00	0.00 ± 0.00
ABCD1	2.19 ± 1.04	12.84 ± 1.51 [h***; r**]	9.39 ± 2.51

Table 1. *Cont.*

ABC Transporter Genes	Human (Mean ± SD)	Mouse (Mean ± SD)	Rat (Mean ± SD)
ABCD2	0.53 ± 0.11	11.41 ± 1.92 [h**]	11.61 ± 3.44 [h**]
ABCD3	4.49 ± 1.70	39.19 ± 4.07 [h****]	49.20 ± 3.48 [h****; m*]
ABCD4	21.42 ± 11.18	18.38 ± 1.36	6.57 ± 0.60
ABCE1	4.30 ± 1.77	33.30 ± 0.94 [h****; r*]	22.97 ± 4.96 [h***]
ABCF1	8.61 ± 3.62	60.72 ± 1.79 [h****]	63.48 ± 5.19 [h****]
ABCF2	5.54 ± 2.98	54.18 ± 6.08 [h**]	55.27 ± 21.99 [h**]
ABCF3	19.77 ± 9.53	63.67 ± 4.50 [h***; r*]	42.97 ± 2.88 [h**]
ABCG1	3.57 ± 2.96	29.25 ± 5.94 [h**]	23.23 ± 6.08 [h**]
ABCG2	10.01 ± 12.63	185.35 ± 38.42 [h**; r*]	83.22 ± 46.64
Abcg3	Not present	2.24 ± 1.14	18.27 ± 12.99
ABCG4	1.98 ± 1.49	30.69 ± 7.74 [h**; r*]	12.07 ± 4.17
ABCG5	0.08 ± 0.10	0.02 ± 0.02	0.42 ± 0.05 [h**; m***]
ABCG8	0.15 ± 0.08 [m*; r*]	0.01 ± 0.02	0.01 ± 0.02

[§] One-way ANOVA: * $p < 0.05$; ** $p < 0.01$; *** $p < 0.001$; **** $p < 0.0001$; [h]: compare to human; [m]: compare to mouse; [r]: compare to rat. [§] Two-tailed *t*-test: ## $p < 0.01$; ### $p < 0.001$.

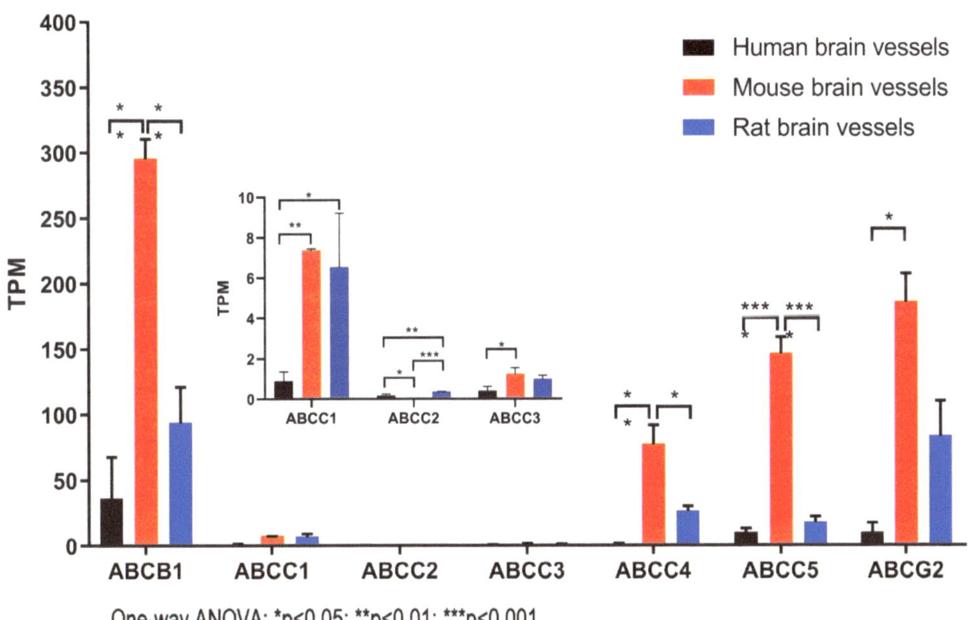

Figure 1. The expression of *ABC* drug transporter genes in brain microvessels of humans, mice and rats analyzed by transcript per million (TPM) across species.

ABCC1/MRP1 has been identified and characterized as one of the major drug efflux transporters in multidrug resistance regarding cancer chemotherapy [3,6]. In this study, the expression levels of the *ABCC1/MRP1*, *ABCC2/MRP2* and *ABCC3/MRP3* genes were found to be generally low in cerebral vessels in all three species examined (Figure 1 insert), although some differences were observed; for example, *Abcc2* in rat cerebral vessels (0.35 ± 0.05) was slightly higher than that in mouse cerebral vessels (0.03 ± 0.01) (one-way ANOVA, ** $p < 0.01$) (Figure 1; Table 1). Interestingly, the expression levels of the *ABCC4/MRP4* and *ABCC5/MRP5* genes were significantly higher than those of *ABCC1-3* in brain vessels in all three species studied (Table 1; Figure 1). Similar to *ABCB1*, the expression levels of both *ABCC4/MRP4* and *ABCC5/MRP5* were significantly higher in the cerebral vessels of mice compared to those in humans or rats (Figure 1; Table 1). These results suggest that *ABCC4/MRP-4* and *ABCC5/MRP-5* may be more important than *ABCC1-3/MRP1-3* for the transport function of brain vessels.

ABCG2/BCRP is a half-transporter in the ABCG subfamily and can form homo- or hetero-dimers; its substrate spectrum mostly overlaps with that of ABCB1 [1–3,6,28]. ABCG2/BCRP may be complementary to ABCB1 and is critical for BBB transport [1–3,6,28]. The RNA-seq data in this study found that *ABCG2/BCRP* expression was higher than *ABCC1-5/MRP1-5*, but lower than *ABCB1* in brain vessels across studied species (Table 1; Figure 1), with the expression levels in mouse brain vessels (185.35 ± 38.42) higher than those in either humans (10.01 ± 12.63) (* $p < 0.05$) or rats (83.22 ± 46.64) (Figure 1A; Table 1). Overall, the expression levels of *ABC* drug transporter genes in cerebral vessels were higher in rodents than in humans (mice > rats > human), with *ABCB1, ABCG2, ABCC4* and *ABCC5* being highly expressed or enriched in cerebral vessels.

The protein expression of three major drug efflux transporters, including ABCB1/Pgp, ABCC1/MRP-1 and ABCG2/BCRP, was then validated in human, mouse and rat cerebral vessels by quantitative Wes^{TM} analysis using species cross-reactive antibodies (Supplementary Figure S1A,B). Protein-level expression analyses showed some discrepancies from the RNA-seq analyses. For example, ABCB1/Pgp protein levels were higher in rat cerebral vessels compared to those in mouse cerebral vessels (Supplementary Figure S1A). The Wes^{TM} analysis results were consistent with the RNA-seq analysis results for *ABCC1/MRP1* and *ABCG2/BCRP* (Supplementary Figure S1), showing the order of protein expression abundance among species similar to what was observed at the mRNA level.

ABCA subfamily transporters, composed of 13 members in humans and 17 members in rodents (Table 1), are mostly involved in lipid and cholesterol transport. ABCA1 is a well-characterized cholesterol transporter that mediates cellular cholesterol efflux in the brain and influences neuroinflammation and neurodegeneration [29,30]. The expression of *ABCA1* was significantly higher in mouse brain vessels (19.45 ± 1.72) compared to rat (5.90 ± 1.47; *** $p < 0.001$) or human brain vessels (0.69 ± 0.66; ** $p < 0.01$) (Table 1). The expression of *ABCA1* in rat brain vessels was also significantly higher than that in human brain vessels (*** $p < 0.001$) (Table 1). *ABCA2* and *A3* were found to be highly expressed in cerebral vessels across species (Table 1; Figure 2). The level of *ABCA2* in human cerebral vessels (208.09 ± 70.27) was higher than that in either rat (71.63 ± 25.12; one-way ANOVA, * $p = 0.0217$) or mouse (107.57 ± 15.42) cerebral vessels; while the level of *ABCA3* was significantly lower in humans (10.85 ± 6.45) compared to either mice (35.98 ± 2.22; ** $p = 0.005$) or rats (26.49 ± 7.73; * $p = 0.0416$) (one-way ANOVA) (Table 1). ABCA2 is a lipid transporter involved in the maintenance of homeostasis of cholesterol and sphingolipids and is highly expressed in brain tissue [31]. *ABCA2* knockout mice display developmental defects similar to aberrant myelination [31]. ABCA3 is involved in phospholipid transport for surfactant production in lung tissue [32] and is overexpressed in childhood acute myeloid leukemia [33]. Nevertheless, the role of ABCA3 in cerebral vessels is still poorly understood. Both ABCA7 and ABCA8 are involved in the transport of cholesterol or/and phospholipids and play roles in sphingomyelin synthesis and the generation of HDL-like particles [34–36]. The two transporter genes were moderately expressed in cerebral vessels across species (Table 1; Figure 2), although the levels of *ABCA8* expression were higher in mice (13.82 ± 3.03) and rats (13.52 ± 2.97) than in humans (4.84 ± 1.54; one-way ANOVA, * $p < 0.05$) (Table 1). The roles of ABCA7 variants in sterol and lipid transport and metabolism makes them one of the top risk factors for Alzheimer's disease development [36–40].

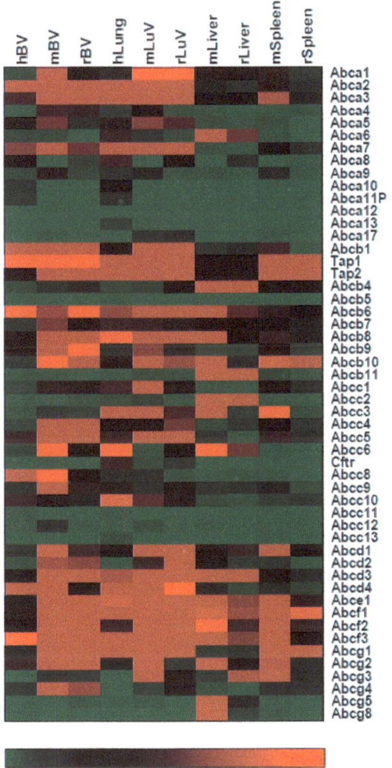

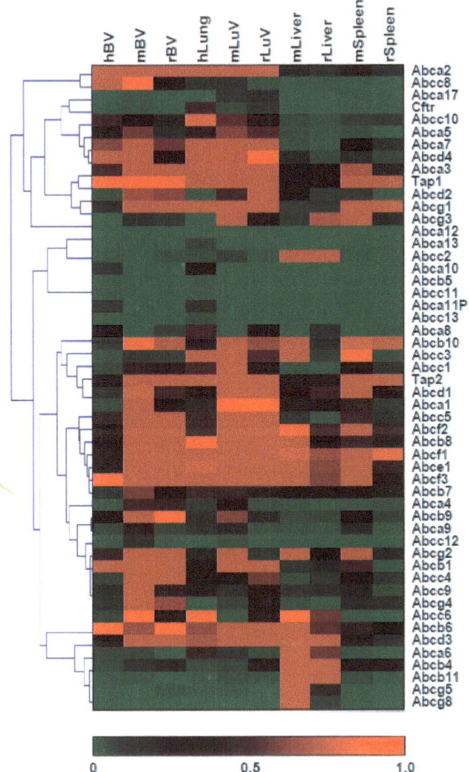

Figure 2. Heatmap for the levels of *ABC* transporter gene expression in humans, mice and rats (h for humans; m for mice and r for rats in the graphs): Three samples were used in RNA-seq analyses for cerebral vessels and peripheral tissues and lung vessels from humans, mice and rats. The RNA-seq data were converted to transcript per million (TPM) counts for comparison across species. The means of three samples (TPM counts) were used in the following analysis. (**A**). Transcript per million counts (TPM) ranking for *ABC* transporter gene expression in human, mouse and rat brain vessels vs. peripheral tissues and lung vessels. (**B**). TPM ranking values of hierarchical clustering for *ABC* transporter gene expression in human, mouse and rat brain vessels vs. peripheral tissues and lung vessels.

There are 11 members of the ABCB subfamily. In addition to the drug efflux transporter ABCB1/Pgp, the other members have diverse or unclear functions. ABCB2/TAP1 (Transporter associated with antigen processing 1) and ABCB3/TAP2 are involved in peptide transport or peptide antigen presentation for immune response [3,6]. These two genes were expressed at moderate to high levels in cerebral vessels across the species (Table 1; Figure 2). The same is true for *ABCB6* (Table 1; Figure 2) which encodes a Fe/S cluster transporter for mitochondrion [6], as well as for ABCB8 and B9, which are both involved in the maintenance of mitochondrial iron homeostasis, maturation of cytosolic Fe/S proteins and transport of peptides and phospholipids [6,41]. ABCB9 is a TAP-like half-transporter associated with lysosomes and may function as a peptide translocase and phosphatidylserine floppase [42,43]. The levels of both *ABCB8* and *ABCB9* expression were significantly lower in human cerebral vessels compared to mouse or rat cerebral vessels (Table 1; Figure 2).

There are 13 members of the ABCC subfamily. In addition to *ABCC1-5*, *ABCC8* was similarly expressed at moderate to low levels in cerebral vessels across species (Table 1; Figure 2). Interestingly, *ABCC6* and *C9* were expressed at moderate levels in mouse cerebral vessels (*ABCC6*: 15.72 ± 4.27; *ABCC9*: 26.48 ± 10.26) but significantly lower in human (*ABCC6*: 1.04 ± 0.74, ** $p < 0.01$; *ABCC9*: 2.40 ± 1.95, ** $p < 0.01$) and rat (*ABCC6*: 4.69 ± 3.00, ** $p < 0.01$; *ABCC9*: 8.14 ± 2.74, * $p < 0.05$) cerebral vessels (one-way ANOVA) (Table 1; Figure 2). ABCC6 is a GS-X pump involved in the transport of glutathione and peptides, while ABCC8 and C9 are known as sulfonylurea receptors and regulate ATP-sensitive K+ channels [6,44] (Table S1).

There are four, one and three members of the ABCD, ABCE and ABCF subfamilies, respectively, in all three species (Table 1). All the members were expressed at low levels in human cerebral vessels except *ABCD4* and *ABCF3* (Table 1; Figure 2). However, all the members were expressed at moderate to relatively high levels in mouse and rat cerebral vessels (Table 1; Figure 2). The four *ABCD* subfamily members are half-transporters and may be involved in peroxisome-related functions [2,6]. The functions of ABCE and ABCF1-3 are not well understood.

ABCG subfamily members, except for ABCG2 and Abcg3, are involved in sterol or/and phospholipid transport [1–3,6,45,46]. *ABCG1* and *G4* were expressed at moderate levels in rodent cerebral vessels but at lower levels in human cerebral vessels (Table 1; Figure 2). Both genes are known to be highly expressed in human brain tissue [3,6] and their function in brain vessels may involve the transport of sterol and phospholipids across the BBB.

3.3. The Expression of ABC Transporter Genes in Brain Vessels vs. Lung Vessels, Liver and Spleen

The second and broader objective of the study was to evaluate the relative enrichment of *ABC* transporter gene expression in brain vessels vs. lung vessels, lung, liver and spleen tissues. As shown in Figure 2 and Tables 1–4, the expression levels of *ABC* transporter genes and their distribution varied among different tissues and vessels across species. Liver, lung and spleen tissues used in the study are a mixture of different cell types; therefore, the expression data shown reflect tissue-averaged levels of all cell types present in that particular tissue. For example, *ABCB1*, *ABCG2* and *ABCC5/MRP-5* were expressed at high levels in cerebral vessels across the species compared with lung vessels (Figure 3), while *ABCC4/MRP-4* and *ABCC9* were highly expressed in mouse and rat cerebral vessels but not in human cerebral vessels. *ABCC8*, which encodes an ATP-sensitive K$^+$ channel, was highly expressed in human cerebral vessels but not in other human, mouse and rat tissues analyzed in this study (Figures 2 and 3; Tables 1–4). *ABCC8* is typically expressed in pancreatic insulin-secreting cells and pituitary glands (Table S1). In addition to their expression in the liver, *ABCA2/A3* and *ABCB6/B8/B9* were expressed at moderate to high levels in brain vessels (Figure 2; Tables 1–4). A moderate level of *ABCA2* and a high level of *ABCA3* were expressed in lung tissues and lung vessels (Tables 2–4). *ABCC1/MRP1* was moderately expressed in human and mouse lung vessels/tissues but not in other tissues analyzed in this study, while *ABCC2/MRP-2* was highly expressed in mouse and rat liver tissues. *ABCC3/MRP3* was expressed at moderate to low levels in mouse and rat cerebral vessels as well as in human and mouse lung tissues/vessels and mouse liver and spleen, but not in other tissues across the species (Figures 2 and 3; Tables 1–4). *ABCC5/MRP5* was expressed at moderate to high levels in lung tissues/vessels across the species (Figures 2 and 3). These results suggest that drug transporters ABCB1, ABCG2 and ABCC5/MRP5 appear to be important in cerebral vessels across species, whereas the lower expression of ABCC4/MRP4 in human vessels compared to that in mouse and rat vessels suggests its higher functional importance in rodent species (Table S1). Drug transporters ABCC1/MRP-1, ABCC2/MRP-2 and ABCC3/MRP-3 play a dominant role in the liver and lungs (Table S1), while they appear to be functionally less important in brain vessels.

Table 2. The levels of *ABC* transporter gene expression in human brain vessels and lung tissues (RNA-seq analysis by transcripts per million counts (TPM)) *.

ABC Transporter Genes	Human Brain Vessels (Mean ± SD)	Human Lung Tissues (Mean ± SD)
ABCA1	0.69 ± 0.66	2.40 ± 0.32 #
ABCA2	208.09 ± 70.27 ##	12.32 ± 5.17
ABCA3	10.85 ± 6.45	95.24 ± 49.02 #
ABCA4	0.19 ± 0.15	0.75 ± 0.56
ABCA5	4.51 ± 1.34	5.50 ± 0.23
ABCA6	0.88 ± 0.08	6.12 ± 1.13 ##
ABCA7	13.04 ± 3.68	19.18 ± 6.67
ABCA8	4.84 ± 1.54	6.8 ± 0.46
ABCA9	0.81 ± 0.91	1.24 ± 0.22
ABCA10	1.93 ± 1.26	3.17 ± 2.47
ABCA12	0.01 ± 0.00	0.03 ± 0.04
ABCA13	0.02 ± 0.01	0.54 ± 0.67
ABCB1	36.62 ± 54.00	4.35 ± 2.84
ABCB2/TAP1	19.84 ± 19.30	48.58 ± 16.21
ABCB3/TAP2	6.99 ± 8.53	20.72 ± 4.66
ABCB4	0.84 ± 0.33	0.38 ± 0.35
ABCB5	0.08 ± 0.03	0.08 ± 0.02
ABCB6	19.99 ± 7.05	7.10 ± 5.68
ABCB7	1.72 ± 0.41	6.94 ± 1.81 ##
ABCB8	7.08 ± 3.07	11.87 ± 1.11
ABCB9	5.26 ± 2.20 #	1.70 ± 0.30
ABCB10	1.88 ± 1.37	4.42 ± 2.15
ABCB11	0.31 ± 0.25	0.10 ± 0.04
ABCC1	0.90 ± 0.80	7.28 ± 1.60 ##
ABCC2	0.18 ± 0.11	0.32 ± 0.20
ABCC3	0.41 ± 0.37	21.07 ± 3.14 ###
ABCC4	1.00 ± 1.02	4.44 ± 2.86
ABCC5	10.00 ± 5.01	5.91 ± 2.24
ABCC6	1.04 ± 0.74	13.82 ± 1.98 ###
ABCC7	0.13 ± 0.07	8.30 ± 6.06
ABCC8	43.31 ± 37.99	1.78 ± 1.56
ABCC9	2.40 ± 1.95	2.74 ± 0.16
ABCC10	10.30 ± 4.29	9.85 ± 3.57
ABCC11	0.23 ± 0.10	0.06 ± 0.07
ABCC12	0.24 ± 0.20	0.03 ± 0.03
ABCC13	0.01 ± 0.01	0.00 ± 0.00

Table 2. *Cont.*

ABC Transporter Genes	Human Brain Vessels (Mean ± SD)	Human Lung Tissues (Mean ± SD)
ABCD1	2.19 ± 1.04	3.95 ± 0.33 #
ABCD2	0.53 ± 0.11 ##	0.14 ± 0.07
ABCD3	4.49 ± 1.70	10.13 ± 3.67
ABCD4	21.42 ± 11.18	24.41 ± 4.06
ABCE1	4.30 ± 1.77	11.34 ± 4.15
ABCF1	8.61 ± 3.62	21.40 ± 3.42 #
ABCF2	5.54 ± 2.98	6.56 ± 0.97
ABCF3	19.77 ± 9.53	28.26 ± 2.61
ABCG1	3.57 ± 2.96	31.01 ± 16.02 #
ABCG2	10.01 ± 12.63	0.03 ± 0.02
ABCG4	1.98 ± 1.49	0.04 ± 0.03
ABCG5	0.08 ± 0.10	0.02 ± 0.01
ABCG8	0.15 ± 0.08	0.00 ± 0.00

* Two-tailed *t*-test: # $p < 0.05$; ## $p < 0.01$; ### $p < 0.001$.

Table 3. The levels of *ABC* transporter gene expression in mouse brain vessels vs. peripheral vessels and tissues (RNA-seq analysis by transcripts per million counts (TPM)) [#].

ABC Transporter Genes	Brain Vessels (Mean ± SD)	Liver Tissue (Mean ± SD)	Lung Vessels (Mean ± SD)	Spleen Tissues (Mean ± SD)
Abca1	19.45 ± 1.72 [LT, LV, ST**]	4.29 ± 4.03	7.18 ± 3.59	5.02 ± 2.74
Abca2	107.57 ± 15.42 [LT, LV, ST***]	2.02 ± 1.40	15.62 ± 5.47	4.81 ± 2.73
Abca3	35.98 ± 2.22 [LT, ST***]	3.04 ± 3.04	37.80 ± 8.16 [LT, ST***]	7.90 ± 4.63
Abca4	8.22 ± 1.19 [LT, LV, ST****]	0.19 ± 0.27	2.54 ± 0.63 [LT, ST*]	0.14 ± 0.09
Abca5	9.77 ± 1.16 [LT, ST**; LV*]	0.32 ± 0.21	5.08 ± 2.65 [LT, ST*]	0.29 ± 0.11
Abca6	0.67 ± 0.10	18.76 ± 9.39 [BV, LV, ST**]	0.76 ± 0.24	0.16 ± 0.14
Abca7	19.29 ± 1.78 [LT, ST**]	0.27 ± 0.12	24.35 ± 7.14 [LT, ST***]	4.22 ± 1.23
Abca8	13.82 ± 3.03	14.47 ± 5.09	82.39 ± 16.70 [BV, LT, ST****]	3.44 ± 0.96
Abca9	7.12 ± 1.23 [LT****; LV, ST***]	0.24 ± 0.19	1.47 ± 0.46	1.70 ± 1.16
Abca10	0.00 ± 0.00	0.00 ± 0.00	0.00 ± 0.00	0.00 ± 0.00
Abca12	0.03 ± 0.01	0.00 ± 0.00	0.02 ± 0.02	0.00 ± 0.00
Abca13	0.03 ± 0.01	0.00 ± 0.00	0.01 ± 0.02	0.01 ± 0.01
Abca14	0.00 ± 0.01	0.06 ± 0.04	2.02 ± 1.30 [LT, ST*]	0.03 ± 0.03
Abca15	0.00 ± 0.00	0.00 ± 0.00	0.00 ± 0.00	0.00 ± 0.00
Abca16	0.00 ± 0.00	0.00 ± 0.00	0.01 ± 0.02	0.00 ± 0.00
Abca17	0.20 ± 0.13	0.02 ± 0.04	0.81 ± 0.44	0.10 ± 0.08
Abcb1a	290.67 ± 26.70 [LT, LV, ST****]	0.27 ± 0.07	26.81 ± 7.34	3.37 ± 0.99
Abcb1b	4.67 ± 0.45 [LT***]	0.80 ± 0.13	3.39 ± 1.12 [LT*]	2.85 ± 0.81 [LT*]
Abca1a + b1b	295.34 ± 26.58 [LT, LV, ST****]	1.06 ± 0.09	30.19 ± 7.63	6.21 ± 0.13
Abcb2/Tap1	14.98 ± 1.72	4.12 ± 1.59	81.16 ± 24.53 [BV, LT**]	58.37 ± 19.49 [BV, LT*]
Abcb3/Tap2	47.64 ± 4.68 [LT**]	4.00 ± 2.28	123.39 ± 20.66 [LT****; BV, ST***]	58.10 ± 8.97 [LT**]
Abcb4	1.71 ± 0.04	26.94 ± 18.47 [BV, LV*]	0.45 ± 0.13	4.10 ± 0.87

Table 3. *Cont.*

ABC Transporter Genes	Brain Vessels (Mean ± SD)	Liver Tissue (Mean ± SD)	Lung Vessels (Mean ± SD)	Spleen Tissues (Mean ± SD)
Abcb5	0.00 ± 0.00	0.00 ± 0.00	0.00 ± 0.00	0.00 ± 0.00
Abcb6	10.93 ± 1.61	34.06 ± 9.07 [BV, LV, ST****]	20.95 ± 2.10 [BV***; ST****]	6.76 ± 0.87
Abcb7	9.95 ± 1.09 [LT, LV, ST**]	3.46 ± 2.07	3.94 ± 0.81	3.99 ± 0.61
Abcb8	33.10 ± 9.46 [ST*]	12.75 ± 2.54	28.88 ± 13.28 [ST*]	6.89 ± 1.83
Abcb9	21.47 ± 4.52 [LT, ST****; LV***]	0.42 ± 0.20	5.32 ± 0.57	3.24 ± 1.17
Abcb10	16.11 ± 1.24	12.51 ± 4.56	15.02 ± 1.10	9.96 ± 1.91
Abcb11	0.14 ± 0.21	89.50 ± 23.92 [BV, LV, ST****]	0.28 ± 0.48	0.36 ± 0.53
Abcc1	7.35 ± 0.19 [LT***]	0.19 ± 0.10	9.54 ± 1.78 [LT***; ST**]	4.00 ± 2.05 [LT*]
Abcc2	0.03 ± 0.01	15.76 ± 13.17	0.43 ± 0.18	0.06 ± 0.09
Abcc3	1.21 ± 0.59	16.62 ± 8.84 [BV*]	29.72 ± 4.05 [BV***; LT*; ST**]	11.97 ± 0.21
Abcc4	77.02 ± 25.83 [LT, LV, ST***]	0.36 ± 0.27	2.60 ± 0.97	2.76 ± 1.38
Abcc5	146.28 ± 22.30 [LT, ST****; LV*]	1.10 ± 0.54	96.94 ± 25.04 [LT, ST***]	5.96 ± 1.90
Abcc6	15.72 ± 4.27 [LV, ST***]	10.60 ± 3.70 [LV, ST**]	0.19 ± 0.17	0.05 ± 0.07
Abcc7	0.14 ± 0.02	0.02 ± 0.00	0.25 ± 0.13 [LT, ST*]	0.02 ± 0.02
Abcc8	14.67 ± 2.43 [LT, LV, ST****]	0.05 ± 0.09	1.02 ± 0.28	0.05 ± 0.07
Abcc9	26.48 ± 10.26 [LT, LV**]	1.03 ± 0.72	1.02 ± 0.40	2.07 ± 1.19
Abcc10	5.89 ± 0.27 [LT, ST**]	0.36 ± 0.17	4.64 ± 2.13 [LT, ST**]	0.65 ± 0.19
Abcc11	0.00 ± 0.00	0.00 ± 0.00	0.00 ± 0.00	0.00 ± 0.00
Abcc12	1.99 ± 0.39 [LT, LV, ST****]	0.06 ± 0.04	0.20 ± 0.17	0.01 ± 0.01
Abcc13	0.00 ± 0.00	0.00 ± 0.00	0.00 ± 0.00	0.00 ± 0.00
Abcd1	12.84 ± 1.51	3.18 ± 2.39	26.75 ± 2.48 [BV, ST*; LT***]	17.06 ± 6.80
Abcd2	11.41 ± 1.92 [LT, ST***; LV**]	2.70 ± 2.12	4.23 ± 0.74	1.01 ± 0.12
Abcd3	39.19 ± 4.07 [LV*; ST****]	100.43 ± 65.85 [ST*]	26.40 ± 4.68 [ST***]	4.73 ± 1.73
Abcd4	18.38 ± 1.36 [LT, ST**]	2.05 ± 0.63	19.81 ± 6.34 [LT, ST***]	1.86 ± 0.28
Abce1	33.30 ± 0.94 [LT, ST**]	13.44 ± 6.11	26.15 ± 5.40 [LT*]	16.58 ± 3.04
Abcf1	60.72 ± 1.79 [LT****; ST**]	8.65 ± 4.83	65.19 ± 7.33 [LT****; ST***]	34.83 ± 4.98 [LT**]
Abcf2	54.18 ± 6.08 [LT****; LV**; ST***]	9.84 ± 4.21	30.66 ± 3.54 [LT**]	22.49 ± 3.93 [LT*]
Abcf3	63.67 ± 4.50 [LT, ST****]	16.19 ± 5.45	59.31 ± 9.07 [LT****; ST***]	19.70 ± 0.78
Abcg1	29.25 ± 5.94 [LT**]	0.80 ± 0.45	27.65 ± 2.70 [LT**]	19.93 ± 9.74 [LT*]
Abcg2	185.35 ± 38.42 [LT, LV, ST****]	28.53 ± 15.63	24.84 ± 5.87	9.38 ± 1.22
Abcg3	2.24 ± 1.14	0.84 ± 0.21	9.07 ± 1.59 [LT*]	25.01 ± 5.93 [BV, LT****; LV**]
Abcg4	30.69 ± 7.74 [LT, LV, ST****]	0.33 ± 0.35	0.45 ± 0.23	1.64 ± 0.34
Abcg5	0.02 ± 0.02	13.96 ± 8.20 [LT, LV, ST*]	0.15 ± 0.19	0.09 ± 0.10
Abcg8	0.01 ± 0.02	10.87 ± 8.46	0.01 ± 0.02	0.12 ± 0.14

One-way ANOVA: * $p < 0.05$; ** $p < 0.01$; *** $p < 0.001$; **** $p < 0.0001$; BV: compare to brain vessels; LT: compare to liver tissues; LV: compare to lung vessels; ST: compare to spleen tissues.

Table 4. The levels of *ABC* transporter gene expression in rat brain vessels vs. peripheral vessels and tissues (RNA-seq analysis by transcripts per million counts (TPM)) #.

ABC Transporter Genes	Brain Vessels (Mean ± SD)	Liver Tissues (Mean ± SD)	Lung Vessels (Mean ± SD)	Spleen Tissues (Mean ± SD)
Abca1	5.90 ± 1.47	11.81 ± 2.86	13.32 ± 6.25 [ST*]	3.92 ± 0.90
Abca2	71.63 ± 25.12 [LT***; LV, ST**]	4.19 ± 1.92	16.56 ± 5.44	5.97 ± 0.30
Abca3	26.49 ± 7.73 [LT, ST*]	9.58 ± 2.57	113.33 ± 6.40 [BV, LT, ST****]	9.57 ± 0.29

Table 4. Cont.

ABC Transporter Genes	Brain Vessels (Mean ± SD)	Liver Tissues (Mean ± SD)	Lung Vessels (Mean ± SD)	Spleen Tissues (Mean ± SD)
Abca4	7.85 ± 4.64 LT, LV, ST*	0.20 ± 0.04	0.36 ± 0.08	0.97 ± 0.21
Abca5	2.60 ± 0.72 LT*	0.54 ± 0.05	7.89 ± 2.98 LT, ST**	0.11 ± 0.03
Abca6	0.03 ± 0.01	21.79 ± 4.20 BV, LV, ST****	0.45 ± 0.07	0.03 ± 0.01
Abca7	9.86 ± 4.96	0.30 ± 0.11	19.32 ± 7.94 LT**; ST*	5.72 ± 0.17
Abca8	13.52 ± 2.10	21.57 ± 2.51 ST**	21.59 ± 9.15 ST**	3.28 ± 0.50
Abca9	0.78 ± 0.26	0.48 ± 0.25	0.24 ± 0.18	0.50 ± 0.07
Abca10	0.00 ± 0.00	0.00 ± 0.00	0.00 ± 0.00	0.00 ± 0.00
Abca12	0.05 ± 0.02	0.00 ± 0.00	0.00 ± 0.01	0.00 ± 0.00
Abca13	0.05 ± 0.02	0.00 ± 0.00	0.00 ± 0.01	0.05 ± 0.01
Abca14	0.00 ± 0.00	0.01 ± 0.02	0.00 ± 0.00	0.00 ± 0.00
Abca15	0.00 ± 0.00	0.00 ± 0.00	0.00± 0.00	0.00 ± 0.00
Abca16	0.04 ± 0.04	0.69 ± 0.33 BV, LV, ST**	0.03 ± 0.04	0.04 ± 0.01
Abca17	0.04 ± 0.02	0.07 ± 0.04	2.12 ± 1.04 BV, LT, ST**	0.06 ± 0.01
Abcb1a	93.50 ± 47.78 LT, LV, ST**	4.18 ± 1.07	2.43 ± 0.90	0.21 ± 0.08
Abcb1b	0.28 ± 0.12	1.97 ± 1.39	9.37 ± 7.72	0.42 ± 0.05
Abca1a + b1b	93.78 ± 47.9078 LT, ST**, LV*	6.15 ± 2.46	11.80 ± 8.56	0.63 ± 0.13
Abcb2/Tap1	18.94 ± 5.33	11.65 ± 3.24	140.59 ± 55.46 BV, LV**; ST*	53.59 ± 8.14
Abcb3/Tap2	12.64 ± 20.52	11.07 ± 4.69	61.88 ± 101.10	50.28 ± 3.97
Abcb4	3.21 ± 1.00	28.83 ± 16.53 BV, LV*	3.62 ± 3.79	15.81 ± 2.31
Abcb5	0.00 ± 0.00	0.00 ± 0.00	0.00 ± 0.00	0.00 ± 0.00
Abcb6	17.39 ± 2.98	20.74 ± 3.91	20.01 ± 3.67	14.43 ± 1.16
Abcb7	5.79 ± 2.54	13.74 ± 1.09 BV**; LV***; ST*	3.01 ± 0.45	9.19 ± 1.38 LV**
Abcb8	40.43 ± 5.68 LT, ST***; LV*	9.37 ± 4.18	25.33 ± 5.35 LT**; ST*	9.66 ± 0.57
Abcb9	17.09 ± 4.28 LT, LV, ST***	1.54 ± 0.77	2.47 ± 0.41	2.77 ± 0.54
Abcb10	14.00 ± 2.80	5.84 ± 2.34	9.59 ± 2.38	58.53 ± 6.4328 BV, LT, LV***
Abcb11	0.03 ± 0.01	103.10 ± 6.5828 BV, LV, ST****	0.33 ± 0.50	0.59 ± 0.13
Abcc1	6.52 ± 4.65	0.35 ± 0.20	3.50 ± 1.47	5.72 ± 0.75
Abcc2	0.35 ± 0.05	72.54 ± 23.86 BV, LV, ST***	0.18 ± 0.02	0.06 ± 0.03
Abcc3	0.98 ± 0.34	2.16 ± 0.56	7.17 ± 1.76 BV, LT, ST***	1.31 ± 0.02
Abcc4	26.29 ± 6.88 LT, ST***; LV**	3.05 ± 1.41	7.92 ± 3.01	3.64 ± 0.70
Abcc5	17.69 ± 7.92 LT**; ST*	0.40 ± 0.07	33.37 ± 24.38	5.19 ± 0.12
Abcc6	4.69 ± 3.00	21.87 ± 1.95 BV, LV, ST****	3.36 ± 2.22	0.08 ± 0.01
Abcc7	0.04 ± 0.04	0.06 ± 0.06	1.20 ± 0.46 BV, LV, ST**	0.01 ± 0.01
Abcc8	4.95 ± 1.34 LT**; LV***	0.51 ± 0.12	0.28 ± 0.04	0.00 ± 0.00
Abcc9	8.14 ± 2.74 ST*	4.05 ± 1.50	4.53 ± 3.51	1.47 ± 0.16
Abcc10	3.48 ± 1.13	0.84 ± 0.47	5.80 ± 1.60 LT**	3.37 ± 0.16
Abcc11	0.00 ± 0.00	0.00 ± 0.00	0.00 ± 0.00	0.00 ± 0.00
Abcc12	0.11 ± 0.02 LT, LV, ST***	0.01 ± 0.01	0.01 ± 0.02	0.01 ± 0.01
Abcc13	0.00 ± 0.00	0.00 ± 0.00	0.00 ± 0.00	0.00 ± 0.00
Abcd1	9.39 ± 2.51	18.47 ± 8.20	17.72 ± 6.65	11.54 ± 0.22
Abcd2	11.61 ± 3.44 LT, ST**	2.64 ± 0.20	20.21 ±25.38	0.58 ± 0.06
Abcd3	49.20 ± 3.48 ST**	93.03 ± 14.82 BV**; LV***; ST****	44.92 ± 8.65 ST**	6.19 ± 0.64
Abcd4	6.57 ± 0.60	1.28 ± 0.57	12.17 ± 4.39 LT, ST**	2.18 ± 0.46
Abce1	22.97 ± 4.96	23.71 ± 0.63	19.48 ± 3.47	22.87 ± 3.26

Table 4. *Cont.*

ABC Transporter Genes	Brain Vessels (Mean ± SD)	Liver Tissues (Mean ± SD)	Lung Vessels (Mean ± SD)	Spleen Tissues (Mean ± SD)
Abcf1	63.48 ± 5.19 [LT**; ST*]	25.22 ± 8.45	60.05 ± 13.24 [LT**]	40.27 ± 4.01
Abcf2	55.27 ± 21.99 [LT*]	19.43 ± 3.85	39.85 ± 12.28	23.77 ± 1.47
Abcf3	43.97 ± 2.88 [LT*; ST**]	23.21 ± 6.41	39.58 ± 9.53 [LT*; ST*]	19.44 ± 1.53
Abcg1	23.23 ± 6.08 [LT*]	3.68 ± 1.21	299.66 ± 357.98	36.56 ± 3.12 [BV*; LT***]
Abcg2	83.22 ± 46.64 [LT, LV, ST*]	13.83 ± 2.39	1.17 ± 0.45	6.75 ± 0.97
Abcg3	18.27 ± 12.99	62.98 ± 15.33 [BV**]	105.58 ± 126.24	35.68 ± 2.39
Abcg4	12.07 ± 4.17 [LT**; LV*]	0.17 ± 0.04	3.34 ± 3.29	5.28 ± 0.97
Abcg5	0.42 ± 0.05	7.38 ± 3.68 [BV, LV, ST**]	0.04 ± 0.01	0.03 ± 0.01
Abcg8	0.01 ± 0.02	2.36 ± 2.06	0.00 ± 0.01	0.02 ± 0.00

[#] One-way ANOVA: * $p < 0.05$; ** $p < 0.01$; *** $p < 0.001$; **** $p < 0.0001$; BV: compare to brain vessels; LT: compare to liver tissues; LV: compare to lung vessels; ST: compare to spleen tissues.

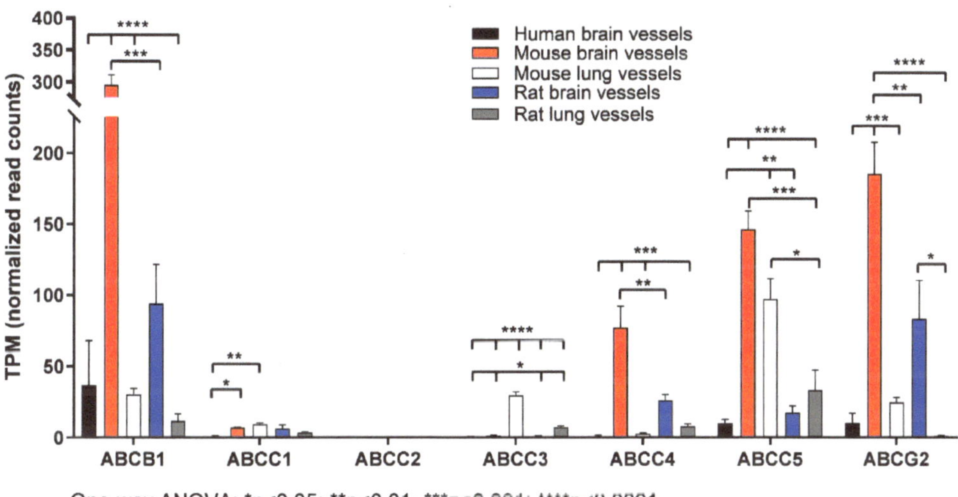

Figure 3. Expression of *ABC* drug transporter genes in cerebral and lung vessels across species. Three samples of cerebral vessels from humans, mice and rats were used in RNA-seq analysis. The RNA-seq data were then converted to transcript per million (TPM) counts for comparison across species. Mean ± SD from three samples was used in the analysis (Tables 1–4). One-way ANOVA was used for statistical analysis (* $p < 0.05$; ** $p < 0.01$; *** $p < 0.001$; **** $p < 0.0001$).

K-means clustering analysis was performed on human, mouse and rat RNA-seq data (TPM counts). Based on the abundance of gene expression, eight clusters of genes were identified from the analysis of brain vessel vs. peripheral tissue expression data (including liver and spleen) (Figure 4). The analysis considered *ABC* transporter gene expression relative to the expression of all the genes detected in the RNA-seq dataset of each species. Cluster 1 shows the *ABC* transporter genes that were expressed at very high levels in brain vessels but at very low levels in peripheral tissues (Figure 4). *ABCA2* was the only gene found among 805 genes in humans (Figure 4A). There were 15 *ABC* transporter genes (among 9884 genes in cluster 1) in mice (Figure 4B) and 12 *ABC* transporter genes in rats (among 6589 genes in cluster 1) (Figure 4C) that showed this profile. Cluster 2 represents the *ABC* transporter genes that were expressed at very high levels in brain vessels but at moderate levels in peripheral tissues. *ABCD2* was the only gene found among 875 genes in humans in cluster 2. There were 6 *ABC* transporter genes in both mice (among 2415 genes)

(Figure 4B) and rats (2808 genes) (Figure 4C) in cluster 2. Cluster 3 includes genes that were expressed at high levels in brain vessels but at moderate to low levels in peripheral tissues. There were 9 *ABC* transporter genes in humans (among 5269 genes in cluster 3) (Figure 4A), 13 *ABC* transporter genes in mice (among 6349 genes) (Figure 4B) and 11 *ABC* transporter genes in rats (among 4885 genes in cluster 3) (Figure 4C). Cluster 4 identifies genes that were highly expressed in both brain vessels and peripheral tissues; none such *ABC* transporter gene was identified in any of the studied species (Figure 4A–C). Cluster 5 includes the genes that were expressed moderately in both brain vessels and peripheral tissues. There were 4 *ABC* transporter genes in both humans (among 4015 genes) (Figure 4A) and rats (among 1897 genes) and none in mice (among 1430 genes) in cluster 5 (Figure 4B). Cluster 6 identifies *ABC* transporter genes that were expressed at low levels in both brain vessels and peripheral tissues (Figure 4). Cluster 7 includes *ABC* transporter genes that were expressed at low levels in brain vessels but moderate to high levels in peripheral tissues. Cluster 8 identifies *ABC* transporter genes that were expressed at low levels in brain vessels but at very high levels in peripheral tissues, as shown in Figure 4. The clustering analysis shows *ABC* transporter gene expression patterns relative to all genes expressed in brain vessels vs. those expressed in peripheral tissues in each of the species. The patterns of *ABC* transporter gene clustering in these subgroups were similar in mice and rats.

Finally, the data generated allowed us to analyze tissue-specific enrichment of *ABC* genes and compare this across preclinical species and humans. Clusters 1 and 8 represent the *ABC* transporter genes that were enriched in either brain vessels or peripheral tissues, respectively. Species differences between humans and rodents in *ABC* transporter gene enrichment were clearly identified through clustering analyses. There were over 10 *ABC* transporter genes enriched in rodent brain vessels but only one in human brain vessels (Figure 4). The opposite was true for the *ABC* transporter genes enriched in peripheral tissues; 11 *ABC* transporter genes were enriched in human peripheral (lung) tissues, but none were enriched in rodent peripheral (liver and spleen) tissues (Figure 4).

Species differences were also found regarding the number of *ABC* transporter genes encoded in human and rodent genomes. In addition to the well-known presence of two isoforms of *Abcb1* (*Abcb1a* and *Abcb1b*) in rodents compared to one *ABCB1* in humans, other species differences were also identified. For *ABCA* subfamily genes, four additional genes (*Abca14* to *Abca17*) were present in the rodent but not in the human genome. However, these four genes were expressed at extremely low levels in both rodent brain vessels and rodent tissues analyzed in our study (Tables 2–4). *Abcg3* was found in rodents but not in humans (Tables 2–4). Bioinformatics analysis suggested that *Abcg3* is closely related to *Abcg2* and was predicted to be a drug efflux transporter, although its function has not been characterized [2,47].

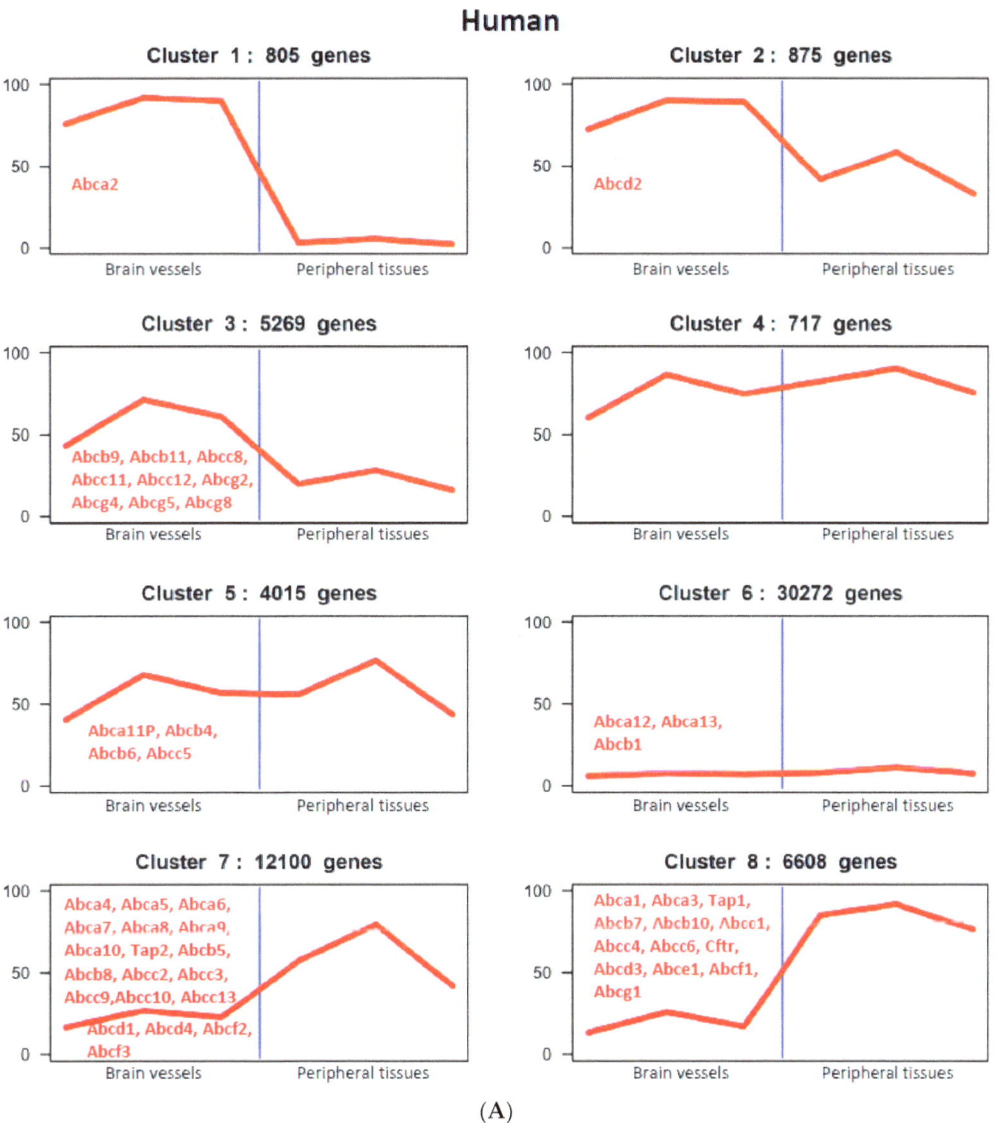

Figure 4. *Cont.*

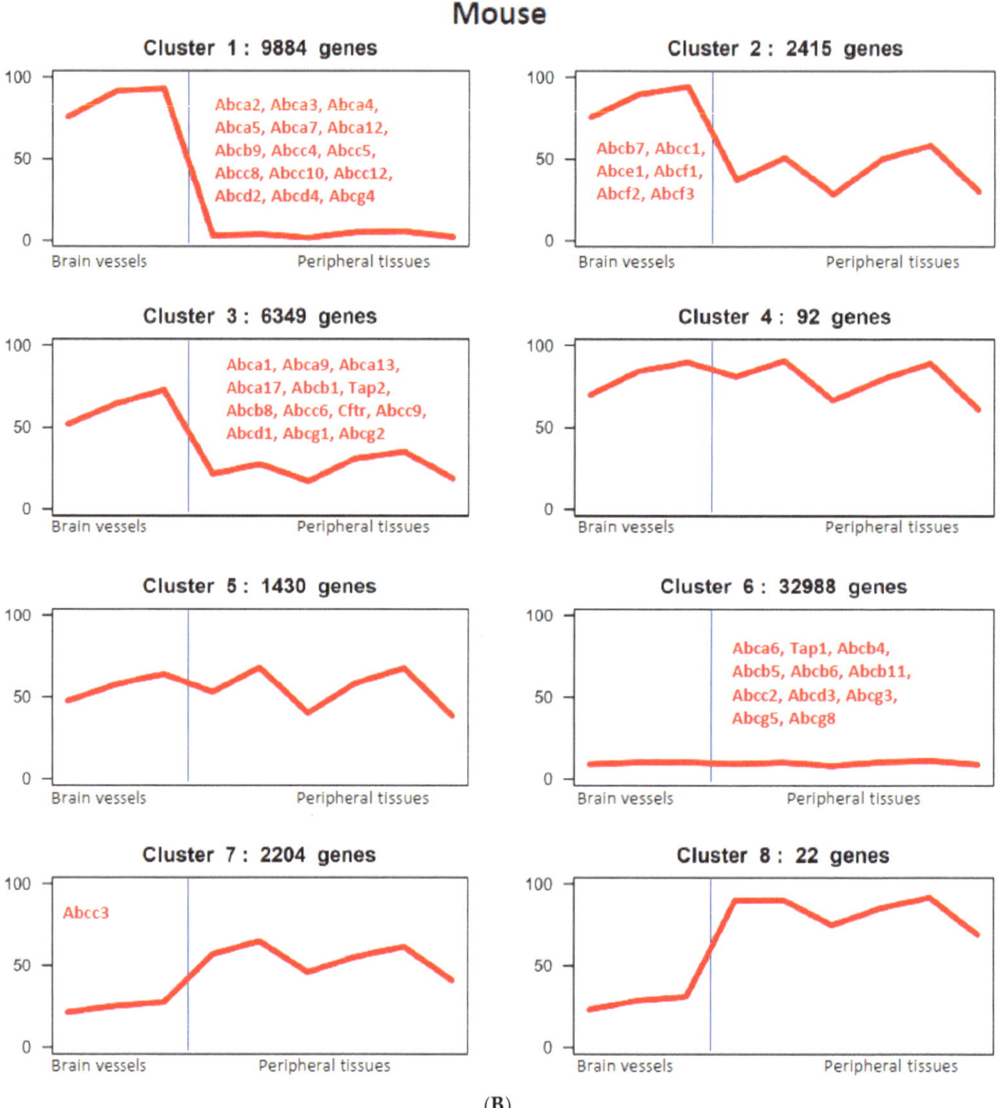

(**B**)

Figure 4. *Cont.*

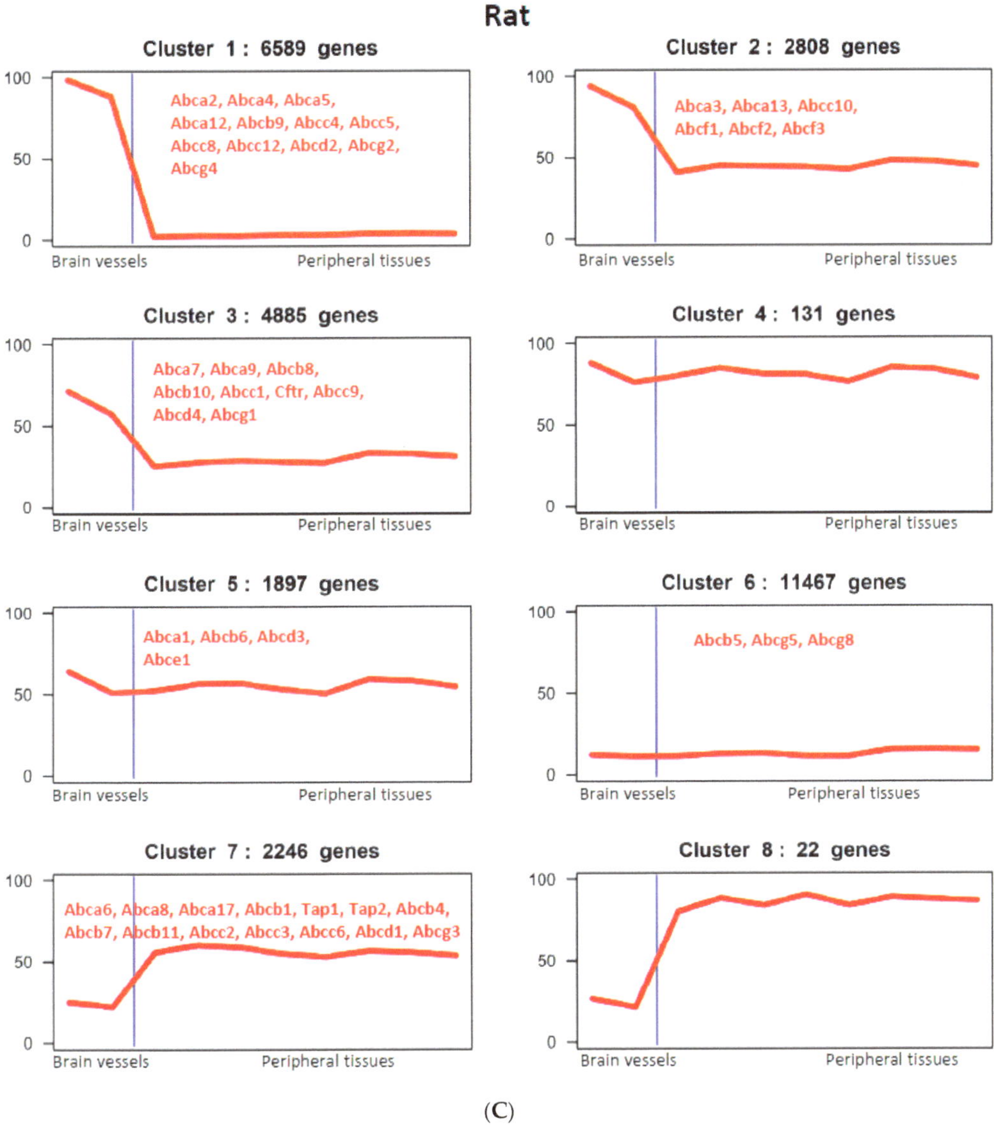

(**C**)

Figure 4. K-means clustering of RNA-seq data (TPM counts) showing classification of *ABC* transporter gene expression in brain vessels vs. peripheral tissues. (**A**). Human brain vessels vs. human lung tissues. (**B**). Mouse brain vessels vs. mouse peripheral tissues (including mouse liver and spleen tissues). (**C**). Rat brain vessels vs. rat peripheral tissues (including rat liver and spleen tissues). The 8 clusters in each species correspond to the following: (1) high in brain vessels and low in peripheral tissues, (2) high in brain vessels and medium in peripheral tissues, (3) medium in brain vessels and low in peripheral tissues, (4) high in brain vessels and high in peripheral tissues, (5) medium in brain vessels and medium in peripheral tissues, (6) low in brain vessels and low in peripheral tissues, (7) low in brain vessels and medium in peripheral tissues, (8) low in brain vessels and high in peripheral tissues.

4. Discussion

ABC transporters play critical roles in transporting a variety of substrates across plasma membranes against substrate gradients, although the function of some of them is still not completely understood [1–3]. This study analyzed the expression and distribution patterns of *ABC* transporter genes and demonstrated marked differences in their tissue abundance, as well as differential expression patterns in three studied species—human, mouse and rat. In particular, the study identified species differences in *ABC* transporter enrichment and expression levels in brain vessels, relevant for the function of BBB.

Several prior studies investigated and quantified the expression of transporters [including some ABC and solute carrier (SLC) transporters] in cerebral vessels and the liver or kidneys of humans, mice, rats and non-human primates using proteomics approaches [19,48–52]. However, none of these studies analyzed the expression patterns of all known *ABC* transporter genes in brain vessels vs. peripheral (lung, liver and spleen) tissues in different species. One of the most important features of cerebral vessels is the cellular and molecular anatomy of brain endothelial and adjacent cells that underlies the BBB. The BBB restricts the passage of blood-borne neurotoxic substances, xenobiotics and drugs into the brain and maintains the homeostasis of the CNS. In addition to physical tightness for molecules > 500 D, the drug transport barrier is achieved mostly by the polarized expression of ABC drug efflux transporters such as ABCB1/Pgp, ABCG2/BCRP and ABCC4/5/MRP4/5 [3,6,53]. RNA-seq analyses performed in this study revealed that *Abcb1a*, *Abcg2*, *Abcc4* or/and *Abcc5* were highly expressed in the brain vessels of mice and rats compared to other *ABC* transporter genes, as well as highly enriched compared to the peripheral tissues analyzed. Similar to mice and rats, *ABCB1*, *ABCG2* and *ABCC5* were also highly abundant and enriched in human cerebral vessels compared to lung tissue.

To complement the RNA-Seq analyses, the protein level expression of major drug efflux transporters ABCB1/Pgp, ABCC1/MRP1 and ABCG2/BCRP was confirmed in the isolated cerebral vessels from humans, mice and rats, showing general congruency between gene and protein expression, with some variations.

It is important to note that the expression levels of *ABC* transporter genes were different in the same tissues between humans and rodents. For example, the expression levels of *ABCB1* and *ABCG2* were significantly higher in mouse and rat brain vessels than that in human brain vessels. Given that both ABCB1 and ABCG2 are major drug efflux transporters expressed at the BBB and additional transporter genes are present in rodents (such as Abcb1b), there are different levels of gene expressions/abundance in brain vessels, and since the high sensitivity of humans to drug toxicity, the direct extrapolation of drug testing/transport assays in animal in vitro and in vivo BBB models should be cautious.

Furthermore, *Abcc4/Mrp4* and *Abcc5/Mrp5* were also found to be more abundantly expressed in rodent brain vessels compared to human brain vessels. Hence, the above consideration of 'translatability' should also apply to these GS-X pump drug efflux transporters and their substrate spectrums, which are different from those of ABCB1 and ABCG2.

In addition to *ABCB1*, *ABCG2* and *ABCC5* drug transporter genes, there was only one additional *ABC* transporter gene (*ABCA2*) that was highly enriched/expressed in human brain vessels. In contrast, clustering analyses identified 15 and 12 other *ABC* genes in mice and rats, respectively, that were highly enriched/expressed in brain vessels; the identified genes mostly overlapped between rodent species.

Generally, higher expression levels of these drug transporters were observed in rodent cerebral vessels compared to human cerebral vessels, consistent with previous reports in the literature [48,54,55]. The evolutionary adaptation to the higher exposure to broader sources of foods and environmental toxins in rodents compared to humans has been considered as a potential explanation for this 'enhanced' barrier function through the induced expression of drug efflux transporters. The neurotoxic substances or xenobiotics in the food sources of rodents might induce higher habitual levels of gene expressions [8,12]. Alternatively, genetic changes might have occurred, for example, in regulatory regions of drug transporter genes, during evolution, which led to higher-level expression (or silencing) in different species. For example, there are two isoforms of Abcb1/Pgp in both rodent species and only one isoform of ABCB1/Pgp in humans. The two rodent isoforms may complement each other functionally to achieve a wider substrate spectrum than that of humans. For example, one study found that bisphenol A is likely a substrate for rat Abcb1b/mdr1b based on high ATPase activity assay, but not for human ABCB1/MDR1 or rat Abcb1a/mdr1a [56]. The homology between human ABCB1 and rodent Abcb1a protein sequences is approximately 87%. The sequence difference may determine the affinity of drug substrates to the transporter protein, the spectrum of substrates accepted by the transporter and their transporting efficiency. One study found that the substrate recognition or transport efficiency of ABCB1/Pgp differs between humans and mice for certain antiepileptic drugs [57]. Strong drug efflux efficiency or a wider substrate spectrum ensures that the blood-borne neurotoxic agents and xenobiotics are not accessible to the CNS [55–57]. Similarly, *Abcg3*, predicted to be a drug transporter gene and considered to be closely related to *Abcg2* [47], is present in rodents but not in humans.

Neuro-drug development and testing often employ in vitro and in vivo rodent BBB models. Mice and rats are the most used pre-clinical species for CNS disease models and toxicology studies. It has been observed that many drug candidates found to be efficacious and safe in rodent models have failed in human clinical trials [58,59]. It appears that humans might be more sensitive or have stronger toxic responses to drugs than rodents [60], which might be due to differences between human and rodent drug transporter expression levels or the presence/absence of variant genes. The differences between humans and rodents have been noted in preclinical studies [58–60], and the results obtained from rodent in vitro and in vivo models may not be completely translatable to human models [58–60].

In addition to the enrichment of *ABCB1/Abcb1a* in cerebral vessels, this study also found that *ABCA2* was highly expressed in brain vessels relative to other tissues and other *ABC* transporter genes in both rodent and human species. It was observed that *ABCA2* expression in human brain vessels was higher than that in rodent brain vessels. ABCA2 is a transporter for cholesterol and sphingolipids [31] and is involved in supplying cholesterol-rich brains with essential lipids. *Abca2* knockout in mice resulted in developmental defects, suggesting that ABCA2 is a transporter with more specific roles in brain vessels and high functional significance to brain development [31]. The expression levels of *ABCA1*, *ABCG1* and *ABCG4* cholesterol/lipid transporter genes in the brain [29,30,45,46] were higher in rodent brain vessels than in human brain vessels.

This study provided additional evidence that some *ABC* transporter genes are specifically expressed or enriched in certain tissues and vessels, such as, for example, the liver and lungs. Both *ABCC2/MRP2* and *ABCC3/MRP3*, known to encode GS-X pump drug efflux transporters [1–3,6], were highly expressed in the liver; *ABCC2/MRP2* expression in rodent liver was comparable to or higher than that of *ABCB1* in brain vessels. Another liver-specific transporter detected in the study was *Abcb11*, which mediates bile acid export.

In contrast, *ABCC1/MRP1*, *ABCC3/MRP3* and *ABCC5/MRP5* were found to be expressed at higher levels in human and rodent lung vessels relative to their expression in brain vessels. ABCC4 and ABCC5, both transporters involved in the cellular export of cyclic nucleotides such as cAMP and cGMP, may contribute to the elimination pathway for cyclic nucleotides in the regulation of signal transduction [2]. Higher levels of *ABCC4/MRP4* and *ABCC5/MRP5* in rodent lung vessels, as well as higher levels of *ABCC1/MRP1* and *ABCC3/MRP3* in human lungs, suggest that the functions of these transporters are needed for essential lung function. The high expression of *ABCC4* in airway epithelia and its role in the regulation of CFTR/ABCC7 ion channel activity have been documented [61]. ABCA3 is involved in phospholipid transport and implicated in surfactant production in the lungs [6,32]. *ABCA3* was also moderately expressed in rodent brain vessels, although its functional role in brain vessels remains unclear.

The limitations of this study include the use of three human brain samples of older donors (>73 years, two females and one male) with underlying non-brain diseases likely requiring medications, some of which may had altered the expression of *ABC* transporter genes. In addition, there was a scarcity of clinical information available for the three patients' lung tissues. It is noted that the age, gender, underlying diseases, medications and ethnicities of the patients could have impacted *ABC* transporter expression. Human brain and lung tissues were frozen and stored at $-80\,^\circ$C before vessel isolation and RNA extraction, while mouse and rodent vessels and RNA were isolated from fresh tissues. Moreover, adult male mice and rats were used; gender/sex hormones may have affected *ABC* transporter expression [62]. Given that the limitations, as well as the sample size, selection and preparation, may have impacted the *ABC* transporter gene expression observed in the datasets, generalization of the findings observed in this study should be cautioned and made within the context of this particular study. Some of these deficiencies were mitigated by the extensive comparisons and high correlative alignment of the generated datasets with those available in public repositories. Although multicellular tissues were used for sequencing, the impact of the overall tissue expression of *ABC* transporter genes in different species on drug biodistribution and potential tissue toxicity could be inferred from the dataset. Furthermore, many of the transporters analyzed in this study were less studied in the literature, providing useful insights about their potential functional importance in specific tissues. Since gene expression at the mRNA level may not be completely translated at the protein level, future investigations on species differences at the protein levels of ABC transporters (such as proteomics analysis) and even at the functional levels (such as drug pharmacokinetics) are warranted.

5. Conclusions

In summary, our study provides the first overview of *ABC* transporter gene expression patterns in brain vessels vs. peripheral tissues (lung, liver and spleen) and lung vessels across species. RNA-seq and Wes^{TM} analyses on brain vessels demonstrated that *ABC* drug transporter genes, including *ABCB1*, *ABCG2/BCRP*, *ABCC4/MRP4* and *ABCC5/MRP5*, were highly expressed in both human and rodent brain vessels. These transporters cover a wide spectrum of drug substrates, preventing blood-borne neurotoxic substances from entering the CNS. It is noted that the expression levels of *ABC* drug transporter genes were generally higher in rodent cerebral vessels compared to human cerebral vessels. Some of the other *ABC* transporter genes were expressed more abundantly or specifically in peripheral organs, such as the liver and lungs. In each case, there were significant differences in patterns of expression or relative enrichment between rodents and humans. Therefore, the interpretation (and translation) of drug biodistribution and toxicity studies with substrates of these transporters performed in rodents must take into account differences in the abundance and patterns of expression of ABC transporters in both peripheral tissues and brain vessels with specialized barriers, such as the BBB.

Supplementary Materials: The following supporting information can be downloaded at https://www.mdpi.com/article/10.3390/pharmaceutics15051563/s1: Figure S1: Wes^{TM} analysis of ABCB1, ABCC1 and ABCG2 expression in human, mouse and rat brain vessels at the protein level; Table S1: ABC transporters and their functions.

Author Contributions: W.Z. conceived of the research idea, analyzed the data, designed some of the experiments, drafted figures and wrote and revised the manuscript; Q.Y.L. designed the experiments, performed some data analysis and drafted part of the manuscript; A.S.H. designed some of the experiments, performed some data analysis and drafted some figures; Z.L. normalized the RNAseq datasets and converted data into TPM counts; C.S. performed Wes analyses; S.L. isolated RNA, prepared RNA-seq libraries and performed RNA sequencing; E.B. (Ewa Baumann) isolated brain vessels, dissected animal tissues and performed immunohistochemistry on brain vessels; C.E.D. isolated brain vessels and dissected animal tissues; E.B. (Eric Brunette) dissected mice and collected tissues; D.B.S. conceived of the research idea and contributed to writing and revising the manuscript. All authors have read and agreed to the published version of the manuscript.

Funding: This research was funded intramurally by the National Research Council of Canada.

Institutional Review Board Statement: The use of human tissues and animals in this study was approved by the Research Ethics Board of the National Research Council Canada (#2013-38 approved in 2014; #2006-03 approved in 2008) and Animal Care Committee (Animal User Protocol #2016-04 approved in 2016), respectively.

Informed Consent Statement: For human lung tissues, all patients signed informed consent as per CDHA-RS/2013-271. Human post-mortem brain tissues were obtained from the Human Brain and Spinal Fluid Resource Center, VAMC (Los Angeles, CA, USA) sponsored by NINDS/NIMN, National Multiple Sclerosis Society, VA Greater Los Angeles Healthcare System and Veterans Health Services and Research Administration, Department of Veteran Affairs. All patients had signed informed consent.

Data Availability Statement: The data presented in this study are available upon request.

Conflicts of Interest: The authors declare no conflict of interest.

References

1.   Dean, M.; Rzhetsky, A.; Allikmets, R. The human ATP-binding cassette (ABC) transporter superfamily. *Genome Res.* **2001**, *11*, 1156–1166. [CrossRef] [PubMed]
2.   Dean, M.; Moitra, K.; Allikmets, R. The human ATP-binding cassette (ABC) transporter superfamily. *Hum. Mutat.* **2022**, *43*, 1162–1182. [CrossRef] [PubMed]
3.   Zhang, W.; Bamji-Mirza, M.; Chang, N.; Haqqani, A.; Stanimirovic, D.B. Expression and Function of ABC Transporters at the Blood-Brain Barrier. In *The Blood-Brain Barrier in Health and Disease*; Dorovini-Zis, K., Ed.; CRC press/Taylor & Francis Group: New York, NY, USA, 2015; Volume one, pp. 172–214.
4.   Dezi, M.; Fribourg, P.F.; Di Cicco, A.; Arnaud, O.; Marco, S.; Falson, P.; Di Pietro, A.; Lévy, D. The multidrug resistance half-transporter ABCG2 is purified as a tetramer upon selective extraction from membranes. *Biochim. Biophys. Acta* **2010**, *1798*, 2094–2101. [CrossRef] [PubMed]
5.   Shimabuku, A.M.; Nishimoto, T.; Ueda, K.; Komano, T. P-glycoprotein. ATP hydrolysis by the N-terminal nucleotide-binding domain. *J. Biol. Chem.* **1992**, *267*, 4308–4311.
6.   Shen, S.; Zhang, W. ABC transporters and drug efflux at the blood-brain barrier. *Rev. Neurosci.* **2010**, *21*, 29–53. [CrossRef]
7.   Redzic, Z. Molecular biology of the blood-brain and the blood-cerebrospinal fluid barriers: Similarities and differences. *Fluids Barriers CNS* **2011**, *8*, 3. [CrossRef]
8.   Miller, D.S. Regulation of ABC transporters at the blood-brain barrier. *Clin. Pharmacol. Ther.* **2015**, *97*, 395–403. [CrossRef]
9.   Zhang, W.; Stanimirovic, D.B. The transport systems of the blood-brain barrier. In *The Blood-Brain Barrier and Its Microenvironment: Basic Physiology to Neurological Disease*; De Vries, E., Prat, A., Eds.; Taylor & Francis Group: New York, NY, USA, 2005; pp. 103–142.
10. Han, L.W.; Gao, C.; Mao, Q. An update on expression and function of P-gp/ABCB1 and BCRP/ABCG2 in the placenta and fetus. *Expert Opin. Drug Metab. Toxicol.* **2018**, *14*, 817–829. [CrossRef]
11. Abbott, N.J.; Patabendige, A.A.; Dolman, D.E.; Yusof, S.R.; Begley, D.J. Structure and function of the blood-brain barrier. *Neurobiol. Dis.* **2010**, *37*, 13–25. [CrossRef]
12. Miller, D.S. Regulation of P-glycoprotein and other ABC drug transporters at the blood-brain barrier. *Trends Pharmacol. Sci.* **2010**, *31*, 246–254. [CrossRef]
13. Terstappen, G.; Meyer, A.; Bell, R.; Zhang, W. Strategies for delivering central nervous system therapeutics across the blood-brain barrier. *Nat. Rev. Drug Discov.* **2021**, *20*, 362–383. [CrossRef] [PubMed]
14. Schinkel, A.H.; Mayer, U.; Wagenaar, E.; Mol, C.A.; van Deemter, L.; Smit, J.J.; van der Valk, M.A.; Voordouw, A.C.; Spits, H.; van Tellingen, O.; et al. Normal viability and altered pharmacokinetics in mice lacking mdr1-type (drug-transporting) P-glycoproteins. *Proc. Natl. Acad. Sci. USA* **1997**, *94*, 4028–4033. [CrossRef] [PubMed]
15. Wijnholds, J.; deLange, E.C.; Scheffer, G.L.; van den Berg, D.J.; Mol, C.A.; van der Valk, M.; Schinkel, A.H.; Scheper, R.J.; Breimer, D.D.; Borst, P. Multidrug resistance protein 1 protects the choroid plexus epithelium and contributes to the blood-cerebrospinal fluid barrier. *J. Clin. Investig.* **2000**, *105*, 279–285. [CrossRef]
16. Zhang, W.; Xiong, H.; Callaghan, D.; Jones, A.; Pei, K.; Stanimirovic, D. Blood-brain barrier transport of amyloid beta peptides in efflux pump knock-out animals evaluated by in vivo optical imaging. *Fluids Barriers CNS* **2013**, *10*, 13. [CrossRef]
17. Wang, J.; Gan, C.; Retmana, I.A.; Sparidans, R.W.; Li, W.; Lebre, M.C.; Beijnen, J.H.; Schinkel, A.H. P-glycoprotein (MDR1/ABCB1) and Breast Cancer Resistance Protein (BCRP/ABCG2) limit brain accumulation of the FLT3 inhibitor quizartinib in mice. *Int. J. Pharm.* **2019**, *556*, 172–180. [CrossRef] [PubMed]
18. Mak, I.W.; Evaniew, N.; Ghert, M. Lost in translation: Animal models and clinical trials in cancer treatment. *Am. J. Transl. Res.* **2014**, *6*, 114–118.
19. Morris, M.E.; Rodriguez-Cruz, V.; Felmlee, M.A. SLC and ABC Transporters: Expression, Localization, and Species Differences at the Blood-Brain and the Blood-Cerebrospinal Fluid Barriers. *AAPS J.* **2017**, *19*, 1317–1331. [CrossRef]

20. Zhang, W.; Liu, Q.Y.; Haqqani, A.S.; Leclerc, S.; Baumann, E.; Ly, D.; Delaney, C.E.; Liu, Z.-Y.; Star, A.T.; Brunette, E.; et al. Differential expression of receptors mediating receptor-mediated transcytosis (RMT) in brain microvessels, brain parenchyma and peripheral tissues of mouse and human. *Fluids Barriers CNS* **2020**, *17*, 47. [CrossRef]
21. Stanimirovic, D.B.; Bani-Yaghoub, M.; Perkins, M.; Haqqani, A.S. Blood–brain barrier models: In vitro to in vivo translation in preclinical development of CNS-targeting biotherapeutics. *Expert Opin. Drug Discov.* **2015**, *10*, 141–155. [CrossRef]
22. Dobin, A.; Davis, C.A.; Schlesinger, F.; Drenkow, J.; Zaleski, C.; Jha, S.; Batut, P.; Chaisson, M.; Gingeras, T.R. STAR: Ultrafast universal RNA-seq aligner. *Bioinformatics* **2013**, *29*, 15–21. [CrossRef]
23. Li, B.; Dewey, C.N. RSEM: Accurate transcript quantification from RNA-Seq data with or without a reference genome. *BMC Bioinform.* **2011**, *12*, 323. [CrossRef] [PubMed]
24. Love, M.I.; Huber, W.; Anders, S. Moderated estimation of fold change and dispersion for RNA-seq data with DESeq2. *Genome Biol.* **2014**, *15*, 550. [CrossRef] [PubMed]
25. Leinonen, R.; Sugawara, H.; Shumway, M. International nucleotide sequence database collaboration. The sequence read archive. *Nucleic Acids Res.* **2011**, *39*, D19–D21. [CrossRef] [PubMed]
26. Grossman, R.L.; Heath, A.; Murphy, M.; Patterson, M.; Wells, W. A case for data commons: Toward data science as a service. *Comput. Sci. Eng.* **2016**, *18*, 10–20. [CrossRef] [PubMed]
27. Gomez-Zepeda, D.; Taghi, M.; Scherrmann, J.M.; Decleves, X.; Menet, M.C. ABC Transporters at the Blood-Brain Interfaces, Their Study Models, and Drug Delivery Implications in Gliomas. *Pharmaceutics* **2019**, *12*, 20. [CrossRef] [PubMed]
28. Zhang, W.; Mojsilovic-Petrovic, J.; Andrade, M.; Zhang, H.; Ball, M.; Stanimirovic, D. Expression and Functional Characterization of ABCG2 in Brain Endothelial Cells and Vessels. *FASEB J.* **2003**, *17*, 2085–2087. [CrossRef]
29. Hirsch-Reinshagen, V.; Zhou, S.; Burgess, B.L.; Bernier, L.; McIsaac, S.A.; Chan, J.Y.; Tansley, G.; Cohn, J.S.; Hayden, M.R.; Wellington, C.L. Deficiency of ABCA1 impairs apolipoprotein E metabolism in brain. *J. Biol. Chem.* **2004**, *279*, 41197–41207. [CrossRef] [PubMed]
30. Karasinska, J.M.; de Haan, W.; Franciosi, S.; Ruddle, P.; Fan, J.; Kruit, J.K.; Stukas, S.; Lütjohann, D.; Gutmann, D.H.; Wellington, C.L.; et al. ABCA1 influences neuroinflammation and neuronal death. *Neurobiol. Dis.* **2013**, *54*, 445–455. [CrossRef]
31. Davis, W., Jr.; Tew, K.D. ATP-binding cassette transporter-2 (ABCA2) as a therapeutic target. *Biochem. Pharmacol.* **2018**, *151*, 188–200. [CrossRef]
32. Stahlman, M.T.; Besnard, V.; Wert, S.E.; Weaver, T.E.; Dingle, S.; Xu, Y.; von Zychlin, K.; Olson, S.J.; Whitsett, J.A. Expression of ABCA3 in developing lung and other tissues. *J. Histochem. Cytochem.* **2007**, *55*, 71–83. [CrossRef]
33. Steinbach, D.; Gillet, J.P.; Sauerbrey, A.; Gruhn, B.; Dawczynski, K.; Bertholet, V.; de Longueville, F.; Zintl, F.; Remacle, J.; Efferth, T. ABCA3 as a possible cause of drug resistance in childhood acute myeloid leukemia. *Clin. Cancer Res.* **2006**, *12 Pt 1*, 4357–4363. [CrossRef] [PubMed]
34. Wang, N.; Lan, D.; Gerbod-Giannone, M.; Linsel-Nitschke, P.; Jehle, A.W.; Chen, W.; Martinez, L.O.; Tall, A.R. ATP-binding cassette transporter A7 (ABCA7) binds apolipoprotein A-I and mediates cellular phospholipid but not cholesterol efflux. *J. Biol. Chem.* **2003**, *278*, 42906–42912. [CrossRef] [PubMed]
35. Trigueros-Motos, L.; van Capelleveen, J.C.; Torta, F.; Castaño, D.; Zhang, L.H.; Chai, E.C.; Kang, M.; Dimova, L.G.; Schimmel, A.W.M.; Tietjen, I.; et al. ABCA8 Regulates Cholesterol Efflux and High-Density Lipoprotein Cholesterol Levels. *Arterioscler. Thromb. Vasc. Biol.* **2017**, *37*, 2147–2155. [CrossRef] [PubMed]
36. Iqbal, J.; Suarez, M.D.; Yadav, P.K.; Walsh, M.T.; Li, Y.; Wu, Y.; Huang, Z.; James, A.W.; Escobar, V.; Mokbe, A.; et al. ATP-binding cassette protein ABCA7 deficiency impairs sphingomyelin synthesis, cognitive discrimination, and synaptic plasticity in the entorhinal cortex. *J. Biol. Chem.* **2022**, *298*, 102411. [CrossRef] [PubMed]
37. Bamji-Mirza, M.; Li, Y.; Najem, D.; Liu, Q.-Y.; Stupak, J.; Chen, K.; Li, J.; Walker, D.; Lue, L.-F.; Yang, Z.; et al. Genetic variations in ABCA7 can increase secreted levels of Aβ_{1-42} and ABCA7 transcription in cell culture models. *J. Alzheimer's Dis.* **2016**, *53*, 875–892; Erratum in *J. Alzheimer's Dis.* **2018**, *66*, 853–854. [CrossRef]
38. Ma, F.C.; Wang, H.F.; Cao, X.P.; Tan, C.C.; Tan, L.; Yu, J.T. Meta-Analysis of the Association between Variants in ABCA7 and Alzheimer's Disease. *J. Alzheimer's Dis.* **2018**, *63*, 1261–1267. [CrossRef]
39. Sakae, N.; Liu, C.C.; Shinohara, M.; Frisch-Daiello, J.; Ma, L.; Yamazaki, Y.; Tachibana, M.; Younkin, L.; Kurti, A.; Carrasquillo, M.M.; et al. ABCA7 Deficiency Accelerates Amyloid-β Generation and Alzheimer's Neuronal Pathology. *J. Neurosci.* **2016**, *36*, 3848–3859. [CrossRef]
40. Abe-Dohmae, S.; Yokoyama, S. ABCA7 links sterol metabolism to the host defense system: Molecular background for potential management measure of Alzheimer's disease. *Gene* **2021**, *768*, 145316. [CrossRef]
41. Ichikawa, Y.; Bayeva, M.; Ghanefar, M.; Potini, V.; Sun, L.; Mutharasan, R.K.; Wu, R.; Khechaduri, A.; Jairaj Naik, T.; Ardehali, H. Disruption of ATP-binding cassette B8 in mice leads to cardiomyopathy through a decrease in mitochondrial iron export. *Proc. Natl. Acad. Sci. USA* **2012**, *109*, 4152–4157. [CrossRef]
42. Zhang, F.; Zhang, W.; Liu, L.; Fisher, C.L.; Hui, D.; Childs, S.; Dorovini-Zis, K.; Ling, V. Characterization of ABCB9, an ATP binding cassette protein associated with lysosomes. *J. Biol. Chem.* **2000**, *275*, 23287–23294. [CrossRef]
43. Park, J.G.; Kim, S.; Jang, E.; Choi, S.H.; Han, H.; Ju, S.; Kim, J.W.; Min, D.S.; Jin, M.S. The lysosomal transporter TAPL has a dual role as peptide translocator and phosphatidylserine floppase. *Nat. Commun.* **2022**, *13*, 5851. [CrossRef] [PubMed]
44. Bryan, J.; Muñoz, A.; Zhang, X.; Düfer, M.; Drews, G.; Krippeit-Drews, P.; Aguilar-Bryan, L. ABCC8 and ABCC9: ABC transporters that regulate K+ channels. *Pflug. Arch.* **2007**, *453*, 703–718. [CrossRef] [PubMed]

45. Tansley, G.H.; Burgess, B.L.; Bryan, M.T.; Su, Y.; Hirsch-Reinshagen, V.; Pearce, J.; Chan, J.Y.; Wilkinson, A.; Evans, J.; Naus, K.E.; et al. The cholesterol transporter ABCG1 modulates the subcellular distribution and proteolytic processing of beta-amyloid precursor protein. *J. Lipid Res.* **2007**, *48*, 1022–1034. [CrossRef]
46. Wang, N.; Yvan-Charvet, L.; Lütjohann, D.; Mulder, M.; Vanmierlo, T.; Kim, T.W.; Tall, A.R. ATP-binding cassette transporters G1 and G4 mediate cholesterol and desmosterol efflux to HDL and regulate sterol accumulation in the brain. *FASEB J.* **2008**, *22*, 1073–1082. [CrossRef] [PubMed]
47. Mickley, L.; Jain, P.; Miyake, K.; Schriml, L.M.; Rao, K.; Fojo, T.; Bates, S.; Dean, M. An ATP-binding cassette gene (ABCG3) closely related to the multidrug transporter ABCG2 (MXR/ABCP) has an unusual ATP-binding domain. *Mamm. Genome* **2001**, *12*, 86–88. [CrossRef]
48. Uchida, Y.; Ohtsuki, S.; Katsukura, Y.; Ikeda, C.; Suzuki, T.; Kamiie, J.; Terasaki, T. Quantitative targeted absolute proteomics of human blood-brain barrier transporters and receptors. *J. Neurochem.* **2011**, *117*, 333–345. [CrossRef]
49. Uchida, Y.; Zhang, Z.; Tachikawa, M.; Terasaki, T. Quantitative targeted absolute proteomics of rat blood-cerebrospinal fluid barrier transporters: Comparison with a human specimen. *J. Neurochem.* **2015**, *134*, 1104–1115. [CrossRef]
50. Ito, K.; Uchida, Y.; Ohtsuki, S.; Aizawa, S.; Kawakami, H.; Katsukura, Y.; Kamiie, J.; Terasaki, T. Quantitative membrane protein expression at the blood-brain barrier of adult and younger cynomolgus monkeys. *J. Pharm. Sci.* **2011**, *100*, 3939–3950. [CrossRef]
51. Wang, L.; Prasad, B.; Salphati, L.; Chu, X.; Gupta, A.; Hop, C.E.; Evers, R.; Unadkat, J.D. Interspecies variability in expression of hepatobiliary transporters across human, dog, monkey, and rat as determined by quantitative proteomics. *Drug Metab. Dispos.* **2015**, *43*, 367–374. [CrossRef]
52. Fallon, J.K.; Smith, P.C.; Xia, C.Q.; Kim, M.S. Quantification of Four Efflux Drug Transporters in Liver and Kidney across Species Using Targeted Quantitative Proteomics by Isotope Dilution NanoLC-MS/MS. *Pharm. Res.* **2016**, *33*, 2280–2288. [CrossRef]
53. Qosa, H.; Miller, D.S.; Pasinelli, P.; Trotti, D. Regulation of ABC efflux transporters at blood-brain barrier in health and neurological disorders. *Brain Res.* **2015**, *1628 Pt B*, 298–316. [CrossRef]
54. Al Feteisi, H.; Al-Majdoub, Z.M.; Achour, B.; Couto, N.; Rostami-Hodjegan, A.; Barber, J. Identification and quantification of blood-brain barrier transporters in isolated rat brain microvessels. *J. Neurochem.* **2018**, *146*, 670–685. [CrossRef] [PubMed]
55. Verscheijden, L.F.M.; Koenderink, J.B.; de Wildt, S.N.; Russel, F.G.M. Differences in P-glycoprotein activity in human and rodent blood-brain barrier assessed by mechanistic modelling. *Arch. Toxicol.* **2021**, *95*, 3015–3029. [CrossRef] [PubMed]
56. Mazur, C.S.; Marchitti, S.A.; Dimova, M.; Kenneke, J.F.; Lumen, A.; Fisher, J. Human and rat ABC transporter efflux of bisphenol a and bisphenol a glucuronide: Interspecies comparison and implications for pharmacokinetic assessment. *Toxicol. Sci.* **2012**, *128*, 317–325. [CrossRef] [PubMed]
57. Baltes, S.; Gastens, A.M.; Fedrowitz, M.; Potschka, H.; Kaever, V.; Löscher, W. Differences in the transport of the antiepileptic drugs phenytoin, levetiracetam and carbamazepine by human and mouse P-glycoprotein. *Neuropharmacology* **2007**, *52*, 333–346. [CrossRef] [PubMed]
58. Van Norman, G.A. Limitations of Animal Studies for Predicting Toxicity in Clinical Trials: Is it Time to Rethink Our Current Approach? *JACC Basic Transl. Sci.* **2019**, *4*, 845–854. [CrossRef]
59. Van Norman, G.A. Limitations of Animal Studies for Predicting Toxicity in Clinical Trials: Part 2: Potential Alternatives to the Use of Animals in Preclinical Trials. *JACC Basic Transl. Sci.* **2020**, *5*, 387–397. [CrossRef]
60. Atkins, J.T.; George, G.C.; Hess, K.; Marcelo-Lewis, K.L.; Yuan, Y.; Borthakur, G.; Khozin, S.; LoRusso, P.; Hong, D.S. Pre-clinical animal models are poor predictors of human toxicities in phase 1 oncology clinical trials. *Br. J. Cancer* **2020**, *123*, 1496–1501. [CrossRef]
61. Rider, C.F.; Newton, R.; Bear, C.; Carlsten, C.; Hirota, J.A. The ATP Binding Cassette C4 (ABCC4) Transporter Regulates Extracellular cAMP Transport, Intracellular PKA Activity, and CFTR Channel Activity in Human Airway Epithelial Cells. *Am. J. Respir. Crit. Care Med.* **2016**, *193*, A5557.
62. Mares, L.; Vilchis, F.; Chávez, B.; Ramos, L. Expression and regulation of ABCG2/BCRP1 by sex steroids in the Harderian gland of the Syrian hamster (*Mesocricetus auratus*). *Comptes Rendus Biol.* **2019**, *342*, 279–289. [CrossRef]

pharmaceutics

Article

Applicability of MDR1 Overexpressing Abcb1KO-MDCKII Cell Lines for Investigating In Vitro Species Differences and Brain Penetration Prediction

Emőke Sóskuti [1,2,†], Nóra Szilvásy [1,†], Csilla Temesszentandrási-Ambrus [1], Zoltán Urbán [1], Olivér Csíkvári [1], Zoltán Szabó [3], Gábor Kecskeméti [3], Éva Pusztai [4] and Zsuzsanna Gáborik [1,*]

1   Charles River Laboratories Hungary, H-1117 Budapest, Hungary; emoke.soskuti@crl.com (E.S.); nora.szilvasy@crl.com (N.S.); csilla.temesszentandrasi-ambrus@crl.com (C.T.-A.); zoltan.urban@crl.com (Z.U.); oliver.csikvari@crl.com (O.C.)
2   Doctoral School of Semmelweis University, Molecular Medicine Division, H-1085 Budapest, Hungary
3   Department of Medical Chemistry, Albert Szent-Györgyi Medical School, University of Szeged, H-6720 Szeged, Hungary; szabo.zoltan@med.u-szeged.hu (Z.S.); kecskemeti.gabor@med.u-szeged.hu (G.K.)
4   Department of Chemical and Environmental Process Engineering, Faculty of Chemical Technology and Biotechnology, Budapest University of Technology and Economics, H-1111 Budapest, Hungary; pusztai.eva@vbk.bme.hu
*   Correspondence: zsuzsanna.gaborik@crl.com
†   These authors contributed equally to this work.

Abstract: Implementing the 3R initiative to reduce animal experiments in brain penetration prediction for CNS-targeting drugs requires more predictive in vitro and in silico models. However, animal studies are still indispensable to obtaining brain concentration and determining the prediction performance of in vitro models. To reveal species differences and provide reliable data for IVIVE, in vitro models are required. Systems overexpressing MDR1 and BCRP are widely used to predict BBB penetration, highlighting the impact of the in vitro system on predictive performance. In this study, endogenous Abcb1 knock-out MDCKII cells overexpressing MDR1 of human, mouse, rat or cynomolgus monkey origin were used. Good correlations between ERs of 83 drugs determined in each cell line suggest limited species specificities. All cell lines differentiated CNS-penetrating compounds based on ERs with high efficiency and sensitivity. The correlation between in vivo and predicted $K_{p,uu,brain}$ was the highest using total ER of human MDR1 and BCRP and optimized scaling factors. MDR1 interactors were tested on all MDR1 orthologs using digoxin and quinidine as substrates. We found several examples of inhibition dependent on either substrate or transporter abundance. In summary, this assay system has the potential for early-stage brain penetration screening. IC_{50} comparison between orthologs is complex; correlation with transporter abundance data is not necessarily proportional and requires the understanding of modes of transporter inhibition.

Keywords: BBB penetration; $K_{p,uu,brain}$ prediction; MDR1; BCRP; ABCB1 knock-out; preclinical MDR1; species differences; MDR1 inhibitors

Citation: Sóskuti, E.; Szilvásy, N.; Temesszentandrási-Ambrus, C.; Urbán, Z.; Csíkvári, O.; Szabó, Z.; Kecskeméti, G.; Pusztai, É.; Gáborik, Z. Applicability of MDR1 Overexpressing Abcb1KO-MDCKII Cell Lines for Investigating In Vitro Species Differences and Brain Penetration Prediction. *Pharmaceutics* **2024**, *16*, 736. https://doi.org/10.3390/pharmaceutics16060736

Academic Editors: Gert Fricker and Elena Puris

Received: 30 April 2024
Revised: 22 May 2024
Accepted: 24 May 2024
Published: 29 May 2024

1. Introduction

Determining the brain exposure of drugs, especially of those with a central nervous system (CNS) target, is crucial for de-risking efficacy and toxicity issues early in drug discovery [1]. The brain is separated from the systemic circulation by the blood–brain barrier (BBB), a complex and dynamic interface with multiple roles. It is composed of brain microvascular endothelial cells connected by tight junctions, and supported by microglial cells, astrocytes, pericytes and the capillary basement membrane, adding up to a cellular membrane with constant thickness, limited pinocytotic activity and negative surface charge [2]. The transport of compounds across the BBB is tightly regulated, resulting in distinct pharmacokinetic properties of certain drugs in the brain compared to those in the blood. It involves ATP-binding cassette (ABC) transporter, multidrug resistance 1 (ABCB1/P-gp or MDR1), breast cancer resistance protein (ABCG2/BCRP) and multidrug resistance-associated protein 4 (ABCC4/MRP4). The major quantifiable solute carrier (SLC) transporters in brain microvessels are OAT3 (SLC22A8), GLUT1 (SLC2A1), LAT1 (SLC7A5) and MCT1 (SLC16A1) [3–9]. Of these, MDR1 is the most studied and most relevant gatekeeper [10,11].

Tools are available to determine the BBB penetration of drugs in development ranging from preclinical animal models to various in vitro and in silico tools, but every model has its limitations and translation to human is still challenging. Preclinical animals, mainly rodents, are used in neurotoxicity studies and to determine drug concentration in the brain [1]. The advantage of animal models is that they comprise all factors that influence the transport across the BBB, but because of substantial differences in the abundance and nature of transporters, translation to humans requires further investigation [1,12]. Despite the highly conserved nature of MDR1 [13], controversial data have been reported previously in vivo [14–16], underpinning the necessity for a set of in vitro assays enabling direct comparisons of MDR1 interactions across species. Hence, we aimed to create an in vitro test system to collect data about MDR1 from human and preclinical species on quantitative transport, affinity to various transporters and transporter abundance [1,4–7,12,17–19].

The major pharmacological factor that needs to be optimized in CNS drug discovery is $K_{p,uu,brain}$ (unbound brain-to-plasma partition coefficient), since it is the unbound brain concentration that drives target binding and subsequent pharmacological response, according to the free drug hypothesis [10,20–24]. Higher $K_{p,uu,brain}$ values of CNS drugs are preferable during drug development, as this will result in lower systemic toxicity concerns [25]. Considering recent efforts of the pharmaceutical industry to implement the 3R initiative to reduce preclinical animal experiments with more predictive in vitro tools, the improvement of in vitro in vivo extrapolation (IVIVE) is of high importance. Because of the complexity of the BBB, primary cell-based in vitro models are cost- and work-intense, and therefore less ideal for screening purposes. Systems overexpressing the two most important efflux transporters, MDR1 and BCRP, in all combinations have been evaluated previously by numerous groups. We learnt from these studies about the importance of the assay systems; the parental cell line and the transporter expression level have a significant influence on the predictive performance [17,26–31]. Our aim was to generate an in vitro system utilizing endogenous canine Abcb1-knock-out (KO) MDCKII cells overexpressing the MDR1 transporter of either human, mouse, rat or cynomolgus monkey origin to enable direct correlation of human and preclinical in vitro data. Bidirectional permeability and efflux ratio (ER) in the presence or absence of zosuquidar were determined for 83 drugs. In vivo rodent $K_{p,uu,brain}$ data from the literature were compared to investigate whether the in vitro ERs determined in screening setup can be used for CNS permeability classification and quantitative $K_{p,uu,brain}$ prediction. Our MCDKII-BCRP cell line, which has previously been shown to be useful in BBB penetration prediction [32,33], has been added to the screen, and combined MDR1 and BCRP substrate data were used for $K_{p,uu,brain}$ prediction [17,28–31,34]. To compare data across species, transporter abundance was determined to calculate relative expression factors (REF), which are important input data for PBPK models as well. Inhibition was assessed in all cell lines with 21 compounds using two representative MDR1 substrates, digoxin and quinidine, to identify potential species differences in transporter specificity and sensitivity.

2. Materials and Methods

2.1. Materials

Reagents and non-radiolabeled chemicals were purchased from Merck/Sigma-Aldrich (St. Louis, MO, USA), Cayman Europe OÜ (Tallinn, Estonia), Selleck Chemicals (Houston, TX, USA), Biosynth (Compton, UK), Thermo Fisher Scientific (Waltham, MA, USA), Toronto Research Chemicals (Toronto, ON, Canada), Biogal Rt. (Debrecen, Hungary), VWR (Debrecen, Hungary) and MedChemExpress (Monmouth Junction, NJ, USA). All chemicals were of analytical grade. Ultima Gold XR scintillation fluid was obtained from PerkinElmer (Waltham, MA, USA). ^{3}H-digoxin ([^{3}H(G)], 23.8 Ci/mmol) and ^{3}H-quinidine ([9-^{3}H], 20 Ci/mmol) were from American Radiolabeled Chemicals (St. Louis, MO, USA). Stable isotope (^{13}C and ^{15}N) labeled proteospecific peptide fragments common for all orthologs of MDR1 (IATEAIENFR), ATP1A1 (IVEIPFNSTNK) and dog specific MDR1 (FYDPLAGSVLIDGK) were ordered from JPT Peptide Technologies (Berlin, Germany) in the form of tagged SpikeTides™ TQL peptides.

2.2. Cell Line Generation and Culture Conditions

Cell lines were generated as described previously in [PMID: 36901890]. In brief, sequence-verified cDNA encoding human MDR1 (NCBI Reference Sequence: NM_000927.4), rat Mdr1a (NM_133401.1), mouse Mdr1a (NM_011076.3) and cynomolgus monkey Mdr1 (NM_001287322.1) was synthesized by GenScript. Transduced and antibiotic-selected Abcb1KO-MDCKII cells were subjected to single cell cloning by calcein-AM-based FACS, and amplified clones were functionally tested for transporter-specific efflux activity. The best-performing clones were selected for continued validation, and are hereafter referred to as the hMDR1, mMDR1, rMDR1, and cyMDR1 cell lines. Empty vector-transduced Abcb1KO-MDCKII-Mock cells were used as control. Wild type parental MDCKII cells were transduced with sequence-verified cDNA (GenScript) encoding human BCRP (NCBI Reference Sequence: NM_004827.1). The maintenance and seeding of cells are described in detail [35]. Briefly, Abcb1KO-MDCKII-MDR1 and Mock, and MDCKII-BCRP, cells were cultured in DMEM with high glucose, supplemented with 10% fetal bovine serum, 2 mM GlutaMAX™, 100 units/mL penicillin and 100 µg/mL streptomycin at 37 °C, 5% CO$_2$, and 90% relative humidity. For transport experiments, cells were seeded on Millicell™ high-pore-density 0.4 µm PCF 96 well cell culture plate inserts (Millipore, Merck KGaA, Darmstadt, Germany) at a density of 25,000 cells/well and grown for 5 days at 37 °C in an atmosphere of 5% CO$_2$ and 95% relative humidity. The culture medium was changed once, the day before the experiment.

2.3. Bidirectional Transport Assays

Transport assays were performed as previously described [36] with minor modifications. After washing and preincubation for 15 min with prewarmed Hanks' balanced salt solution (HBSS) at pH 7.4, the experiment was started at t = 0 by replacing HBSS in the donor compartment (either apical or basolateral) with HBSS containing the substrate (1 µM) or the mixture of substrate and inhibitors (zosuquidar, 1 or 5 µM for MDR1 or Ko143, 1 µM for BCRP). The final concentration of DMSO was ≤0.1% in all transport buffers. Samples were taken from both the receiver and donor sides at 120 min to determine the recovery in addition to the transport activity. Samples and dosing solutions were diluted twofold in methanol and analyzed by LC-MS/MS. In inhibitory studies, A-B and B-A permeabilities of ^{3}H-digoxin (1 µM, 0.17 µCi/mL) and ^{3}H-quinidine (0.1 or 1 µM, 0.17 µCi/mL) probe substrates were applied in the absence or presence of increasing concentrations of the selected inhibitors at predetermined timepoints by cell lines (digoxin—120 min; quinidine—30 min (rMDR1); hMDR1, mMDR1 and cyMDR1—60 min). To determine the amounts of radiolabeled substrates (digoxin or quinidine) transported, samples mixed with Ultima Gold XR liquid scintillation cocktail were measured with a MicroBeta2 microplate counter (PerkinElmer). The tightness of the cell monolayer was controlled via the permeability of Lucifer Yellow (LY, 40 µg/mL). Experiments showing LY permeation higher than P_{app} 2 × 10^{-6} cm/s were rejected.

2.4. Analytical Measurements

Sample analysis was performed on an LS-I autosampler (Sound Analytics, Niantic, CT, USA) equipped with Agilent 1260 HPCL pumps coupled to a Sciex 6500+ Triple Quadrupole Mass Spectrometer (AB SCIEX, Framingham, MA, USA). Chromatographic separation was achieved on a Phenomenex Kinetex F5 column (30 × 2.1 mm, 2.6 µm,

Phenomenex Inc., Torrance, CA, USA) with gradient elution starting from 2% B eluent with a flow rate of 0.7 mL/min, holding for 0.12 min, followed by a linear gradient from 2 to 95% B in 0.48 min, then holding at 95% B for 0.1 min with a flow rate of 1 mL/min, then a re-equilibration of the column with 2% B with 1 mL/min for 0.3 min. Eluent A consisted of 0.1% formic acid in water, eluent B consisted of 0.1% formic acid in acetonitrile. Then, 10 μL samples were injected into the HPLC-MS/MS system. Mass spectrometric detection was performed in SRM mode, and ion transitions for each analyte were optimized during method development. Samples were analyzed in pools, with each pool containing three to four analytes.

2.5. Quantitative Targeted Absolute Proteomics (QTAP)

Absolute protein expression levels (pmol/mg membrane protein) of target transporters were determined via a PRM analysis in the NanoLC–MS/MS using the peptide probes to quantify the target molecules. MDR1 was quantified from filter-grown hMDR1, mMDR1, rMDR1a and cyMDR1 cell lines, and BCRP from MDCKII-BCRP cell monolayers. Cell culturing and seeding conditions were equivalent to those in bidirectional transport assays. The membrane protein fractions of the cells were enriched using ProteoExtract™ Native Membrane Protein Extraction kit (Merck) according to the manufacturer's protocol. The membrane-enriched buffer II fractions were used for further analysis. The protein content of the samples was determined using BCA Protein Assay kit (Thermo Fisher Scientific) following the manufacturer's acetone precipitation protocol. For all the samples, 10 μg protein was processed with an On Pellet Digestion protocol. The samples were reduced with 20 mmol DTT at 60 °C for 30 min and alkylated with 40 mmol IAA in the dark at room temperature for 30 min. The protein content was precipitated by adding seven volumes of ice-cold acetone and incubated at −20 °C overnight. After centrifugation at 15,000× g for 10 min, at 4 °C, the supernatant was discarded. The protein pellet was washed three times with 0.5 mL acetone/water (85/15, v/v) mixture. After centrifugation at 14,000× g for 10 min at 4 °C, the protein pellet was dissolved in 15 μL RapiGest SF Surfactant (Waters, Milford, MA, USA) and was incubated at 100 °C for 5 min. After cooling to room temperature, 55 μL 100 mmol AmBic (pH = 8), 10 μL IS peptide mix (1 pmol/peptide) and 0.25 μg/5 μL trypsin were added to the mixtures. The samples were incubated at 37 °C for 30 min and another 0.25 μg/5 μL trypsin was added, and the mixture was digested at 37 °C for 5.5 h. Digestion was stopped by the addition of 1 μL concentrated FA. The resulting peptide samples were purified using Pierce™ C18 Spin Tips (Thermo Fisher Scientific) according to the manufacturer's instructions, with an additional detergent removal step. This was achieved by washing the tips with 2 × 50 μL DCE after the prescribed desalting step. The samples were then evaporated under vacuum, then resolved in 90 μL of the initial eluent. The samples were centrifuged with 10,000× g for 10 min at 4 °C and 5 μL of the supernatant was injected into NanoLC–MS/MS. NanoLC-MS/MS analysis was carried out on a Waters ACQUITY UPLC M-Class LC system (Waters) coupled with an Orbitrap Exploris™ 240 mass spectrometer (Thermo Fisher Scientific). A Symmetry® C18 (100 Å, 5 μm, 180 μm × 20 mm) trap column was used for trapping and desalting the samples. The chromatographic separation of peptides was accomplished on an ACQUITY UPLC® M-Class Peptide BEH C18 analytical column (130 Å, 1.7 μm, 75 μm × 250 mm) at 45 °C by gradient elution. Water (solvent A) and acetonitrile (solvent B), both containing 0.1% FA, were used as mobile phases at a flow rate of 350 nL/min. The sample temperature was maintained at 5 °C. The mass spectrometer was operated using the equipped Nanospray Flex Ion Source. The parallel reaction monitoring (PRM) method was used to monitor the m/z transitions for the + 2 charged peptides precursors of interest. To reach the maximum sensitivity, precursor ions were fragmented using optimal collision energies. The automatic gain control (AGC) setting was defined as 1×10^5 charges, the maximum injection time was set to auto, and resolution was set to 15,000. Data acquisition was performed using Xcalibur™ 4.6 (Thermo Fisher Scientific), and Skyline 22.2.1.278 [37] was used for data evaluation. The ratio of the peptides to stable isotope labeled internal standard was used for protein quantification. Final quantitative results are shown as the ratio to total protein content injected into NanoLC-MS/MS (pmol protein/mg total membrane protein).

2.6. Data Analysis

(a)   Bidirectional transport assays

All the experiments were performed in three biological replicates and repeated two or three times; each data point corresponds to the mean of at least six values.

Apparent permeability (P_{app}), ER and mass balance (recovery) were determined as has been published [36]. Compounds with lower than 60% recovery were excluded from the data analysis. The half-maximal inhibitory concentration (IC_{50}) is used as a measure of inhibitory drug potency. If inhibition did not exceed 50% at the highest inhibitor concentration tested IC_{50} calculations were not performed, and the highest applied concentration was used for correlation. IC_{50} values were derived from a four-parametric logistic equation (log(inhibitor) vs. response–variable slope); the curve was fitted to the ER vs. inhibitor concentration plot using non-linear regression in GraphPad Prism version 9.0 (GraphPad, La Jolla, CA, USA). In the comparison studies, using linear regression, the coefficient of determination (R^2) was determined with GraphPad Prism version 9.0. The difference from the line of identity was quantified with residual standard error (RSE), which was calculated as follows:

$$\text{RSE} = \sqrt{\frac{\sum_1^n (x_i - y_i)^2}{n}} \tag{1}$$

where x_i and y_i represent the calculated ER or the IC_{50} of the component i in substrate or inhibition studies, respectively, and n represents the number of data points.

(b)   Predictive Performance Metrics

MDR1 and/or BCRP substrate properties were investigated by binary classification analysis on CNS+ and CNS− compounds. The possible outcomes are the following: True positive (TP), when the CNS− compound showed impaired brain distribution due to MDR1 and/or BCRP efflux mechanisms. True negative (TN), where there was no efflux and the CNS+ compound crossed the BBB. False negative (FN), when a compound showed impaired brain penetration, but not due to an efflux mechanism. False positive (FP) compounds are efflux transporter substrates that do not show impaired brain distribution. The sensitivity and specificity of the estimation of brain penetration along with MDR1 and BCRP activity were calculated from predictive performance metrics.

(c)   Calculations of predicted unbound brain-to-plasma partition coefficient ($K_{p,uu,brain}$)

First, NET ER was calculated and used for human predictions,

$$\text{NET ER} = \text{ER}_{(-\text{inh})} - \text{ER}_{(+\text{inh})} \tag{2}$$

where $\text{ER}_{(-\text{inh})}$ is ER in the absence of inhibitor, and $\text{ER}_{(+\text{inh})}$ is ER in the presence of inhibitor, for hMDR1 (ER_{MDR1}) or hBCRP (ER_{BCRP}).

Total ER comprising both hMDR1 and hBCRP activities is

$$\text{Total ER} = (\text{NET ER}_{\text{MDR1}}) + (\text{NET ER}_{\text{BCRP}}) + 1 \tag{3}$$

REF was calculated as follows [38]:

$$\text{REF} = \frac{\text{transporter abundance in brain capillaries}(\text{pmol/mg protein})}{\text{transporter abundance in transporter overexpressing cell line}(\text{pmol/mg protein})} \tag{4}$$

The data used for REF calculation are shown in Table S2.

Equation (3) modified with REF:

$$\text{Total ER}_{\text{REF corrected}} = (\text{NET ER}_{\text{MDR1}}) \times \text{REF}_{\text{MDR1}} + (\text{NET ER}_{\text{BCRP}}) \times \text{REF}_{\text{BCRP}} + 1 \tag{5}$$

Predicted $K_{p,uu,brain}$ calculated with four different methods, listed here (Equations (6)–(9)).

1. Predicted $K_{p,uu,brain}$ based on hMDR1 ER was calculated as follows:

$$K_{p,uu,brain} = \frac{1}{NET\ ER \times REF + 1} \tag{6}$$

2. Predicted $K_{p,uu,brain}$ calculated from total ER is as follows:

$$K_{p,uu,brain} = \frac{1}{Total\ ER} \tag{7}$$

3. Predicted $K_{p,uu,brain}$ calculated from total ER corrected with relative expression levels of MDR1 and BCRP is as follows:

$$K_{p,uu,brain} = \frac{1}{REF_{MDR1} \times (NET\ ER_{MDR1}) + REF_{BCRP} \times (NET\ ER_{BCRP}) + 1} \tag{8}$$

4. $K_{p,uu,brain}$ calculated with α and β scaling factors, with a modification of Equation (8), is as follows:

$$K_{p,uu,brain} = \frac{1}{\alpha \times (NET\ ER_{MDR1}) + \beta \times (NET\ ER_{BCRP}) + 1} \tag{9}$$

where α and β represent scaling factors determined with non-linear least squares regression using the Levenberg–Marquardt algorithm, detailed below.

(d) Estimation of α and β scaling factors for $K_{p,uu,brain}$ prediction

Non-linear least squares regression using the Levenberg–Marquardt algorithm was applied to estimate the α and β parameters of the model. The regression was performed in RStudio (version 2023.12.0) using the nls_multstart() function from the nls.multstart package [39], and the best fit was determined based on the best Akaike information criterion (AIC). Residual plots were used to assess the goodness of fit for each model (Figures S1 and S2).

The dataset of 55 compounds (atenolol, mannitol, sumatriptan have been excluded) was used for the first fit. On the residual plots (Figure S1), four compounds had extremely high residuals compared to the others (standardized residuals were 2 or higher). Investigating these outliers, we observed that their $K_{p,uu,brain}$ values were much higher than 1. Since the mathematical formula used for $K_{p,uu,brain}$ calculation is incapable of predicting $K_{p,uu,brain}$ values higher than 1.2, we excluded these compounds from the regression analysis. A new model was fitted omitting these data and the parameters were re-estimated accordingly. Since these new residual plots were appropriate (Figure S2), the re-estimated model parameters were accepted.

3. Results

3.1. Substrate Screen

3.1.1. Comparing the Functional Activity of hMDR1, rMDR1, mMDR1 and cyMDR1 Cell Lines

First, MDR1-overexpressing cells were fully validated using two prototypical MDR1 substrates, digoxin and quinidine, and assay parameters for substrate and inhibitor assesment were defined accordingly. For substrate screen 83 proprietary compounds, including both CNS+ and CNS− drugs, were selected, and in vitro ERs were assessed across hMDR1, rMDR1, mMDR1 and cyMDR1 as well as mock control cell lines (1 µM, 120 min, ± zosuquidar). Using the same conditions across substrates allowed for the proper comparison of ERs determined (Figure 1, Table S1). Among the 83 compounds, 47 were known hMDR1 substrates. Using an ER cut-off value of 2, according to the regulatory guidelines (ICH M12), we identified 47, 43, 40 and 36 substrates of hMDR1, rMDR1, mMDR1 and cyMDR1, respectively. The ER in mock cells for all compounds were close to unity. Famotidine was applied as a permeability control; its passive permeability was comparable across cell lines.

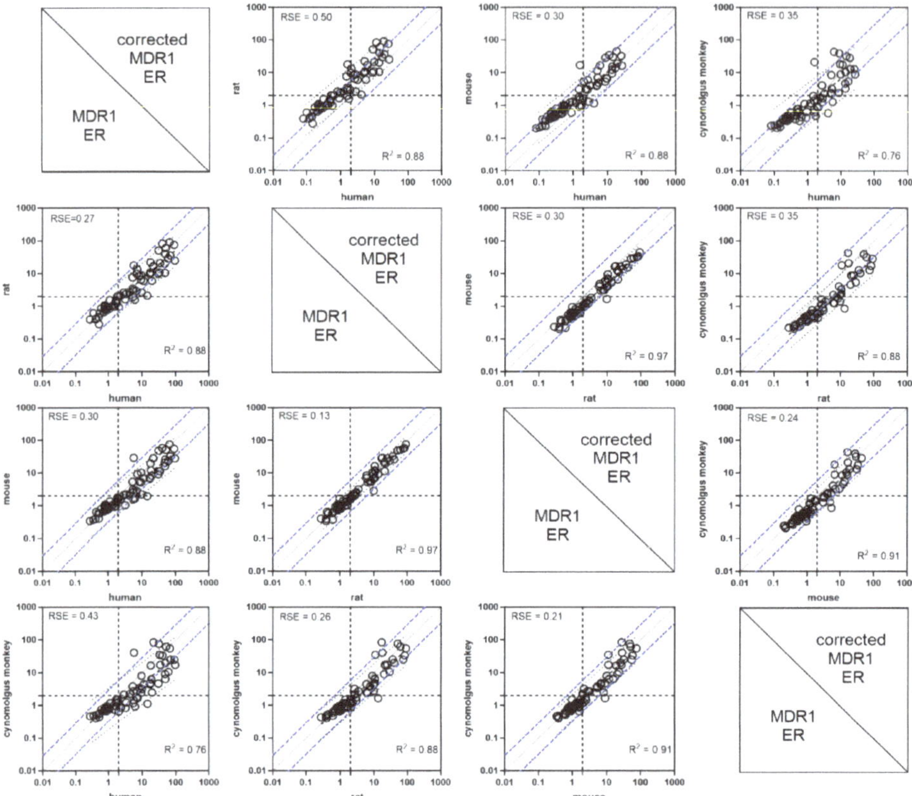

Figure 1. ERs across MDR1-expressing cell lines for a set of selected compounds were positively correlated. Graphs in the upper right triangle show transporter abundance-corrected ERs, while the lower-left triangle represents determined ERs. Solid lines show the line of identity. The densely dotted lines represent three-fold difference from the line of identity. ER values of 2 are used as the cut-off to classify MDR1 substrates (dotted horizontal and vertical lines).

MDR1 ERs calculated for this set of compounds across cell lines were positively correlated: rMDR1 vs. mMDR1 ($R^2 = 0.97$, $p < 0.001$); cyMDR1 vs. mMDR1 ($R^2 = 0.91$, $p < 0.001$); cyMDR1 vs. rMDR1 ($R^2 = 0.88$, $p < 0.001$); hMDR1 vs. rMDR1 ($R^2 = 0.88$, $p < 0.001$); hMDR1 vs. mMDR1 ($R^2 = 0.87$, $p < 0.001$). Surprisingly, the weakest correlation was seen between hMDR1 and cyMDR1 ($R^2 = 0.76$; $p < 0.001$) (Figure 1).

Next, to correct for transporter expression differences between the cell lines, ERs were corrected by transporter abundance (Table S2). Correlation analysis between the transporter abundance-corrected ERs was run between each pair of MDR1 orthologs. (Figure 1, upper right triangle). This correction changed the slope of the fitted line to deviate more from the line of identity. Of note, this analysis was run for all drugs including non-substrates and species-specific substrates. Remarkably, if we analyzed only shared substrates of orthologs, correction with transporter abundance improved the correlation (The data presented in this study are available on request from the corresponding author). To measure standard deviation from the line of identity, residual standard errors (RSE) were calculated.

Remarkably, the dispersion of ER values differs across the four orthologs, so we calculated interquartile ranges (IQR) for each dataset. Human MDR1 has, by far, the largest IQR (16.57), and hence, the widest dynamic range, suggesting the highest sensitivity for substrate recognition. Rat and mouse MDR1 data had similar interquartile ranges

(8.92 and 8.09, respectively), and cyMDR1 (4.85) showed the smallest dispersion of ER values. The human MDR1 cell line showed the highest MDR1 activity for most tested compounds, which aligns well with the highest transporter expression level. This shows that transporter protein abundance is not proportionate in all cases with the ER; still, human MDR1 proved to be the best in revealing weak MDR1 substrates. However, the ER of cimetidine and trimethoprim was the highest in rat, paclitaxel and ritonavir ER were higher in rat and mouse MDR1, daunorubicin was higher in cyMDR1, mitoxantrone and saquinavir ER were the lowest in hMDR1, and the ER of vinblastine and ondansetron were comparable in all cells, suggesting potential species differences.

3.1.2. Brain Permeability Classification Based on In Vitro ER

To identify the correlation between the in vivo rodent $K_{p,uu,brain}$ and ERs, 58 compounds with known $K_{p,uu,brain}$ were selected. The correlation between the $K_{p,uu,brain}$ in rodents and in vitro ERs is shown in Figure 2. An ER of 2 was used as a cut-off to classify MDR1 substrates and a $K_{p,uu,brain}$ of 0.3 was used as a threshold for CNS penetrant compounds. The resulting graphs are thus divided into four quadrants. Brain penetration ($K_{p,uu,brain}$) was used as the logical value true condition, with compounds having $K_{p,uu,brain} \leq 0.3$ considered positive. Q2 and Q3 comprise molecules with impaired brain penetration. Drugs in Q2 (true positives, TP) represent the subset of MDR1 substrates, where efflux activity likely limits brain penetration. Drugs in Q3 (false negatives, FN) have limited brain penetration but are unlikely to be MDR1 substrates. Drugs in Q1 (false positives, FP) are MDR1 substrates, but their brain penetration is not limited, probably due to other mechanisms, such as uptake or high passive permeability overwrite efflux, e.g., for trimethoprim and ondansetron [40,41]. Molecules in Q4 (true negatives, TN) can freely access the brain (Figure 2). Predictive performance metrics were calculated to show the efficacy of the established cell lines in BBB permeability classification (Table 1). Sensitivity refers to the ability of the model to correctly classify CNS-restricted molecules based on MDR1 substrate nature. We found that the human and rat MDR1 cell lines have the highest resolution to differentiate compounds with a sensitivity $\geq$80% (Figure 2A,B, respectively), while the sensitivity of the mouse and cynomolgus monkey MDR1 cell lines are lower; 71% and 65%, respectively (Figure 2C,D). Specificity was above 80% for all tested cell lines.

Subsequently, for a more reliable prediction of brain penetration, data generated in the hBCRP cell line were included in the analysis. hBCRP assay conditions and evaluations were identical to that of MDR1 cell lines, which allowed proper comparisons (Figure 2E,F). Total ER was calculated by combining ERs determined in BCRP and hMDR1 cell lines based on Equation (3) [42], with an ER cut-off of 3. The predictive performance of this total ER scored better than hMDR1 alone with respect to sensitivity, as out of the five false negatives, only three remained in this category—atenolol, mannitol and sumatriptan—revealing low passive permeability and subsequent limited brain penetration. Entacapone and indomethacin were first identified as false negatives using the hMDR1 screen alone, but as they are BCRP substrates, here they turned into true positives. However, the specificity of the total ER prediction is lower (79%), resulting in five false positives (ondansetron, trimethoprim, warfarin, etoricoxib, guanabenz). Eventually, total ERs were also assessed with the Relative Expression Factor (REF) (Equation (4); Figure 2E,F). REF is an in vitro and in vivo correlation scaler, calculated from transporter protein abundances in brain microvessels versus overexpressing cell lines [4,5]. Remarkably, metrics calculated without REF showed better prediction in all aspects than those with transporter abundance correction. Taken together, among all approaches, the data clearly show that total ER without REF correction had the highest sensitivity (91%) in identifying CNS compounds.

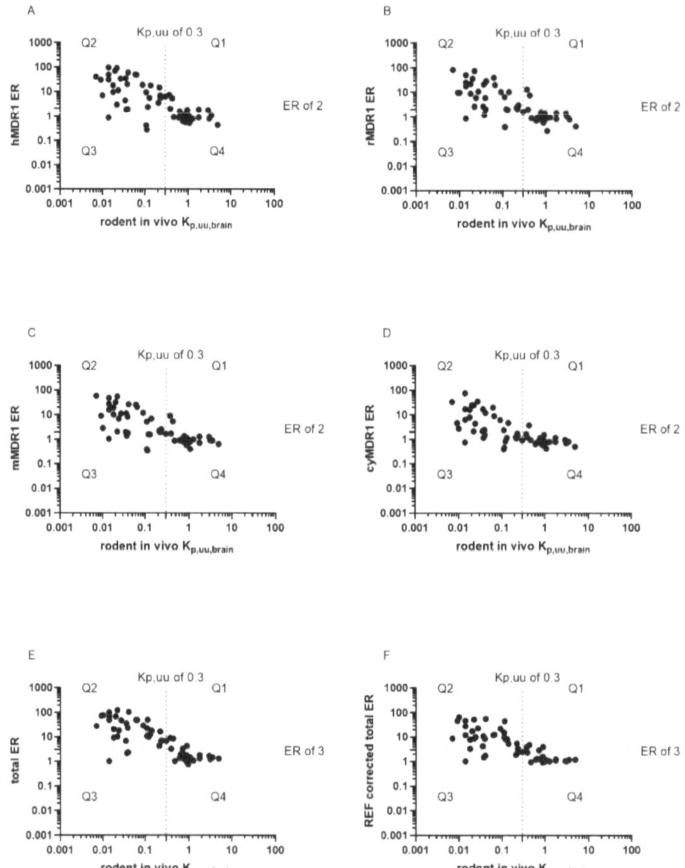

Figure 2. Correlation of rodent $K_{p,uu,brain}$ with ER for 58 selected compounds. ER values were determined in hMDR1 (**A**), rMDR1 (**B**), mMDR1 (**C**) and cyMDR1 (**D**) cell lines or calculated as total ER (**E**) and REF-corrected total ER (**F**) of hMDR1 and hBCRP. Cut-offs of 2 and 3 were used for MDR1 ERs and for total ERs, respectively. A threshold of 0.3 was used for $K_{p,uu,brain}$. The quadrants illustrate true negative (Q4), true positive (Q2), false positive (Q1) and false negative (Q3) predictions.

Table 1. Predictive performance metrics (ER cut-off 2 or 3; in vivo $K_{p,uu,brain}$ cut-off 0.3).

	hMDR1	rMDR1	mMDR1	cyMDR1	Total ER	REF-Corrected Total ER
False Negative (n)	5	6	10	12	3	6
False Positive (n)	2	3	2	1	5	3
True Negative (n)	22	21	22	23	19	21
True Positive (n)	29	28	24	22	31	28
PPV	94%	90%	92%	96%	86%	90%
NPV	81%	78%	69%	66%	86%	78%
Sensitivity	85%	82%	71%	65%	91%	82%
Specificity	92%	88%	92%	96%	79%	88%

Table 1. *Cont.*

	hMDR1	rMDR1	mMDR1	cyMDR1	Total ER	REF-Corrected Total ER
Accuracy	88%	84%	79%	78%	86%	84%
FN Rate	15%	18%	29%	35%	9%	18%

Positive predictive values (PPV) were calculated by dividing the number of true positives by the total number of true and false positives. Similarly, negative predictive values (NPV) were calculated as the number of true negatives divided by the total number of true and false negatives. Assay sensitivity was calculated by dividing true positives by the total number of false negatives and true positives, whereas specificity was determined as true negatives divided by the total number of false positives and true negatives. Assay accuracy was calculated by dividing true positives and true negatives by the total number of studies.

3.1.3. Comparison of In Vitro Predicted $K_{p,uu,brain}$ and In Vivo Rodent $K_{p,uu,brain}$

The quantitative prediction of BBB penetration from physicochemical parameters is challenging even with the use of in silico models, and needs further refinement, e.g., by using in vitro efflux transporter data [10,43,44]. Numerous methods have been published on $K_{p,uu,brain}$ prediction based on in vitro data, of which MDR1 ER is indispensable [10,17,28–30]. It needs to be emphasized that the goodness of predictions depends on in vitro assay properties [10,45,46]. Therefore, we were eager to see to what extent our cell lines would predict in vivo brain penetration data.

For this quantitative prediction, we calculated in vitro $K_{p,uu,brain}$ with four different methods using ERs obtained in our hMDR1 and hBCRP cell lines. First, we calculated $K_{p,uu,brain}$ from hMDR1 ERs according to Equation (5). Next, $K_{p,uu,brain}$ values were calculated from either total ERs or REF-corrected total ERs according to Equations (6) and (7), to account for the difference between in vitro in vivo transporter abundance. Finally, to further improve the prediction performance of our data and achieve the best fit to in vivo $K_{p,uu,brain}$, we used selected compounds to define α and β factors according to Equation (9) [28,29]. Three compounds (atenolol, mannitol and sumatriptan) with relatively low passive permeability and ER < 2 were excluded from the correlation, since one known limitation of these simple prediction models is that they do not account for the passive permeability of substances. BDDCS classification can be incorporated into the pipeline in a multi-step approach, as published previously [46]. The calculated in vitro $K_{p,uu,brain}$ were compared against the corresponding in vivo rodent $K_{p,uu,brain}$ data. The least promising correlation was seen using only the human MDR1 data (Figure 3A), where R^2 was 0.61. The prediction was more accurate using both hMDR1 and hBCRP-derived total ER, as it improved the fitting to $R^2 = 0.73$ (Figure 3B). The REF-corrected prediction was similar, with $R^2 = 0.75$ (Figure 3C). The strongest correlation was observed between the predicted and in vivo $K_{p,uu,brain}$ using the estimated α and β factors in the equation (Table 2), with R^2 being 0.83 (Figure 3D,E).

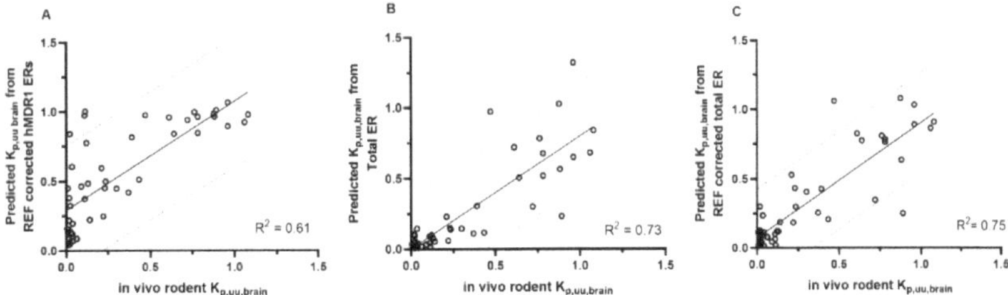

Figure 3. *Cont.*

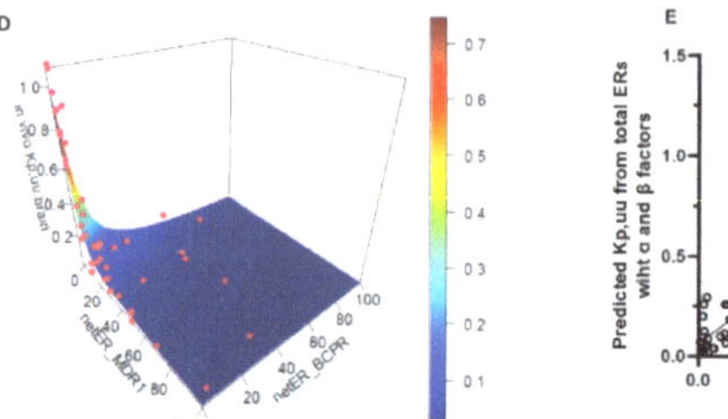

Figure 3. Prediction of $K_{p,uu,brain}$ using ERs of hMDR1 and hBCRP cell lines. The solid lines represent best fit from simple linear regression analysis, while the dotted lines show the 95% prediction band of the best fit line. Atenolol, mannitol, sumatriptan and compounds with in vivo $K_{p,uu,brain} > 1.2$ were excluded from analysis. (**A**) Correlation between in vivo rodent $K_{p,uu,brain}$ and predicted $K_{p,uu,brain}$ from REF-corrected hMDR1 ERs. (**B**) Correlation between in vivo rodent $K_{p,uu,brain}$ and predicted $K_{p,uu,brain}$ from total ER. (**C**) Correlation between in vivo rodent $K_{p,uu,brain}$ and predicted $K_{p,uu,brain}$ from REF-corrected total ER. (**D**) Visual representation of data used for the estimation of α and β parameters. (**E**) Correlation between $K_{p,uu,brain}$ calculated according to Equation (9) with the estimated α and β scaling factors and in vivo $K_{p,uu,brain}$.

Table 2. Estimated α and β parameters with confidence intervals using Equation (9).

Parameter	Estimate	Standard Error	Lower 95% CI	Upper 95% CI
α	0.52	0.103	0.315	0.731
β	0.29	0.080	0.126	0.450

3.1.4. Species- and Substrate-Specific Differences in IC_{50} Values

First, kinetic parameters in hMDR1, rMDR1, mMDR1 and cyMDR1 cell lines were determined for both substrates, digoxin and quinidine, and accordingly the final concentrations of digoxin and quinidine were set as 1 µM, except for cyMDR1, for which a lower Km was estimated, necessitating a quinidine concentration of 0.1 µM. Inhibition potencies were assessed for 21 known MDR1 interactors (Table S3), including the two probe substrates, digoxin and quinidine, and species- and substrate-specific differences were analyzed. IC_{50} values were calculated from the ERs [47] determined at seven concentrations and normalized to solvent control. To investigate whether observed differences among species are due to different abundances of MDR1 proteins, IC_{50} values were normalized, and protein-corrected and non-corrected IC_{50} values were systematically compared. Of the 21 compounds, 18 inhibited the MDR1-mediated transport of probe substrates in a consistent and reproducible manner, and were analyzed; ritonavir and talinolol had no inhibitory effect on either substrate. Also, consistently with previous results, digoxin did not inhibit the transport of quinidine. In cases where the IC_{50} values were given as "greater than" due to solubility issues, the highest tested concentration was used as a surrogate for inhibition potential comparison. The IC_{50} values obtained cover three orders of magnitude demonstrating a wide dynamic range for the assays.

The species-specific MDR1 transport activity for digoxin and quinidine were assessed by ERs calculated and averaged from all the vehicle controls of the inhibition studies. Interestingly, the efflux activity of human MDR1 toward quinidine was significantly higher than for digoxin (47.3 ± 16.6 vs. 17.7 ± 5.9), while such substrate-specific differences were not seen in the other cell lines (cyMDR1, 10.7 ± 1.4 vs. 13.3 ± 4.4; rMDR1, 16.7 ± 4.8 vs. 17.2 ± 4.9; mMDR1, 15.9 ± 3.8 vs. 13.9 ± 3.6).

The correlation between IC_{50} values determined for each ortholog for digoxin and quinidine are shown in Figure 4A,B, respectively. Based on the linear regression, the strongest correlation ($R^2 = 0.78$) was found between rat and cynomolgus MDR1 IC_{50} values for digoxin. To quantify differences, residual standard errors (RSE) were calculated against the line of identity, where lower RSE values correspond to smaller differences. The lowest RSE (0.47) was found between mouse and rat MDR1 with digoxin as the substrate. Interestingly, while IC_{50} values from human and cynomolgus MDR1 correlate well using quinidine, this correlation is weaker with digoxin. Overall, data indicate that the inhibition of digoxin resulted in higher IC_{50} values for human MDR1 than for the other investigated orthologs. No such pattern was observed using quinidine as a substrate. Correction with the transporter abundance has no effect on R^2 but on the slope of the regression line. Protein correction resulted in lower RSE for all comparisons using digoxin, in contrast to other orthologs. Using quinidine as a probe, correction with transporter abundance resulted in higher RSE values in all cases. Comparing IC_{50} values between digoxin and quinidine for each MDR1 ortholog highlighted substrate-specific differences, especially for hMDR1 (Figure 5), where digoxin transport inhibition resulted in much higher IC_{50} values than that of quinidine. This tendency was also observed for mouse and rat MDR1, but to a lesser extent. The IC_{50} values determined in cyMDR1 cells correlated with an R^2 of 0.77.

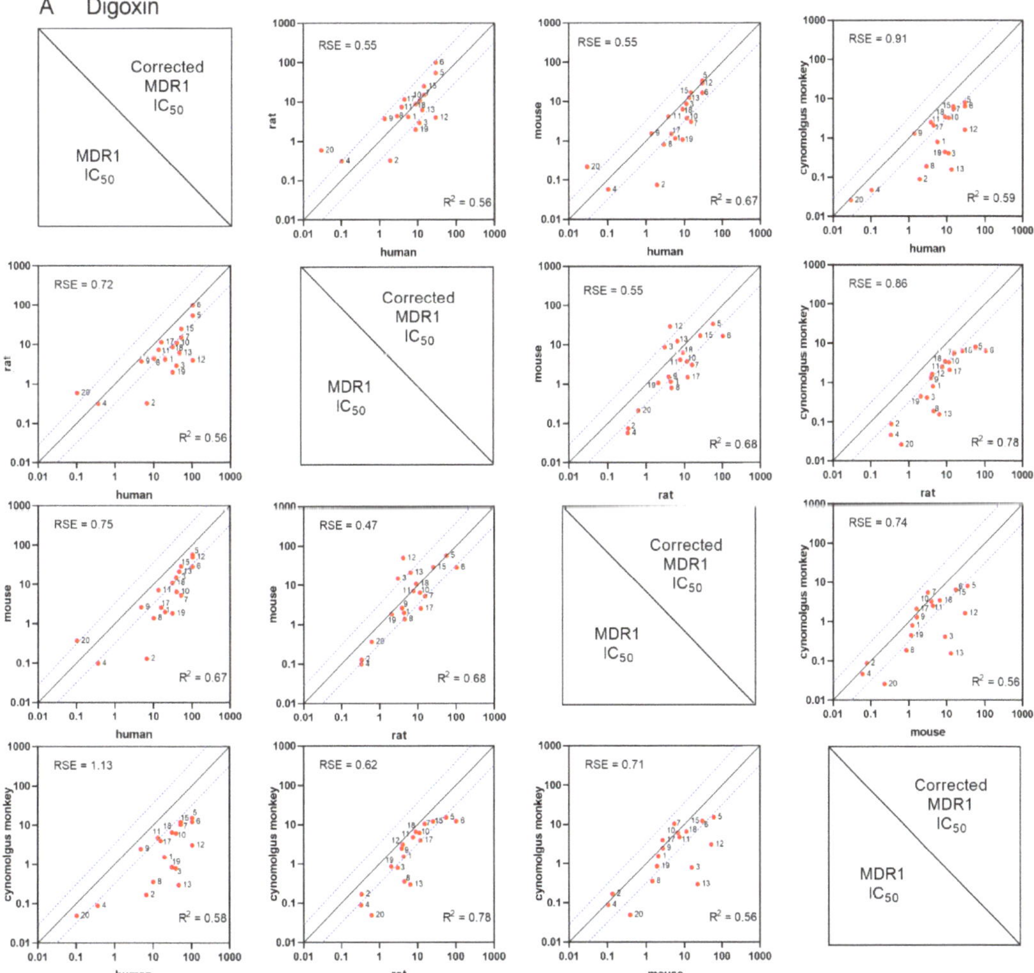

Figure 4. *Cont.*

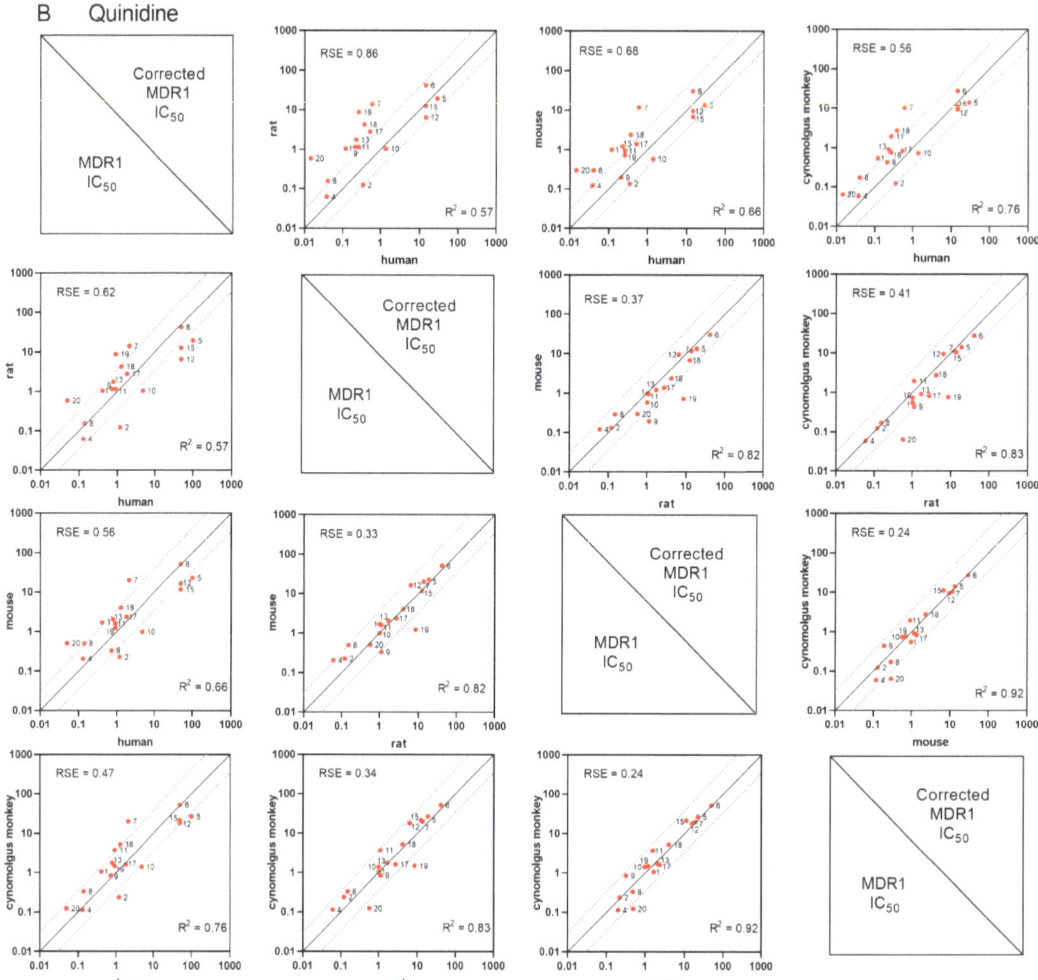

Figure 4. Correlation of IC_{50} values from human, rat, mouse or cynomolgus monkey MDR1 using digoxin (**A**) and quinidine (**B**) as probe substrates. R^2 values have been determined for each pair using the logarithmically transformed IC_{50} values. Considering all correlations, all *p*-values of Pearson correlation were lower than 0.0079. RSE values were calculated against the line of identity and are shown on each graph. Red dots represent the means of at least two independent experiments, each with three technical parallels. Densely dotted lines show the 3-fold difference from the line of identity. Compounds are marked with numbers (Table S3). Graphs in the upper right triangle show transporter abundance-corrected IC_{50}, while the lower left triangle represents uncorrected IC_{50} values.

Next, we examined trends in IC_{50} values across compounds between orthologs. A $\geq$3-fold ratio between any two IC_{50} values was considered to be a real difference. The same set of IC_{50} comparisons were run after normalization for transporter expression. For elacridar, ketoconazole and zosuquidar IC_{50}s are in the same range for all MDR1 orthologs using either digoxin or quinidine, independent of transporter abundance. Similarly, zosuquidar's IC_{50} values are comparable, with the highest potency for human and cynomolgus MDR1 in both substrates.

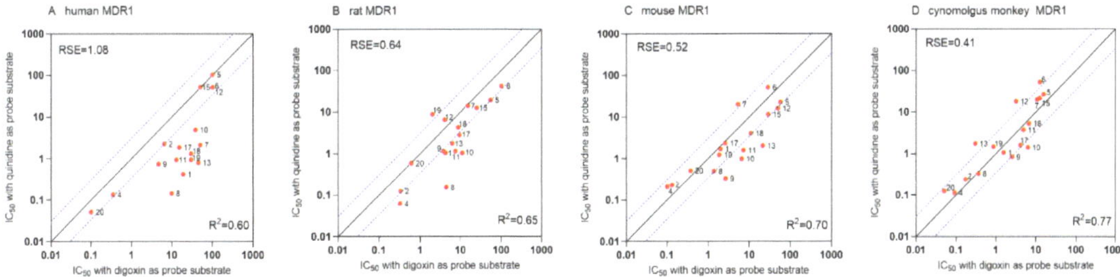

Figure 5. Correlation of IC$_{50}$ values determined with either digoxin or quinidine as probe substrates in human (**A**), rat (**B**), mouse (**C**) or cynomolgus monkey (**D**) MDR1 cells. R^2 values have been determined for each pair using the logarithmically transformed IC$_{50}$ values. Considering all correlations, all p-values of Pearson correlation were lower than 0.0030. RSE values were calculated against the line of identity and are shown on each graph. Red dots represent the means of at least two independent experiments, each with three technical parallels. Densely dotted lines show the 3-fold difference from the line of identity. Compounds are marked with numbers (Table S3).

For some compounds, such as the high-affinity MDR1 substrate verapamil, there were notable differences in IC$_{50}$s for digoxin across the cell lines. The verapamil IC$_{50}$ in the hMDR1 cell line was 15–35-fold higher than in rat, mouse and cyMDR1 cell lines. With quinidine, there was no difference in verapamil IC$_{50}$ values, except for rMDR1, where the IC$_{50}$ value was 6–9 times higher than for other orthologs. While protein-corrected IC$_{50}$ values resulted in even more pronounced differences for quinidine, with digoxin, these differences diminished. Carvedilol, isradipine and tolvaptan inhibited digoxin and quinidine transport with the same pattern as verapamil. Digoxin transport inhibition by CSA yielded >20-fold (20–50-fold) higher IC$_{50}$ values for hMDR1 than for other MDR1 orthologs. This difference was within 5–10-fold when using quinidine as the substrate. When comparing CSA IC$_{50}$ values across cell lines, the differences between the two substrates were less than 3-fold, except for hMDR1, which showed a 5-fold difference. For high-affinity human MDR1 substrate loperamide, higher IC$_{50}$ values were measured for both substrates in human MDR1. Correction with transporter abundance diminished these differences. IC$_{50}$ values for human, rat and mouse MDR1-mediated digoxin or quinidine transport inhibition could not be calculated for etoricoxib (hMDR1), felodipine (hMDR1, mMDR1, rMDR1), saquinavir (hMDR1) or isradipine (hMDR1).

4. Discussion

This report is the first to investigate MDR1 activity in human as well as three preclinical species in the context of CNS penetration using transporter-overexpressing endogenous MDR1-knockout MDCKII cell lines, thus allowing side-by-side study of individual MDR1 orthologs. Despite recent advances in in silico and in vitro models, K$_{p,uu,brain}$ prediction is still challenging, making preclinical studies unavoidable. Although in silico models using molecular descriptors and in vitro parameters exist, their predictive performance is moderate [44]. One challenge is to account for the role of transporters at the BBB [43]; further improvement of these models requires a better understanding of transporter–drug interactions and their impact on BBB penetration [10,44,48]. Inherent differences in cell lines lead to confounding results [10,45,46], likely because parental cells differ in tightness and transporter abundance; factors that make MDCK cells and higher transporter expression preferable [10,29,49–53]. Our cell lines fulfill these requirements, in that endogenous MDR1 activity is abolished, and relatively high MDR1 expression is achieved.

For the majority of the 83 compounds analyzed, comparable ERs were found across cell lines, with no notable outliers. The best correlation was observed between rMDR1 and mMDR1 (R^2 = 0.96), in accordance with the published data [54]. Correlations between human vs. rat and human vs. mouse MDR1 (R^2 = 0.88 for both) are comparable to previous results [49]. The weakest correlation we found, between cynomolgus and human MDR1, is still relatively good ($R^2 = 0.75$), confirming previous findings that overall substrate recognition is conserved across species [15,49,55–58]. A few studies, however, showed moderate differences between human, rat and mouse [14,55,59–61]. Examples include differences in substrate susceptibility between human and mouse for phenytoin and levetiracetam [16,62] or a GSK discovery compound being a rat but not human MDR1 substrate [10].

Our results indicate that the dynamic range is different across cell lines; therefore, resolution efficiency for weaker substrates also varies [63,64]. The largest dynamic range and transporter abundance were found in hMDR1 cells [62,63]. However, there is no apparent correlation between rank order in MDR1 quantity and resolution efficiency in the other three cell lines (Table S1). This was apparent after normalization for transporter abundance, which changed the slope of correlation as well as deviation from the line of identity [54,62], highlighting that although transport rate is linked to MDR1 abundance, the kinetic parameters need to be taken into account in in vitro measurements. A few compounds deviated from the correlation line, suggesting potential species differences. Since the usefulness of an in vitro system depends on its predictive performance and translatability [53], ERs in our cells were compared to rodent $K_{p,uu,brain}$ data. The best sensitivity (85%) was achieved with the hMDR1 cells, likely because it has the highest transporter expression, in agreement with studies comparing the two widely used MDCK-MDR1 cell lines, NIH and Borst [26,42,63,65]. Our hMDR1 cells excel because of (a) the lack of endogenous MDR1 expression, and (b) their high human transporter expression and corresponding ability to recognize weak substrates [29,42,66,67]. Using total ERs derived from MDR1 and BCRP cells further increased predictive power, even compared to other cell lines evaluated for CNS penetration classification [42]. This holds up even when compared to double-transfected cells, likely due to the relatively high but comparable expression level of the two transporters in our cell lines [27,68–70]. It follows that a combination screening of MDR1 and BCRP can improve prediction performance, but dual overexpression at high levels in the same cell is technically challenging. In addition to BBB penetration classification, in vitro data can be used for the quantitative prediction of $K_{p,uu,brain}$ [28,29,42,71]. We compared calculations using hMDR1 and total ERs, REF-corrected total ERs, and a mathematically parametrized calculation. The predictive performance was increased accordingly, with the best correlation ($R^2 = 0.83$) between in vivo and predicted $K_{p,uu,brain}$ achieved with scaling factors.

Studies on species-related differences in MDR1 inhibition are scarce [15,72,73]; therefore, we aimed to further investigate this phenomenon in our MDR1 cells. The selection of substrates is a key consideration when establishing an in vitro inhibition assay [74,75]. As almost the entire inner surface of MDR1 can interact with the ligand, it is difficult to identify a genuine affinity site. Digoxin is the most common clinical probe for MDR1 DDI studies, and a sensitive in vitro probe with high ER but relatively low affinity towards MDR1 (ICH, M12), [76–79]. As digoxin is polar, active uptake and partitioning into the membrane might be rate-limiting for its interaction with MDR1 [80]; its IC_{50} therefore represents multiple transport processes [76]. In contrast, quinidine is amphiphilic with moderate permeability and higher affinity [81], and an MDR1 inhibitor as well [82,83]. Quinidine is thought to be transported across the cells solely by MDR1 and passive permeability [84,85]. In our cells, higher hMDR1 expression likely contributed to the larger dynamic range; still, digoxin—but not quinidine—ERs were comparable across cell lines, with quinidine ER values 3–4-fold higher in hMDR1 than in orthologs. This discrepancy was also seen when comparing IC_{50}s across orthologs and substrates. IC_{50} values of non- human orthologs with digoxin were generally much lower compared to those of hMDR1, while this was not typical with quinidine. Although hMDR1 featured higher digoxin IC_{50}s, digoxin DDI risk was correctly predicted using this cell line when implementing the regulatory agency-recommended static model [35]. Despite several notable differences in the absolute IC_{50}s between the two substrates, rank order profiles were generally similar. Testing the inhibition potential of the two substrates against each other resulted in low IC_{50} for cynomolgus and rat, and high IC_{50} values for human and mouse MDR1 when using quinidine as the inhibitor [72]. However, digoxin did not inhibit quinidine transport, confirming its classification as a substrate but not an inhibitor of MDR1 [86–88]. All inhibitors interacted with the MDR1, except talinolol and ritonavir. Although talinolol is a known MDR1 substrate [89], in accordance with others [35], our results support the lack of clinical interaction with digoxin [90]. Despite its known effect on digoxin pharmacokinetics [91,92], ritonavir's low permeability prevents studying its interaction with MDR1 in vectorial transport assays [93]. However, ritonavir DDI potential was shown in another assay [94]. To reveal the IC_{50}'s dependence on substrate, transporter abundance or orthologs, patterns in IC_{50}s were recognized. The first group of inhibitors, comprising elacridar, ketoconazole and zosuquidar, had comparable IC_{50}s in all MDR1 assays with both substrates, independent of protein abundance. This might be explained by physicochemical properties and the nature of MDR1 interaction, since elacridar and zosuquidar can uncouple ATPase activity from transport [95–99], while ketoconazole is thought to inhibit MDR1 by binding to a central modulatory site [100]. In a second group (etoricoxib, loperamide, nitrendipine, saquinavir and CSA), IC_{50}s are proportional to transporter abundance, independently of substrate [101]. Thus, transporter abundance-corrected IC_{50}s correlate better. As an example, CSA IC_{50}s are proportional to transporter expression, but less dependent on the substrate, consistent with previous findings that CSA [102] is a substrate of MDR1, suggesting a competitive inhibitory mechanism.

In the third group (tolvaptan, isradipine, carvedilol and verapamil), dependence on transporter abundance was apparent only with digoxin and not with quinidine, suggesting different inhibitory mechanisms with the two substrates. Differences in IC_{50}s across cell lines with varying MDR1 expression were reported for verapamil [101]. Studies support the simultaneous binding of verapamil and digoxin to MDR1 [78,103], suggesting competitive inhibition. In contrast, verapamil likely displays noncompetitive interaction with quinidine [104].

The mechanism of MDR1 inhibition is still poorly understood, involving multiple binding sites and inhibitory mechanisms. Although our study does not provide information on the mechanism of inhibition per se, it gives new insights into substrate and transporter abundance-dependent inhibition of MDR1. We confirmed that the inhibitory potency of many drugs is highly substrate-dependent, resulting in generally lower IC_{50} with quinidine, despite the good overlap in rank order of inhibitors. Rank order was similar between species as well; however, differences of an order of magnitude were observed in IC_{50}s. In some cases, IC_{50} difference was resolved by transporter abundance correction, but this correlation is not always proportional and dependent on substrate-inhibitor interaction, suggesting that the correlation between IC_{50} and MDR1 abundance is dependent on the mode of inhibition.

In summary, species differences in DDI potential can be investigated in this system, but transporter abundance and mode of inhibition need to be considered. Our in vitro assays are ready to be integrated into CNS drug discovery screening programs to predict BBB penetration or reveal potential species differences in transporter susceptibility, improving the translatability of in vivo preclinical data.

Supplementary Materials: The following supporting information can be downloaded at: https://www.mdpi.com/article/10.3390/pharmaceutics16060736/s1, Figure S1: Residual plots of the fitted model (Equation (9)) for 55 tested compounds according to Table S1; Figure S2: Residual plots of the fitted model (Equation (9)) with exclusion of compounds with $K_{p,uu,brain}$ >1.2 according to Table S1; Table S1: ERs across the six cell lines for 83 compounds (1 μM, 120 min); Table S2: Summary of transporter protein abundance in isolated brain capillaries (mouse, rat, cynomolgus monkey and human) in Abcb1KO-MDCKII-MDR1 and in MDCKII-BCRP cell lines; Table S3: IC_{50} values when investigating inhibition of hMDR1, rMDR1, mMDR1 and cyMDR1 transport of either digoxin or quinidine by a range of marketed compounds

Author Contributions: Conceptualization, C.T.-A. and Z.G.; methodology, C.T.-A., Z.G., E.S., N.S., Z.U., O.C., G.K. and Z.S.; validation, C.T.-A., E.S. and N.S.; investigation, C.T.-A., E.S., N.S., Z.U., O.C. and G.K.; data curation, C.T.-A., E.S., N.S., Z.U., O.C., Z.S. and É.P.; writing—original draft preparation, C.T.-A., Z.G., E.S. and N.S.; writing—review and editing, C.T.-A., Z.G., E.S. and N.S.; visualization, E.S., N.S. and É.P.; supervision, Z.G. All authors have read and agreed to the published version of the manuscript.

Funding: Project no. 2023-2.1.2-KDP-2023-00016 has been implemented with the support provided by the Ministry of Culture and Innovation of Hungary from the National Research, Development and Innovation Fund, financed under the KDP-2023 funding scheme.

Institutional Review Board Statement: Not applicable.

Informed Consent Statement: Not applicable.

Data Availability Statement: The data presented in this study are available on request from the corresponding author. The data are not publicly available due to company participation.

Acknowledgments: Special thanks to Gábor Nagy for his help in reviewing and revising the manuscript. We would like to thank Flóra Gréta Petényi for her assistance in the systematic literature review.

Conflicts of Interest: Charles River Laboratories Hungary, as the employer of some of the authors, develops and commercializes reagents and assays to study membrane transporters. The authors declare no conflicts of interest.

References

1. Murata, Y.; Neuhoff, S.; Rostami-Hodjegan, A.; Takita, H.; Al-Majdoub, Z.M.; Ogungbenro, K. In Vitro to In Vivo Extrapolation Linked to Physiologically Based Pharmacokinetic Models for Assessing the Brain Drug Disposition. *AAPS J.* **2022**, *24*, 28. [CrossRef]
2. Jagtiani, E.; Yeolekar, M.; Naik, S.; Patravale, V. In vitro blood brain barrier models: An overview. *J. Control. Release* **2022**, *343*, 13–30. [CrossRef]
3. Kamiie, J.; Ohtsuki, S.; Iwase, R.; Ohmine, K.; Katsukura, Y.; Yanai, K.; Sekine, Y.; Uchida, Y.; Ito, S.; Terasaki, T. Quantitative atlas of membrane transporter proteins: Development and application of a highly sensitive simultaneous LC/MS/MS method combined with novel in-silico peptide selection criteria. *Pharm. Res.* **2008**, *25*, 1469–1483. [CrossRef] [PubMed]
4. Uchida, Y.; Ohtsuki, S.; Katsukura, Y.; Ikeda, C.; Suzuki, T.; Kamiie, J.; Terasaki, T. Quantitative targeted absolute proteomics of human blood-brain barrier transporters and receptors. *J. Neurochem.* **2011**, *117*, 333–345. [CrossRef]
5. Ito, K.; Uchida, Y.; Ohtsuki, S.; Aizawa, S.; Kawakami, H.; Katsukura, Y.; Kamiie, J.; Terasaki, T. Quantitative membrane protein expression at the blood-brain barrier of adult and younger cynomolgus monkeys. *J. Pharm. Sci.* **2011**, *100*, 3939–3950. [CrossRef]
6. Hoshi, Y.; Uchida, Y.; Tachikawa, M.; Inoue, T.; Ohtsuki, S.; Terasaki, T. Quantitative atlas of blood-brain barrier transporters, receptors, and tight junction proteins in rats and common marmoset. *J. Pharm. Sci.* **2013**, *102*, 3343–3355. [CrossRef]
7. Shawahna, R.; Uchida, Y.; Declèves, X.; Ohtsuki, S.; Yousif, S.; Dauchy, S.; Jacob, A.; Chassoux, F.; Daumas-Duport, C.; Couraud, P.-O.; et al. Transcriptomic and quantitative proteomic analysis of transporters and drug metabolizing enzymes in freshly isolated human brain microvessels. *Mol. Pharm.* **2011**, *8*, 1332–1341. [CrossRef]
8. Uchida, Y.; Tachikawa, M.; Obuchi, W.; Hoshi, Y.; Tomioka, Y.; Ohtsuki, S.; Terasaki, T. A study protocol for quantitative targeted absolute proteomics (QTAP) by LC-MS/MS: Application for inter-strain differences in protein expression levels of transporters, receptors, claudin-5, and marker proteins at the blood-brain barrier in ddY, FVB, an. *Fluids Barriers CNS* **2013**, *10*, 21. [CrossRef] [PubMed]
9. Braun, C.; Sakamoto, A.; Fuchs, H.; Ishiguro, N.; Suzuki, S.; Cui, Y.; Klinder, K.; Watanabe, M.; Terasaki, T.; Sauer, A. Quantification of Transporter and Receptor Proteins in Dog Brain Capillaries and Choroid Plexus: Relevance for the Distribution in Brain and CSF of Selected BCRP and P-gp Substrates. *Mol. Pharm.* **2017**, *14*, 3436–3447. [CrossRef] [PubMed]
10. Liu, H.; Dong, K.; Zhang, W.; Summerfield, S.G.; Terstappen, G.C. Prediction of brain:blood unbound concentration ratios in CNS drug discovery employing in silico and in vitro model systems. *Drug Discov. Today* **2018**, *23*, 1357–1372. [CrossRef]
11. Sarkadi, B.; Homolya, L.; Szakács, G.; Váradi, A. Human multidrug resistance ABCB and ABCG transporters: Participation in a chemoimmunity defense system. *Physiol. Rev.* **2006**, *86*, 1179–1236. [CrossRef]
12. Shawahna, R.; Decleves, X.; Scherrmann, J.-M. Hurdles with using in vitro models to predict human blood-brain barrier drug permeability: A special focus on transporters and metabolizing enzymes. *Curr. Drug Metab.* **2013**, *14*, 120–136. [CrossRef]
13. Shen, H.; Yang, Z.; Rodrigues, A.D. Cynomolgus Monkey as an Emerging Animal Model to Study Drug Transporters: In Vitro, In Vivo, In Vitro-to-In Vivo Translation. *Drug Metab. Dispos.* **2022**, *50*, 299–319. [CrossRef]
14. Katoh, M.; Suzuyama, N.; Takeuchi, T.; Yoshitomi, S.; Asahi, S.; Yokoi, T. Kinetic analyses for species differences in P-glycoprotein-mediated drug transport. *J. Pharm. Sci.* **2006**, *95*, 2673–2683. [CrossRef]

15. Chu, X.; Bleasby, K.; Evers, R. Species differences in drug transporters and implications for translating preclinical findings to humans. *Expert Opin. Drug Metab. Toxicol.* **2013**, *9*, 237–252. [CrossRef]
16. Baltes, S.; Gastens, A.M.; Fedrowitz, M.; Potschka, H.; Kaever, V.; Löscher, W. Differences in the transport of the antiepileptic drugs phenytoin, levetiracetam and carbamazepine by human and mouse P-glycoprotein. *Neuropharmacology* **2007**, *52*, 333–346. [CrossRef]
17. Trapa, P.E.; Troutman, M.D.; Lau, T.Y.; Wager, T.T.; Maurer, T.S.; Patel, N.C.; West, M.A.; Umland, J.P.; Carlo, A.A.; Feng, B.; et al. In Vitro-In Vivo Extrapolation of Key Transporter Activity at the Blood-Brain Barrier. *Drug Metab. Dispos.* **2019**, *47*, 405–411. [CrossRef]
18. Al Feteisi, H.; Al-Majdoub, Z.M.; Achour, B.; Couto, N.; Rostami-Hodjegan, A.; Barber, J. Identification and quantification of blood-brain barrier transporters in isolated rat brain microvessels. *J. Neurochem.* **2018**, *146*, 670–685. [CrossRef]
19. Al-Majdoub, Z.M.; Al Feteisi, H.; Achour, B.; Warwood, S.; Neuhoff, S.; Rostami-Hodjegan, A.; Barber, J. Proteomic Quantification of Human Blood-Brain Barrier SLC and ABC Transporters in Healthy Individuals and Dementia Patients. *Mol. Pharm.* **2019**, *16*, 1220–1233. [CrossRef]
20. Visser, S.A.G.; Wolters, F.L.C.; Gubbens-Stibbe, J.M.; Tukker, E.; Van Der Graaf, P.H.; Peletier, L.A.; Danhof, M. Mechanism-based pharmacokinetic/pharmacodynamic modeling of the electroencephalogram effects of GABAA receptor modulators: In vitro-in vivo correlations. *J. Pharmacol. Exp. Ther.* **2003**, *304*, 88–101. [CrossRef]
21. Kalvass, J.C.; Olson, E.R.; Cassidy, M.P.; Selley, D.E.; Pollack, G.M. Pharmacokinetics and pharmacodynamics of seven opioids in P-glycoprotein-competent mice: Assessment of unbound brain EC50,u and correlation of in vitro, preclinical, and clinical data. *J. Pharmacol. Exp. Ther.* **2007**, *323*, 346–355. [CrossRef] [PubMed]
22. Liu, X.; Vilenski, O.; Kwan, J.; Apparsundaram, S.; Weikert, R. Unbound brain concentration determines receptor occupancy: A correlation of drug concentration and brain serotonin and dopamine reuptake transporter occupancy for eighteen compounds in rats. *Drug Metab. Dispos.* **2009**, *37*, 1548–1556. [CrossRef] [PubMed]
23. Watson, J.; Wright, S.; Lucas, A.; Clarke, K.L.; Viggers, J.; Cheetham, S.; Jeffrey, P.; Porter, R.; Read, K.D. Receptor occupancy and brain free fraction. *Drug Metab. Dispos.* **2009**, *37*, 753–760. [CrossRef] [PubMed]
24. Bundgaard, C.; Sveigaard, C.; Brennum, L.T.; Stensbøl, T.B. Associating in vitro target binding and in vivo CNS occupancy of serotonin reuptake inhibitors in rats: The role of free drug concentrations. *Xenobiotica* **2012**, *42*, 256–265. [CrossRef] [PubMed]
25. Hammarlund-Udenaes, M.; Bredberg, U.; Fridén, M. Methodologies to assess brain drug delivery in lead optimization. *Curr. Top. Med. Chem.* **2009**, *9*, 148–162. [CrossRef] [PubMed]
26. Feng, B.; West, M.; Patel, N.C.; Wager, T.; Hou, X.; Johnson, J.; Tremaine, L.; Liras, J. Validation of Human MDR1-MDCK and BCRP-MDCK Cell Lines to Improve the Prediction of Brain Penetration. *J. Pharm. Sci.* **2019**, *108*, 2476–2483. [CrossRef] [PubMed]
27. Colclough, N.; Alluri, R.V.; Tucker, J.W.; Gozalpour, E.; Li, D.; Du, H.; Li, W.; Harlfinger, S.; O'Neill, D.J.; Sproat, G.G.; et al. Utilizing a Dual Human Transporter MDCKII-MDR1-BCRP Cell Line to Assess Efflux at the Blood Brain Barrier. *Drug Metab. Dispos.* **2024**, *52*, 95–105. [CrossRef] [PubMed]
28. Sato, S.; Matsumiya, K.; Tohyama, K.; Kosugi, Y. Translational CNS Steady-State Drug Disposition Model in Rats, Monkeys, and Humans for Quantitative Prediction of Brain-to-Plasma and Cerebrospinal Fluid-to-Plasma Unbound Concentration Ratios. *AAPS J.* **2021**, *23*, 81. [CrossRef] [PubMed]
29. Sato, S.; Tohyama, K.; Kosugi, Y. Investigation of MDR1-overexpressing cell lines to derive a quantitative prediction approach for brain disposition using in vitro efflux activities. *Eur. J. Pharm. Sci.* **2020**, *142*, 105119. [CrossRef]
30. Storelli, F.; Anoshchenko, O.; Unadkat, J.D. Successful Prediction of Human Steady-State Unbound Brain-Plasma Concentration Ratio of P-gp Substrates Using the Proteomics-Informed Relative Expression Factor Approach. *Clin. Pharmacol. Ther.* **2021**, *110*, 432–442. [CrossRef]
31. Liu, S.; Kosugi, Y. Human Brain Penetration Prediction Using Scaling Approach from Animal Machine Learning Models. *AAPS J.* **2023**, *25*, 86. [CrossRef]
32. Rowbottom, C.; Pietrasiewicz, A.; Tuczewycz, T.; Grater, R.; Qiu, D.; Kapadnis, S.; Trapa, P. Optimization of dose and route of administration of the P-glycoprotein inhibitor, valspodar (PSC-833) and the P-glycoprotein and breast cancer resistance protein dual-inhibitor, elacridar (GF120918) as dual infusion in rats. *Pharmacol. Res. Perspect.* **2021**, *9*, e00740. [CrossRef] [PubMed]
33. Sun, K.; Mikule, K.; Wang, Z.; Poon, G.; Vaidyanathan, A.; Smith, G.; Zhang, Z.-Y.; Hanke, J.; Ramaswamy, S.; Wang, J. A comparative pharmacokinetic study of PARP inhibitors demonstrates favorable properties for niraparib efficacy in preclinical tumor models. *Oncotarget* **2018**, *9*, 37080–37096. [CrossRef] [PubMed]
34. Liu, H.; Huang, L.; Li, Y.; Fu, T.; Sun, X.; Zhang, Y.-Y.; Gao, R.; Chen, Q.; Zhang, W.; Sahi, J.; et al. Correlation between Membrane Protein Expression Levels and Transcellular Transport Activity for Breast Cancer Resistance Protein. *Drug Metab. Dispos.* **2017**, *45*, 449–456. [CrossRef] [PubMed]
35. Lazzaro, S.; West, M.A.; Eatemadpour, S.; Feng, B.; Varma, M.V.S.; Rodrigues, A.D.; Temesszentandrási-Ambrus, C.; Kovács-Hajdu, P.; Nerada, Z.; Gáborik, Z.; et al. Translatability of in vitro Inhibition Potency to in vivo P-Glycoprotein Mediated Drug Interaction Risk. *J. Pharm. Sci.* **2023**, *112*, 1715–1723. [CrossRef] [PubMed]
36. Temesszentandrási-Ambrus, C.; Nagy, G.; Bui, A.; Gáborik, Z. A Unique In Vitro Assay to Investigate ABCB4 Transport Function. *Int. J. Mol. Sci.* **2023**, *24*, 4459. [CrossRef] [PubMed]

37. MacLean, B.; Tomazela, D.M.; Shulman, N.; Chambers, M.; Finney, G.L.; Frewen, B.; Kern, R.; Tabb, D.L.; Liebler, D.C.; MacCoss, M.J. Skyline: An open source document editor for creating and analyzing targeted proteomics experiments. *Bioinformatics* **2010**, *26*, 966–968. [CrossRef] [PubMed]

38. Uchida, Y. Quantitative Proteomics-Based Blood-Brain Barrier Study. *Biol. Pharm. Bull.* **2021**, *44*, 465–473. [CrossRef] [PubMed]

39. T. Package. Package 'nls.multstart'. 2023. Available online: https://cran.r-project.org/web/packages/nls.multstart/ (accessed on 26 April 2024).

40. Redeker, K.-E.M.; Jensen, O.; Gebauer, L.; Meyer-Tönnies, M.J.; Brockmöller, J. Atypical Substrates of the Organic Cation Transporter 1. *Biomolecules* **2022**, *12*, 1664. [CrossRef]

41. Nagaya, Y.; Nozaki, Y.; Takenaka, O.; Watari, R.; Kusano, K.; Yoshimura, T.; Kusuhara, H. Investigation of utility of cerebrospinal fluid drug concentration as a surrogate for interstitial fluid concentration using microdialysis coupled with cisternal cerebrospinal fluid sampling in wild-type and Mdr1a(-/-) rats. *Drug Metab. Pharmacokinet.* **2016**, *31*, 57–66. [CrossRef]

42. Cui, Y.; Lotz, R.; Rapp, H.; Klinder, K.; Himstedt, A.; Sauer, A. Muscle to Brain Partitioning as Measure of Transporter-Mediated Efflux at the Rat Blood-Brain Barrier and Its Implementation into Compound Optimization in Drug Discovery. *Pharmaceutics* **2019**, *11*, 595. [CrossRef] [PubMed]

43. Dolgikh, E.; Watson, I.A.; Desai, P.V.; Sawada, G.A.; Morton, S.; Jones, T.M.; Raub, T.J. QSAR Model of Unbound Brain-to-Plasma Partition Coefficient, K(p,uu,brain): Incorporating P-glycoprotein Efflux as a Variable. *J. Chem. Inf. Model.* **2016**, *56*, 2225–2233. [CrossRef] [PubMed]

44. Gupta, M.; Feng, J.; Bhisetti, G. Experimental and Computational Methods to Assess Central Nervous System Penetration of Small Molecules. *Molecules* **2024**, *29*, 1264. [CrossRef]

45. Summerfield, S.G.; Zhang, Y.; Liu, H. Examining the Uptake of Central Nervous System Drugs and Candidates across the Blood-Brain Barrier. *J. Pharmacol. Exp. Ther.* **2016**, *358*, 294–305. [CrossRef] [PubMed]

46. Broccatelli, F.; Larregieu, C.A.; Cruciani, G.; Oprea, T.I.; Benet, L.Z. Improving the prediction of the brain disposition for orally administered drugs using BDDCS. *Adv. Drug Deliv. Rev.* **2012**, *64*, 95–109. [CrossRef] [PubMed]

47. Kalvass, J.C.; Pollack, G.M. Kinetic considerations for the quantitative assessment of efflux activity and inhibition: Implications for understanding and predicting the effects of efflux inhibition. *Pharm. Res.* **2007**, *24*, 265–276. [CrossRef] [PubMed]

48. Montanari, F.; Ecker, G.F. Prediction of drug-ABC-transporter interaction--Recent advances and future challenges. *Adv. Drug Deliv. Rev.* **2015**, *86*, 17–26. [CrossRef] [PubMed]

49. Feng, B.; Mills, J.B.; Davidson, R.E.; Mireles, R.J.; Janiszewski, J.S.; Troutman, M.D.; de Morais, S.M. In vitro P-glycoprotein assays to predict the in vivo interactions of P-glycoprotein with drugs in the central nervous system. *Drug Metab. Dispos.* **2008**, *36*, 268–275. [CrossRef] [PubMed]

50. Lundquist, S.; Renftel, M.; Brillault, J.; Fenart, L.; Cecchelli, R.; Dehouck, M.-P. Prediction of drug transport through the blood-brain barrier in vivo: A comparison between two in vitro cell models. *Pharm. Res.* **2002**, *19*, 976–981. [CrossRef]

51. Faassen, F.; Vogel, G.; Spanings, H.; Vromans, H. Caco-2 permeability, P-glycoprotein transport ratios and brain penetration of heterocyclic drugs. *Int. J. Pharm.* **2003**, *263*, 113–122. [CrossRef]

52. Poirier, A.; Cascais, A.-C.; Bader, U.; Portmann, R.; Brun, M.-E.; Walter, I.; Hillebrecht, A.; Ullah, M.; Funk, C. Calibration of in vitro multidrug resistance protein 1 substrate and inhibition assays as a basis to support the prediction of clinically relevant interactions in vivo. *Drug Metab. Dispos.* **2014**, *42*, 1411–1422. [CrossRef] [PubMed]

53. Kikuchi, R.; de Morais, S.M.; Kalvass, J.C. In vitro P-glycoprotein efflux ratio can predict the in vivo brain penetration regardless of biopharmaceutics drug disposition classification system class. *Drug Metab. Dispos.* **2013**, *41*, 2012–2017. [CrossRef] [PubMed]

54. Takeuchi, T.; Yoshitomi, S.; Higuchi, T.; Ikemoto, K.; Niwa, S.; Ebihara, T.; Katoh, M.; Yokoi, T.; Asahi, S. Establishment and characterization of the transformants stably-expressing MDR1 derived from various animal species in LLC-PK1. *Pharm. Res.* **2006**, *23*, 1460–1472. [CrossRef] [PubMed]

55. Syvänen, S.; Lindhe, O.; Palner, M.; Kornum, B.R.; Rahman, O.; Långström, B.; Knudsen, G.M.; Hammarlund-Udenaes, M. Species differences in blood-brain barrier transport of three positron emission tomography radioligands with emphasis on P-glycoprotein transport. *Drug Metab. Dispos.* **2009**, *37*, 635–643. [CrossRef] [PubMed]

56. Hsiao, P.; Unadkat, J.D. P-glycoprotein-based loperamide-cyclosporine drug interaction at the rat blood-brain barrier: Prediction from in vitro studies and extrapolation to humans. *Mol. Pharm.* **2012**, *9*, 629–633. [CrossRef]

57. Chen, C.; Liu, X.; Smith, B.J. Utility of Mdr1-gene deficient mice in assessing the impact of P-glycoprotein on pharmacokinetics and pharmacodynamics in drug discovery and development. *Curr. Drug Metab.* **2003**, *4*, 272–291. [CrossRef] [PubMed]

58. Jain, S.; Grandits, M.; Ecker, G.F. Interspecies comparison of putative ligand binding sites of human, rat and mouse P-glycoprotein. *Eur. J. Pharm. Sci.* **2018**, *122*, 134–143. [CrossRef] [PubMed]

59. Molden, E.; Christensen, H.; Sund, R.B. Extensive metabolism of diltiazem and P-glycoprotein-mediated efflux of desacetyl-diltiazem (M1) by rat jejunum in vitro. *Drug Metab. Dispos.* **2000**, *28*, 107–109.

60. Adachi, Y.; Suzuki, H.; Sugiyama, Y. Comparative studies on in vitro methods for evaluating in vivo function of MDR1 P-glycoprotein. *Pharm. Res.* **2001**, *18*, 1660–1668. [CrossRef] [PubMed]

61. Lin, J.H.; Yamazaki, M. Role of P-glycoprotein in pharmacokinetics: Clinical implications. *Clin. Pharmacokinet.* **2003**, *42*, 59–98. [CrossRef]

62. Yamazaki, M.; Neway, W.E.; Ohe, T.; Chen, I.; Rowe, J.F.; Hochman, J.H.; Chiba, M.; Lin, J.H. In vitro substrate identification studies for p-glycoprotein-mediated transport: Species difference and predictability of in vivo results. *J. Pharmacol. Exp. Ther.* **2001**, *296*, 723–735.

63. Jiang, L.; Kumar, S.; Nuechterlein, M.; Reyes, M.; Tran, D.; Cabebe, C.; Chiang, P.; Reynolds, J.; Carrier, S.; Sun, Y.; et al. Application of a high-resolution in vitro human MDR1-MDCK assay and in vivo studies in preclinical species to improve prediction of CNS drug penetration. *Pharmacol. Res. Perspect.* **2022**, *10*, e00932. [CrossRef] [PubMed]

64. Li, N.; Kulkarni, P.; Badrinarayanan, A.; Kefelegn, A.; Manoukian, R.; Li, X.; Prasad, B.; Karasu, M.; McCarty, W.J.; Knutson, C.G.; et al. P-glycoprotein Substrate Assessment in Drug Discovery: Application of Modeling to Bridge Differential Protein Expression Across In Vitro Tools. *J. Pharm. Sci.* **2021**, *110*, 325–337. [CrossRef] [PubMed]

65. Patel, N.C.; Feng, B.; Hou, X.; West, M.A.; Trapa, P.E.; Sciabola, S.; Verhoest, P.; Liras, J.L.; Maurer, T.S.; Wager, T.T. Harnessing Preclinical Data as a Predictive Tool for Human Brain Tissue Targeting. *ACS Chem. Neurosci.* **2021**, *12*, 1007–1017. [CrossRef] [PubMed]

66. Simoff, I.; Karlgren, M.; Backlund, M.; Lindström, A.-C.; Gaugaz, F.Z.; Matsson, P.; Artursson, P. Complete Knockout of Endogenous Mdr1 (Abcb1) in MDCK Cells by CRISPR-Cas9. *J. Pharm. Sci.* **2016**, *105*, 1017–1021. [CrossRef] [PubMed]

67. Chen, E.C.; Broccatelli, F.; Plise, E.; Chen, B.; Liu, L.; Cheong, J.; Zhang, S.; Jorski, J.; Gaffney, K.; Umemoto, K.K.; et al. Evaluating the Utility of Canine Mdr1 Knockout Madin-Darby Canine Kidney I Cells in Permeability Screening and Efflux Substrate Determination. *Mol. Pharm.* **2018**, *15*, 5103–5113. [CrossRef] [PubMed]

68. Poller, B.; Wagenaar, E.; Tang, S.C.; Schinkel, A.H. Double-transduced MDCKII cells to study human P-glycoprotein (ABCB1) and breast cancer resistance protein (ABCG2) interplay in drug transport across the blood-brain barrier. *Mol. Pharm.* **2011**, *8*, 571–582. [CrossRef] [PubMed]

69. Trapa, P.E.; Belova, E.; Liras, J.L.; Scott, D.O.; Steyn, S.J. Insights From an Integrated Physiologically Based Pharmacokinetic Model for Brain Penetration. *J. Pharm. Sci.* **2016**, *105*, 965–971. [CrossRef] [PubMed]

70. Patel, N.C. Methods to optimize CNS exposure of drug candidates. *Bioorg. Med. Chem. Lett.* **2020**, *30*, 127503. [CrossRef]

71. Kalvass, J.C.; Maurer, T.S.; Pollack, G.M. Use of plasma and brain unbound fractions to assess the extent of brain distribution of 34 drugs: Comparison of unbound concentration ratios to in vivo p-glycoprotein efflux ratios. *Drug Metab. Dispos.* **2007**, *35*, 660–666. [CrossRef]

72. Suzuyama, N.; Katoh, M.; Takeuchi, T.; Yoshitomi, S.; Higuchi, T.; Asashi, S.; Yokoi, T. Species differences of inhibitory effects on P-glycoprotein-mediated drug transport. *J. Pharm. Sci.* **2007**, *96*, 1609–1618. [CrossRef] [PubMed]

73. Zolnerciks, J.K.; Booth-Genthe, C.L.; Gupta, A.; Harris, J.; Unadkat, J.D. Substrate- and species-dependent inhibition of P-glycoprotein-mediated transport: Implications for predicting in vivo drug interactions. *J. Pharm. Sci.* **2011**, *100*, 3055–3061. [CrossRef] [PubMed]

74. Aller, S.G.; Yu, J.; Ward, A.; Weng, Y.; Chittaboina, S.; Zhuo, R.; Harrell, P.M.; Trinh, Y.T.; Zhang, Q.; Urbatsch, I.L.; et al. Structure of P-glycoprotein reveals a molecular basis for poly-specific drug binding. *Science* **2009**, *323*, 1718–1722. [CrossRef] [PubMed]

75. Grigoreva, T.; Romanova, A.; Sagaidak, A.; Vorona, S.; Novikova, D.; Tribulovich, V. Mdm2 inhibitors as a platform for the design of P-glycoprotein inhibitors. *Bioorg. Med. Chem. Lett.* **2020**, *30*, 127424. [CrossRef] [PubMed]

76. Zamek-Gliszczynski, M.J.; Lee, C.A.; Poirier, A.; Bentz, J.; Chu, X.; Ellens, H.; Ishikawa, T.; Jamei, M.; Kalvass, J.C.; Nagar, S.; et al. ITC recommendations for transporter kinetic parameter estimation and translational modeling of transport-mediated PK and DDIs in humans. *Clin. Pharmacol. Ther.* **2013**, *94*, 64–79. [CrossRef] [PubMed]

77. von Richter, O.; Glavinas, H.; Krajcsi, P.; Liehner, S.; Siewert, B.; Zech, K. A novel screening strategy to identify ABCB1 substrates and inhibitors. *Naunyn-Schmiedeberg's Arch. Pharmacol.* **2009**, *379*, 11–26. [CrossRef] [PubMed]

78. Ledwitch, K.V.; Barnes, R.W.; Roberts, A.G. Unravelling the complex drug-drug interactions of the cardiovascular drugs, verapamil and digoxin, with P-glycoprotein. *Biosci. Rep.* **2016**, *36*, e00309. [CrossRef]

79. Hansen, T.S.; Nilsen, O.G. Echinacea purpurea and P-glycoprotein drug transport in Caco-2 cells. *Phytother. Res.* **2009**, *23*, 86–91. [CrossRef]

80. Seelig, A.; Landwojtowicz, E. Structure-activity relationship of P-glycoprotein substrates and modifiers. *Eur. J. Pharm. Sci.* **2000**, *12*, 31–40. [CrossRef]

81. Sziráki, I.; Erdo, F.; Beéry, E.; Molnár, P.M.; Fazakas, C.; Wilhelm, I.; Makai, I.; Kis, E.; Herédi-Szabó, K.; Abonyi, T.; et al. Quinidine as an ABCB1 probe for testing drug interactions at the blood-brain barrier: An in vitro in vivo correlation study. *J. Biomol. Screen.* **2011**, *16*, 886–894. [CrossRef]

82. Mori, N.; Iwamoto, H.; Yokooji, T.; Murakami, T. Characterization of intestinal absorption of quinidine, a P-glycoprotein substrate, given as a powder in rats. *Pharmazie* **2012**, *67*, 384–388. [PubMed]

83. Kharasch, E.D.; Hoffer, C.; Altuntas, T.G.; Whittington, D. Quinidine as a probe for the role of p-glycoprotein in the intestinal absorption and clinical effects of fentanyl. *J. Clin. Pharmacol.* **2004**, *44*, 224–233. [CrossRef] [PubMed]

84. Acharya, P.; O'Connor, M.P.; Polli, J.W.; Ayrton, A.; Ellens, H.; Bentz, J. Kinetic identification of membrane transporters that assist P-glycoprotein-mediated transport of digoxin and loperamide through a confluent monolayer of MDCKII-hMDR1 cells. *Drug Metab. Dispos.* **2008**, *36*, 452–460. [CrossRef] [PubMed]

85. Gatlik-Landwojtowicz, E.; Äänismaa, P.; Seelig, A. Quantification and characterization of P-glycoprotein-substrate interactions. *Biochemistry* **2006**, *45*, 3020–3032. [CrossRef] [PubMed]

86. Okamura, N.; Hirai, M.; Tanigawara, Y.; Tanaka, K.; Yasuhara, M.; Ueda, K.; Komano, T.; Hori, R. Digoxin-cyclosporin A interaction: Modulation of the multidrug transporter P-glycoprotein in the kidney. *J. Pharmacol. Exp. Ther.* **1993**, *266*, 1614–1619.

87. Taub, M.E.; Podila, L.; Ely, D.; Almeida, I. Functional assessment of multiple P-glycoprotein (P-gp) probe substrates: Influence of cell line and modulator concentration on P-gp activity. *Drug Metab. Dispos.* **2005**, *33*, 1679–1687. [CrossRef] [PubMed]

88. Tanigawara, Y.; Okamura, N.; Hirai, M.; Yasuhara, M.; Ueda, K.; Kioka, N.; Komano, T.; Hori, R. Transport of digoxin by human P-glycoprotein expressed in a porcine kidney epithelial cell line (LLC-PK1). *J. Pharmacol. Exp. Ther.* **1992**, *263*, 840–845. [PubMed]

89. Troutman, M.D.; Thakker, D.R. Efflux ratio cannot assess P-glycoprotein-mediated attenuation of absorptive transport: Asymmetric effect of P-glycoprotein on absorptive and secretory transport across Caco-2 cell monolayers. *Pharm. Res.* **2003**, *20*, 1200–1209. [CrossRef]

90. Westphal, K.; Weinbrenner, A.; Giessmann, T.; Stuhr, M.; Franke, G.; Zschiesche, M.; Oertel, R.; Terhaag, B.; Kroemer, H.K.; Siegmund, W. Oral bioavailability of digoxin is enhanced by talinolol: Evidence for involvement of intestinal P-glycoprotein. *Clin. Pharmacol. Ther.* **2000**, *68*, 6–12. [CrossRef]

91. Ding, R.; Tayrouz, Y.; Riedel, K.-D.; Burhenne, J.; Weiss, J.; Mikus, G.; Haefeli, W.E. Substantial pharmacokinetic interaction between digoxin and ritonavir in healthy volunteers. *Clin. Pharmacol. Ther.* **2004**, *76*, 73–84. [CrossRef]

92. Penzak, S.R.; Shen, J.M.; Alfaro, R.M.; Remaley, A.T.; Natarajan, V.; Falloon, J. Ritonavir decreases the nonrenal clearance of digoxin in healthy volunteers with known MDR1 genotypes. *Ther. Drug Monit.* **2004**, *26*, 322–330. [CrossRef] [PubMed]

93. Brouwer, K.L.R.; Keppler, D.; Hoffmaster, K.A.; Bow, D.A.J.; Cheng, Y.; Lai, Y.; Palm, J.E.; Stieger, B.; Evers, R. In vitro methods to support transporter evaluation in drug discovery and development. *Clin. Pharmacol. Ther.* **2013**, *94*, 95–112. [CrossRef] [PubMed]

94. Ambrus, C.; Bakos, É.; Sarkadi, B.; Özvegy-Laczka, C.; Telbisz, Á. Interactions of anti-COVID-19 drug candidates with hepatic transporters may cause liver toxicity and affect pharmacokinetics. *Sci. Rep.* **2021**, *11*, 17810. [CrossRef] [PubMed]

95. Palmeira, A.; Sousa, E.; Vasconcelos, M.H.; Pinto, M.M. Three decades of P-gp inhibitors: Skimming through several generations and scaffolds. *Curr. Med. Chem.* **2012**, *19*, 1946–2025. [CrossRef] [PubMed]

96. Nosol, K.; Romane, K.; Irobalieva, R.N.; Alam, A.; Kowal, J.; Fujita, N.; Locher, K.P. Cryo-EM structures reveal distinct mechanisms of inhibition of the human multidrug transporter ABCB1. *Proc. Natl. Acad. Sci. USA* **2020**, *117*, 26245–26253. [CrossRef] [PubMed]

97. Alam, A.; Küng, R.; Kowal, J.; McLeod, R.A.; Tremp, N.; Broude, E.V.; Roninson, I.B.; Stahlberg, H.; Locher, K.P. Structure of a zosuquidar and UIC2-bound human-mouse chimeric ABCB1. *Proc. Natl. Acad. Sci. USA* **2018**, *115*, E1973–E1982. [CrossRef] [PubMed]

98. Urgaonkar, S.; Nosol, K.; Said, A.M.; Nasief, N.N.; Bu, Y.; Locher, K.P.; Lau, J.Y.N.; Smolinski, M.P. Discovery and Characterization of Potent Dual P-Glycoprotein and CYP3A4 Inhibitors: Design, Synthesis, Cryo-EM Analysis, and Biological Evaluations. *J. Med. Chem.* **2022**, *65*, 191–216. [CrossRef]

99. Sauna, Z.E.; Ambudkar, S.V. Characterization of the catalytic cycle of ATP hydrolysis by human P-glycoprotein. The two ATP hydrolysis events in a single catalytic cycle are kinetically similar but affect different functional outcomes. *J. Biol. Chem.* **2001**, *276*, 11653–11661. [CrossRef]

100. Martin, C.; Berridge, G.; Higgins, C.F.; Mistry, P.; Charlton, P.; Callaghan, R. Communication between multiple drug binding sites on P-glycoprotein. *Mol. Pharmacol.* **2000**, *58*, 624–632. [CrossRef]

101. Glavinas, H.; von Richter, O.; Vojnits, K.; Mehn, D.; Wilhelm, I.; Nagy, T.; Janossy, J.; Krizbai, I.; Couraud, P.; Krajcsi, P. Calcein assay: A high-throughput method to assess P-gp inhibition. *Xenobiotica* **2011**, *41*, 712–719. [CrossRef]

102. Sharom, F.J. Complex Interplay between the P-Glycoprotein Multidrug Efflux Pump and the Membrane: Its Role in Modulating Protein Function. *Front. Oncol.* **2014**, *4*, 41. [CrossRef] [PubMed]

103. Safa, A.R. Identification and characterization of the binding sites of P-glycoprotein for multidrug resistance-related drugs and modulators. *Curr. Med. Chem. Anticancer Agents* **2004**, *4*, 1–17. [CrossRef] [PubMed]

104. Shao, Y.M.; Ayesh, S.; Stein, W.D. Mutually co-operative interactions between modulators of P-glycoprotein. *Biochim. Biophys. Acta* **1997**, *1360*, 30–38. [CrossRef] [PubMed]

pharmaceutics

Article

Cu(ATSM) Increases P-Glycoprotein Expression and Function at the Blood-Brain Barrier in C57BL6/J Mice

Jae Pyun [1], HuiJing Koay [2], Pranav Runwal [1], Celeste Mawal [3], Ashley I. Bush [3], Yijun Pan [1], Paul S. Donnelly [2], Jennifer L. Short [4] and Joseph A. Nicolazzo [1,*]

1   Drug Delivery, Disposition and Dynamics, Monash Institute of Pharmaceutical Sciences, Monash University, Parkville, VIC 3052, Australia; jae.pyun@monash.edu (J.P.); pranav.runwal1@monash.edu (P.R.)
2   Bio21 Molecular Science and Biotechnology Institute, University of Melbourne, Parkville, VIC 3052, Australia; pauld@unimelb.edu.au (P.S.D.)
3   Oxidation Biology Lab, Melbourne Dementia Research Centre, Florey Institute of Neuroscience and Mental Health, University of Melbourne, Parkville, VIC 3052, Australia; celeste.mawal@florey.edu.au (C.M.); ashley.bush@florey.edu.au (A.I.B.)
4   Drug Discovery Biology, Monash Institute of Pharmaceutical Sciences, Monash University, Parkville, VIC 3052, Australia; jennifer.short@monash.edu
*   Correspondence: joseph.nicolazzo@monash.edu; Tel.: +61-3-9903-9605

Abstract: P-glycoprotein (P-gp), expressed at the blood-brain barrier (BBB), is critical in preventing brain access to substrate drugs and effluxing amyloid beta (Aβ), a contributor to Alzheimer's disease (AD). Strategies to regulate P-gp expression therefore may impact central nervous system (CNS) drug delivery and brain Aβ levels. As we have demonstrated that the copper complex copper diacetyl bis(4-methyl-3-thiosemicarbazone) (Cu(ATSM)) increases P-gp expression and function in human brain endothelial cells, the present study assessed the impact of Cu(ATSM) on expression and function of P-gp in mouse brain endothelial cells (mBECs) and capillaries in vivo, as well as in peripheral organs. Isolated mBECs treated with Cu(ATSM) (100 nM for 24 h) exhibited a 1.6-fold increase in P-gp expression and a 20% reduction in accumulation of the P-gp substrate rhodamine 123. Oral administration of Cu(ATSM) (30 mg/kg/day) for 28 days led to a 1.5 & 1.3-fold increase in brain microvascular and hepatic expression of P-gp, respectively, and a 20% reduction in BBB transport of [3H]-digoxin. A metallomic analysis showed a 3.5 and 19.9-fold increase in Cu levels in brain microvessels and livers of Cu(ATSM)-treated mice. Our findings demonstrate that Cu(ATSM) increases P-gp expression and function at the BBB in vivo, with implications for CNS drug delivery and clearance of Aβ in AD.

Keywords: blood-brain barrier; P-glycoprotein; copper; bis(thiosemicarbazone); CNS drug delivery

Citation: Pyun, J.; Koay, H.; Runwal, P.; Mawal, C.; Bush, A.I.; Pan, Y.; Donnelly, P.S.; Short, J.L.; Nicolazzo, J.A. Cu(ATSM) Increases P-Glycoprotein Expression and Function at the Blood-Brain Barrier in C57BL6/J Mice. *Pharmaceutics* **2023**, *15*, 2084. https://doi.org/10.3390/pharmaceutics15082084

Academic Editors: Gert Fricker and Elena Puris

Received: 19 June 2023
Revised: 28 July 2023
Accepted: 31 July 2023
Published: 3 August 2023

1. Introduction

The blood-brain barrier (BBB) consists of specialised endothelial cells lining the network of microvessels that perfuse the brain and has an essential role in regulating the interface between the central nervous system (CNS) and the systemic circulation [1]. Brain microvascular endothelial cells express various efflux transporter proteins on their luminal membrane, such as P-glycoprotein (P-gp), which acts as a "gatekeeper" to prevent the CNS entry of a wide range of xenobiotic molecules [2]. P-gp is the most well characterised of the ATP-binding cassette transporter family, encoded by the MDR1 gene and is also expressed in other tissues including the liver, kidneys, small intestines, placenta, and blood-testis barrier [3,4]. The brain uptake of structurally diverse compounds (dexamethasone, digoxin, cyclosporin A, ondansetron, and loperamide) has been shown to increase between twenty to one hundred-fold more in P-gp deficient mice compared with wild-type mice [4–7]. Thus, whilst P-gp has a protective role in keeping toxic substrates from entering the CNS, it also poses a major challenge for the delivery of therapeutic agents to their corresponding target sites.

Efforts to inhibit the P-gp-mediated efflux of drugs have been trialled as a means of enhancing CNS therapeutic access, but these have failed to improve clinical outcomes [8–10]. Therefore, to overcome the impact of P-gp on drug delivery to the brain, novel strategies to regulate P-gp expression and function are needed, in conjunction with developing new techniques to overcome the restrictive nature of the BBB, such as the use of carrier molecules, focused-ultrasound and nanoparticles [11–13]. However, on the other hand, reduced function or expression of P-gp can result in significant neurotoxicity and undesirable pharmacodynamic effects, particularly given that this key efflux transporter is known to be disrupted in certain disease states, leading to changes in the exchange of molecules between the blood and brain.

There is compelling preclinical and clinical evidence to indicate that dysregulation of P-gp at the BBB contributes to the pathogenesis of neurodegenerative diseases, including Alzheimer's disease (AD) [14–17]. Research suggests that the BBB may be disrupted in the earliest stages of AD, even before the onset of cognitive symptoms [18,19]. This early BBB disruption may contribute to the accumulation of amyloid beta (Aβ), which is a hallmark neurotoxic peptide that accumulates in the brain of individuals with AD and leads to disease progression [20,21]. P-gp plays a crucial role in the efflux of Aβ peptides and has been shown in vitro to directly transport the toxic 40 and 42 amino acid isoforms of Aβ (Aβ1-40 and Aβ1-42) [22,23]. Further studies have confirmed that Aβ is indeed a substrate of P-gp and that P-gp-mediated clearance of fluorescent Aβ can be reduced in isolated mouse brain microvessels in the presence of P-gp inhibitors [24,25]. One study demonstrated elevated endogenous Aβ levels in the hippocampal area of P-gp deficient mice indicating reduced clearance, while concurrently, administration of P-gp inhibitors also significantly increased endogenous brain levels in a mouse model of AD, clearly indicating the role of P-gp in the clearance of Aβ [26]. Clinically, there is evidence of an inverse correlation between P-gp expression and brain Aβ deposition in individuals with AD, where decreased P-gp abundance in brain microvessel regions was associated with increased Aβ plaque burden, a finding observed in numerous independent studies comparing AD and non-AD post mortem brains [27–30]. In addition to reduced P-gp expression, the efflux function of P-gp appears diminished in individuals with AD, with positron emission tomography imaging demonstrating an increased brain accumulation of the P-gp substrate [^{11}C]verapamil compared to those without AD [17]. Therefore, given that P-gp mediates Aβ efflux, and P-gp expression and function are reduced in AD, increasing P-gp expression and function could be a promising approach to increase Aβ clearance and attenuate amyloid burden in individuals with AD. Targeting intracellular signalling pathways known to upregulate and restore P-gp function led to a reduction in Aβ accumulation in an AD mouse model, suggesting P-gp as a potential therapeutic target in AD [24]. Therefore, investigating approaches to restore P-gp expression and function, which could modulate the access of drugs into the CNS but also enhance the brain clearance of Aβ in AD, is warranted.

To this end, our laboratory has had an ongoing interest in the role of copper (Cu) in transporter regulation. Cu is an essential biometal that maintains brain health and neuronal integrity and modulates various regulatory pathways within the CNS [31,32]. The biodistribution of Cu is disrupted in AD [33,34], and more recently, Cu levels have been proposed to have a protective effect against cognitive decline in AD clinically [35]. Despite this, the effects of Cu and Cu-complexes on brain endothelial cells, particularly given its need to cross the BBB to reach targets in the CNS, are not well understood. Our laboratory previously demonstrated an association between intracellular levels of Cu and P-gp expression and function in a human brain microvascular endothelial (hCMEC/D3) cell line, a model representing the BBB. A metallochaperone clioquinol was utilised in combination with Cu and zinc (Zn), two essential biometals important in maintaining the health of the CNS, where increased Cu levels were associated with increased P-gp expression [36]. To further ascertain this relationship, we assessed the impact of copper diacetyl bis (4-methyl-3-thiosemicarbazone) (Cu(ATSM)) on P-gp expression and function, as this bis (thiosemicarbazone) complex has been shown to deliver exogenously bound Cu, increasing the levels of the biometal in brain cells [37]. Cu(ATSM) is a low molecular weight, highly lipophilic, charge neutral compound that crosses the BBB [38,39]. We have reported the effects of Cu(ATSM) on hCMEC/D3 cells, where we showed a 6-fold increase in Cu in hCMEC/D3 cells, and this was associated with a 2-fold increase in expression and 1.3-fold increase in function of P-gp [40]. However, as described in the previous study, this increase was not simply a result of increasing intracellular Cu as other compounds known to increase Cu did not upregulate P-gp. The combination of the Cu and the backbone structure of the ATSM complex increased P-gp expression and function and this was shown to be mediated partly through activation (1.4-fold) of the extracellular signal-regulated kinase 1 and 2 (ERK1/2) [40]. Given the previous studies were undertaken in human brain endothelial cells, the aim of the present study was to examine the effects of Cu(ATSM) on P-gp expression and function in an in vivo mouse model. This study is significant because multiple studies have demonstrated discrepancies and a lack of translation between human and rodent systems [41,42]. These studies are the first to assess the effect of Cu(ATSM) at the mouse BBB and the findings have the potential to validate a novel therapeutic approach to restore the function of a key efflux transporter, which may result in enhanced clearance of Aβ from the brain.

2. Materials and Methods

2.1. Materials

Ammonium persulfate (APS), beta-mercaptoethanol (β-ME), dimethyl sulfoxide (DMSO), Dulbecco's phosphate buffer saline (D-PBS), glycerol, glycine, 4% (w/v) paraformaldehyde (PFA), cOmplete™ mini protease inhibitor, phosSTOP™ phosphatase inhibitor cocktail tablets, rhodamine 123 (R123), sodium chloride, sodium dodecyl sulphate (SDS), tetramethylethylenediamine (TEMED), thiazolyl blue tetrazolium bromide (MTT), tris base (TRIZMA® base), Triton® X-100, Tween® 20, and trypan blue were purchased from Sigma-Aldrich (St Louis, MO, USA). GFAP antibody (RRID: AB_1074620), HuC/D pan-neuronal antibody (RRID: AB_221448), Iba1 antibody (RRID: AB_11156585), 0.25% trypsin-EDTA (w/v), Mem-PER™ Plus Membrane Protein Extraction kit, Pierce™ bicinchoninic acid (BCA) protein assay kit, and Pierce™ IP lysis buffer composed of 1 mM EDTA, 150 mM NaCl, 1% (w/v) NP-40 and 25 mM Tris-HCl pH 7.4 were purchased from Thermo Fisher Scientific (Rockford, IL, USA). HyClone bovine serum albumin (BSA) was purchased from GE Healthcare Life Sciences (Little Chalfont, UK). Extra thick blot paper, Mini-Protein® TGX™ Precast gels (4–15% acrylamide), 0.45 μm nitrocellulose membranes, and Precision Plus Protein™ Dual Xtra Standards were purchased from Bio-Rad (Hercules, CA, USA). Primary C219 monoclonal antibody to P-gp (RRID: AB_2565033) and purified anti-mouse CD31 antibody (RRID: AB_312908) were purchased from BioLegend® (San Diego, CA, USA), and the primary mouse antibody for β-actin (RRID: AB_303668) and the sodium-potassium ATPase antibody (ab76020) were purchased from Abcam (Cambridge, UK). Intercept® blocking buffer in PBS, IRDye® 800CW goat anti-mouse (RRID: AB_621842), and IRDye® 680LT donkey anti-rabbit (RRID: AB_10706167) secondary antibodies were purchased from LI-COR (Lincoln, NE, USA). Endothelial basal medium-2 (EBM2) media was used for all primary mouse brain endothelial cell (mBEC) culture experiments, and this growth media was supplemented with 1% (v/v) penicillin/streptomycin and 10 mM HEPES purchased from Sigma-Aldrich (St Louis, MO, USA), and EGM™-2 SingleQuot™ Kit (ascorbic acid, b-slice variant fibroblast growth factor, epidermal growth factor, fetal bovine serum (FBS), gentamicin/amphotericin, hydrocortisone, insulin-like growth factor, vascular endothelial growth factor) purchased from Lonza (Walkersville, MD, USA). T75 flasks, 6 and 96-well plates, rat-tail collagen Type I and all plastic cell culture equipment were purchased from Corning (Corning, NY, USA). Dulbecco's Modified Eagle's Medium (DMEM) GlutaMAX™ and Hanks balanced salt solution (HBSS) were purchased from Life Technologies (Carlsbad, CA, USA). Adult Brain Dissociation kit, MACS® CD31+ MicroBeads, PE CD31+ fluorophore for mouse and all MACS® equipment for the magnetic-activated cell sorting for the isolation of mBECs was purchased from Miltenyi Biotech (Bergish Gladbach, Germany). [3H]-digoxin and [14C]-sucrose were purchased from American Radiolabelled Chemicals (St Louis, MO, USA) and Ultima Gold™ liquid scintillation cocktail was purchased from Perkin-Elmer Life Sciences (Waltham, MA, USA). Copper diacetyl bis(4-methyl-3-thiosemicarbazone) (Cu(ATSM)) was synthesised as previously described [43,44].

2.2. Isolation and Characterisation of mBECs

All experiments were conducted in accordance with the National Health and Medical Research Council of Australia guidelines for the care and use of animals for scientific purposes and were approved by the Monash Institute of Pharmaceutical Sciences Animal Ethics Committee (Protocol MIPS 27467). Female C57BL/6J mice (6–8 weeks old) were acquired from the Monash Animal Research Platform (MARP) and were kept on a 12-h light-dark cycle with food and water available ad libitum.

To obtain highly purified primary mBECs, the Adult Brain Dissociation kit for mice was utilised and a magnetic-activated cell sorting (MACS®) system from Miltenyi Biotech was employed following the manufacturer's protocols, albeit with some specific optimisations. In brief, brains from two female 6–8-week-old C57BL/6J mice (2 brains per sample, $n = 6$ per experimental group, 12 mice for transport experiments and 12 mice for protein quantification experiments) were removed under sterile conditions and washed with cold HBSS buffer containing 0.33 M pyruvate. The brain was rolled over filter paper to remove the meningeal layers and large blood vessels. The cortex was dissected and then homogenised and placed in a C-tube followed by enzymatic digestion (provided in the kit) while being maintained at 37 °C for 30 min with rotation on the gentleMACS™ Octo Dissociator. The homogenate was then strained using a pre-wet 70 μm MACS smart strainer. The resulting cell suspension was centrifuged at $300 \times g$ and the pellet was resuspended in 6.2 mL of HBSS buffer and 1.8 mL of cold Debris Removal Solution and centrifuged at $3000 \times g$ for 10 min at 4 °C. Three separation layers were formed, and the remaining pellet was resuspended in $1 \times$ Red Blood Cell Removal solution and incubated in the refrigerator for 10 min. Following the incubation, 10 mL of D-PBS buffer containing 0.5% (w/v) BSA was added. The resulting solution was centrifuged at $300 \times g$ for 10 min and the cell pellet was resuspended in a 100 μL solution containing 90 μL of cold DPBS/BSA buffer and 10 μL of CD31-specific mouse MACS® MicroBeads and incubated at 4 °C in the dark for 15 min. The cells were washed with 1 mL D-PBS/BSA buffer and centrifuged at $300 \times g$ for 5 min retaining the cell pellet. The cells were resuspended in 500 μL of D-PBS/BSA buffer and then subjected to magnetic separation using the MACS® MS columns and a MACS® separator. The column was washed to ensure purity and maximum yield and finally eluted without the magnetic column to collect CD31 positive cells in 900 μL of EBM-2 growth media. The resulting cell suspension was resuspended and plated at a density of 50,000 cells/cm^2 onto pre-collagenated (Rat-tail collagen Type I) cell culture plates. As an additional purification step, mBECs were then subjected to a 3 μg/mL puromycin (in EBM2 media) treatment for 72 h to eliminate non-endothelial cells. mBECs were maintained at 37 °C in a 5% CO_2 incubator with media change every two days until 80% confluency was reached.

Immunocytochemistry (ICC) and fluorescence-activated cell sorting (FACS) were employed to confirm the absence of potentially contaminating neurons, astrocytes or microglia and to determine the percentage of mBECs. Following the isolation, the cell suspension was seeded into a pre-collagenated 8-well ibiTreat (ibidi cells in focus) plate followed by 72 h of puromycin treatment. mBECs were washed twice in PBS and then incubated in EBM2 growth media for 7 days or until approximately 80% confluent. To prepare the cells for ICC, cells were washed three times for 5 min with 200 μL of PBS. The cells were fixed with 4% (w/v) PFA in PBS for 10 min at room temperature. Excess PFA was washed with 200 μL of PBS three times followed by permeabilisation with 0.1% (v/v) Triton® X-100 in PBS. 200 μL of blocking solution of 1% (w/v) BSA in PBS with 0.1% (v/v) Tween-20 was added for 30 min at room temperature. Following the blocking step, primary antibodies diluted in blocking buffer were co-incubated according to Table 1 overnight at 4 °C. The next day, cells were washed three times for 5 min with 200 μL of PBS followed by incubation with a secondary Alexa Fluor® antibody diluted in blocking buffer (Table 1) and incubated for 60 min protected from light. Following incubation, cells were again washed three times for 5 min with 200 μL of PBS and counter-stained with 4′,6-diamidino-2-phenylindole (DAPI) (25 mg/mL) diluted 1:10,000 for 15 min followed by the final repeated wash step as above. The isolated cells were imaged using the Leica SP8 Lightning confocal microscope (Wetzlar, Germany).

Table 1. Antibody dilutions for ICC imaging of mBECs, neurons, astrocytes, and microglia.

	Primary Antibody	Secondary Alexa Fluor® Antibody
mBEC marker	Rat anti-mouse CD31 (1:400)	Anti-rat Alexa 555 (1:500)
Neuronal marker	Mouse anti-human HuC/D (1:1000)	Anti-mouse Alexa 488 (1:500)
Astrocyte marker	Chicken anti-mouse GFAP (1:2000)	Anti-chicken Alexa 647 (1:500)
Microglial marker	Rabbit anti-mouse Iba1 (1:2000)	Anti-rabbit Alexa 488 (1:500)

Following seeding and 72 h puromycin treatment, cells were detached with 0.25% (w/v) trypsin-EDTA and centrifuged at 650× g for 5 min to collect the pellet. mBECs were resuspended in a FACS sorting buffer consisting of 1 mM EDTA, 25 mM HEPES (pH 7) and 1% (w/v) FBS in Ca^{2+} and Mg^{2+} free PBS. The suspension of cells was filtered through a 40 μm mesh strainer to obtain a single-cell suspension. Cells were sorted using the FACS Canto™ II Flow Cytometer (Franklin Lakes, NJ, USA) and gated for cell debris, single cells followed by a viability V450 marker and finally the phycoerythrin (PE) CD31+ to detect the percentage yield and purity of mBECs isolated using the MACS® system. Controls for the correct detection of the antibodies specific to viability V450, fluorescein isothiocyanate (FITC) CD45+ (peripheral cells) and allophycocyanin (APC) CD11b+ (microglia) were sampled concurrently with PE CD31+.

2.3. Treatment of mBECs with Cu(ATSM) to Assess P-gp Expression and Function

Once 80% confluency was reached, mBECs were treated with either 100 nM Cu(ATSM) or vehicle (0.1% (v/v) DMSO) for 24 h prior to being lysed for Western blot (WB) analysis or for P-gp functional studies as previously described in our laboratory [40]. Uptake studies were conducted using a fluorescent P-gp substrate R123. In brief, accumulation of R123 in mBECs was measured to ascertain P-gp efflux capacity following a pre-incubation of 5 μM R123 in transport buffer (10 mM HEPES in HBSS pH 7.4) for 60 min (Section 3.2 schematic). Following incubation, cells were lysed with 100 μL of 1% (v/v) Triton® X-100 on a plate shaker at 4 °C for 20 min. 50 μL of R123 standards (prepared in 1% (v/v) Triton® X-100 in milliQ water) and 50 μL of the lysate sample from each well were then transferred into a 96-well plate. Fluorescence was detected using the Enspire fluorescence spectrophotometer at excitation and emission wavelengths of 511 and 534 nm, respectively. The fluorescent R123 (nmol) signal was normalised to total cellular protein determined using the Pierce™ BCA protein assay (mg) and the accumulation (nmol/mg) was compared between vehicle and Cu(ATSM)-treated cells.

2.4. Dosing of Cu(ATSM) to C57BL/6J Mice

The dosing regimen of Cu(ATSM) was selected based on disease-modifying doses of 30 mg/kg in disease models of amyotrophic lateral sclerosis and Parkinson's disease, demonstrated to not have any adverse side effects [45–47]. Female C57BL/6J mice (6–8 weeks old), randomly allocated to a treatment or vehicle group, were orally administered a standard suspension vehicle (SSV) consisting of 0.9% (w/v) NaCl, 0.5% (w/v) Na-carboxymethylcellulose, 0.5% (v/v) benzyl alcohol, and 0.4% (v/v) Tween-80 with or without 30 mg/kg of Cu(ATSM). Following daily oral dosing for 28 days, the mice were humanely killed, and their brains were removed to isolate microvessels and the small intestine, liver and kidneys were isolated for P-gp expression and metallomic analysis, or the mice were anaesthetised for functional P-gp functional studies.

2.5. Brain Microvessel Membrane Fraction Isolation and Organ Harvesting

After completing the dosing regimen, mice were anesthetised with 1–5% isoflurane and then euthanised by cervical dislocation. Brains were removed and carefully rolled over filter paper to remove the meningeal layers and larger surface blood vessels. The olfactory bulb, cerebellum and brain stem were dissected out and only the cerebral cortices of mice (vascularised region representing the BBB) were included in the isolation. In order to obtain enough protein for quantification of P-gp using WB, 3–4 mouse cortices were combined to obtain enough microvessel enriched fractions (MEF) using a modified protocol previously reported by our laboratory ($n = 6$, 24 Vehicle (SSV) and 24 Cu(ATSM) (30 mg/kg) group) [48]. Briefly, the pooled brains were homogenised in 10 mL of DMEM GlutaMAX™ using a pre-cooled Dounce homogenizer (Tissue Grinder, Potter-ELV, Wheaton Industries, Millville, NJ, USA) with 10 vertical strokes and 1–2 twisting motions per vertical stroke. The homogenate was centrifuged at $2000 \times g$ at 4 °C for 5 min to obtain a pellet, and the pellet was resuspended to a final concentration of 15% (w/v) BSA in DMEM GlutaMAX™ and further centrifuged at $2000 \times g$ at 4 °C for 30 min. After centrifugation, the resulting supernatant (brain parenchymal fraction) and pellet (MEF) were separated, and the MEF pellet was rinsed with DMEM GlutaMAX™ to remove excess BSA before being filtered through a 100 μm nylon mesh strainer. The suspension was then centrifuged to obtain a pellet that was rinsed twice with PBS, snap-frozen in liquid nitrogen and stored at −80 °C for future membrane fraction isolation. Small intestines, kidneys, and livers of Cu(ATSM) and vehicle-dosed mice were also dissected, snap-frozen, and stored at −80 °C for later use for P-gp protein quantification or inductively-coupled plasma mass spectrometry (ICP-MS) for metallomic analysis.

To determine a more accurate representation of the expression of the functional membrane-bound ATP-binding cassette transporter (rather than investigate total cellular P-gp), membrane fractions of the isolated MEF samples were obtained according to the Mem-PER™ Plus Membrane Protein Extraction kit manufacturer's protocol, albeit with minor modifications. In brief, the capillary pellet isolated from the MEF were thawed on ice and resuspended in 500 μL of cell permeabilisation buffer provided in the kit with cOmplete™ mini protease inhibitor, transferred to a 1.5 mL Eppendorf tube and incubated at 4 °C for 10 min with constant gentle agitation. The permeabilised samples were centrifuged at $16,000 \times g$ for 15 min at 4 °C. The supernatant containing the cytosolic fractions were separated and aliquoted for protein determination and cytosolic metallomic analysis via ICP-MS. The remaining pellet was resuspended in 500 μL of membrane solubilisation buffer provided in the kit with cOmplete™ mini protease inhibitor and incubated at 4 °C for 30 min under gentle agitation. Following incubation, the solubilised membrane fractions were centrifuged at $16,000 \times g$ for 15 min at 4 °C. The supernatant containing membrane fraction proteins was collected and aliquoted for WB analysis.

2.6. ICP-MS of Brain MEFs and Organ Homogenates

The cytosolic fractions (250 μL) of microvessels, as well as peripheral tissue samples (small intestines, liver and kidneys) isolated from Cu(ATSM) (30 mg/kg) and vehicle-treated mice ($n = 6–8$) were lyophilised prior to sample preparation for ICP-MS. The wet weights of organs were determined prior to lyophilisation. Small intestines were approximately 150 mg, livers were approximately 150 mg, and the two kidneys had a combined weight of approximately 250 mg. Nitric acid (HNO_3) (65% Suprapur, Merck, Darmstadt, Germany) was added to each sample and digested overnight. The samples were further digested at 90 °C for 20 min in a heating block and then an equivalent volume of hydrogen peroxide 30% (v/v) (Aristar, BDH) was added to each sample, which was allowed to digest for 30 min. Samples were allowed to stop effervescing for 30 min before being heated again for 15 min at 70 °C. The average reduced volume was determined, and 50 μL aliquots of samples were further diluted 1:20 for MEF cytosolic fractions and 1:400 for tissue samples with 1% HNO_3 (v/v). The ICP-MS measurements were made using an Agilent 7700 series ICP-MS instrument (Santa Clara, CA, USA) with routine multi-element operating conditions using a Helium Reaction Gas Cell. The instrument was calibrated using 0, 5, 10, 50, 100 and 500 ppb of certified multi-element ICP-MS standard calibration solutions (ICP-MS-CAL2-1, ICP-MS-CAL-3 and ICP-MS-CAL-4, Accustandard, New Haven, CT, USA) for a range of elements. Additionally, a certified standard solution containing 200 ppb of Yttrium (Y89) was used as an internal control (ICP-MS-IS-MIX1-1, Accustandard). The elements assessed included sodium, magnesium, phosphorous, calcium, iron, cobalt, nickel, copper, zinc, selenium, and rubidium, and elements with abundances at or below detection limits were excluded from the data analysis.

2.7. WB of Brain MEFs and Organ Homogenates

An aliquot of the cytosolic and membrane fractions (10 µL) obtained from the isolated MEFs was quantified for total protein using the BCA protein assay against known standards of BSA. For the small intestine (jejunum), liver (left lobe) and kidney (left kidney), approximately 20 mg of wet-weight tissue was dissected from each main sample and placed into a 4 mL homogenisation tube containing 0.5 mL of Pierce IP lysis buffer and $7 \times$ concentrated protease inhibitor solution (1:6 ratio). The tissues were briefly homogenised for 10 s per tube using an IKA T25 digital ULTRA-TURRAX tissue homogeniser (Lab Gear Australia, Milton, Queensland, Australia) set at 4000/min. The samples were then left on ice for 30 min, centrifuged at $15,000 \times g$ for 15 min, and the supernatant was collected. The supernatant was then aliquoted for protein content analysis using the BCA protein assay and stored at $-80\,^\circ$C until further use.

Following protein quantification, the appropriate volume to obtain a consistent 20 µg protein load of cytosolic and membrane fractions as well as tissue lysates was mixed with $6 \times$ Laemmli buffer in a 5:1 ratio and incubated at $37\,^\circ$C for 20 min. Electrophoresis was carried out on Mini-Protein® TGX™ Precast gels (4–15% acrylamide) at 100 V for 30 min and then 150 V for 60 min. Gels, extra thick blot paper, and 0.45 µm pore-sized nitrocellulose membranes were equilibrated in a transfer buffer containing 20% (v/v) methanol for 20 min before transfer. The transfer was executed using the Bio-Rad Trans-Blot Turbo transfer system (Hercules, CA, USA) set to 40 min at 25 V and 1.0 A. Following a brief rinse in TBS-T, membranes were incubated in Intercept® blocking buffer at room temperature for 1 h, rinsed again in TBS-T, and then incubated with a 1:500 dilution of the primary C219 monoclonal antibody for P-gp and a 1:500,000 dilution of the primary mouse antibody for β-actin (for peripheral tissues and cytosolic fractions of MEFs) or a 1:25,000 dilution of the Na/K ATPase membrane loading control antibody (for membrane fractions of MEFs) in TBS-T overnight at $4\,^\circ$C. After four $\times$ 5 min membrane washes in TBS-T, the membranes were incubated with secondary IRDye® 800CW goat anti-mouse antibody (1:7500) and IRDye® 680LT donkey anti-rabbit antibody (1:20,000) for 2 h at room temperature followed by another four $\times$ 5 min washes. Membranes were scanned on the GE Amersham™ Typhoon™ instrument (Chicago, IL, USA). Densitometric analysis was performed via the ImageJ software Version 1.53t (National Institutes of Health, Bethesda, MD, USA) and the data were normalised to the β-actin or Na/K ATPase signal and expressed as relative fold-changes compared to controls.

2.8. In situ Transcardiac Brain Perfusion to Assess BBB Transport

An in situ transcardiac brain perfusion technique was utilised to evaluate the transport of the P-gp substrate [3H]-digoxin across the BBB in vehicle and Cu(ATSM)-treated mice (n = 6–7) (Section 3.4. schematic), following a previously described protocol from our laboratory [49]. [14C]-sucrose permeability across the BBB was determined to ensure vascular integrity remained intact following the Cu(ATSM) treatment. Ketamine (133 mg/kg) and xylazine (10 mg/kg) were used to establish surgical anaesthesia of mice (n = 6–7 per treatment group), and the mice were kept warm on a surgical platform during the procedure. The thoracic cavity was exposed, and the descending aorta was ligated to restrict blood flow to peripheral organs and both jugular veins were severed immediately prior to perfusion. A pre-warmed (37 °C) Krebs bicarbonate ringer buffer (128 mM NaCl,

4.2 mM KCl, 1.5 mM CaCl$_2$, 0.9 mM MgSO$_4$, 24 mM NaHCO$_3$, 2.4 mM NaH$_2$PO$_4$, and 9.0 mM glucose carbogenated with 95% O$_2$ 5% CO$_2$, adjusted to pH 7.4) containing [3H]-digoxin (0.1 µCi/mL) and [14C]-sucrose (0.05 µCi/mL) was injected (25G butterfly needle) into the left ventricle of the heart at a rate of 5 mL/min for 2 min using a Harvard infusion pump (Harvard Apparatus, Holliston, MA, USA). After 2 min, the perfusion was halted followed by cervical dislocation and the brain was collected in liquid scintillation vials and digested in 2 mL of Solvable™ at 50 °C overnight. The following day, colour was neutralised by adding 200 µL of hydrogen peroxide (30% *v/v*). The brain sample was mixed with 10 mL of Ultima Gold scintillation cocktail, while 2 mL of Ultima Gold scintillation cocktail was added to 100 µL of the radioactive perfusion fluid to quantify the amount of probe in the spike solution. The radioactivity of each sample was counted using a PerkinElmer 2800TR liquid scintillation analyser (Waltham, MA, USA). The apparent brain distribution volume of the probe compound (brain:perfusate ratio, mL/g) was determined based on previously described methods [49]. The apparent tissue distribution volume of the [3H]-digoxin (B:P, mL/g) was calculated using Q_{brain}/C_p normalised by brain weight (g), where Q_{brain} is the radioactivity in the brain (DPM/g) (radioactivity from the vascular volume subtracted), and C_p is the radioactivity per mL of perfusate (DPM/mL). The vascular volume was determined using the B:P of [14C]-sucrose.

2.9. Statistical Analyses

All figures and data were generated and analysed using GraphPad Prism® software version 8.1.1 (GraphPad Software Incorporated). All data were expressed as mean ± SEM, where all replicates compared were biological replicates. When comparing between two groups, a student's unpaired *t*-test was performed. Levels of significance are indicated for individual experiments, with differences considered to be statistically significant when the *p*-value was less than 0.05 ($p < 0.05$). The number of biological replicates of individual animals or pooled samples are represented by *n* (i.e., *n* = 1 for each pooled sample). Data were assessed for normality using the Sharpiro–Wilk test and all data points were used for statistical analysis. There were no inclusion or exclusion criteria pre-determined, and no blinding was performed, however, animals selected for treatment groups were randomised to allocated groups in the study.

3. Results

3.1. Characterisation and Purity of Isolated Primary mBECs

To validate the purity and yield of primary mBECs isolated with the adapted MACS® isolation method, a qualitative and quantitative approach was utilised. ICC was employed to assess if other CNS cells such as neurons, astrocytes and microglia remained in the suspension of cells after the magnetic separation of CD31-positive endothelial cells. It was clearly observed that there was an abundance of CD31-positive endothelial cells (i.e., mBECs) with no other contaminating cells present following the isolation and 72 h puromycin treatment (Figure 1a). FACS also confirmed that the cell suspension from the isolation method consisted of CD31-positive endothelial cells of 97.3% purity after the sample was gated for cell debris, single-cell suspension and viability (Figure 1b).

3.2. Cu(ATSM) Increases P-gp Expression and Function in Isolated Primary mBECs

Primary mBECs treated with 100 nM Cu(ATSM) for 24 hrs exhibited a 1.6-fold increase in P-gp expression compared to vehicle (0.1% DMSO)-treated cells (Figure 2a,b). The functionality of P-gp efflux was evaluated using a fluorescence-based assay with a known P-gp substrate, R123. A 25% reduction in the accumulation of R123 was observed in mBECs treated with Cu(ATSM) compared to vehicle (0.1% DMSO)-treated mBECs (Figure 2d), which was in line with the increase in P-gp expression.

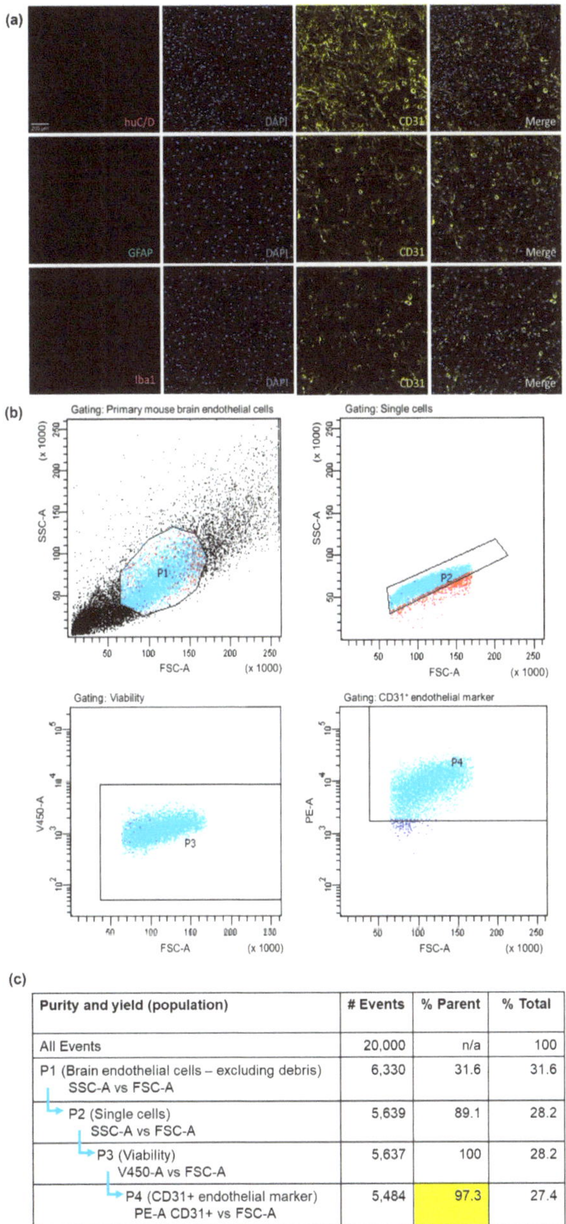

(c)

Purity and yield (population)	# Events	% Parent	% Total
All Events	20,000	n/a	100
P1 (Brain endothelial cells – excluding debris) SSC-A vs FSC-A	6,330	31.6	31.6
P2 (Single cells) SSC-A vs FSC-A	5,639	89.1	28.2
P3 (Viability) V450-A vs FSC-A	5,637	100	28.2
P4 (CD31+ endothelial marker) PE-A CD31+ vs FSC-A	5,484	97.3	27.4

Figure 1. The MACS® isolation method yields pure endothelial cells. Primary mBEC purity and yield assessed by ICC (**a**) indicating no neuronal (huC/D), astrocyte (GFAP) or microglial (Iba1) contamination but positive confirmation of CD31+ brain endothelial cells. Each sample consists of 2 brains pooled from C57BL/6J mice. A 97.3% CD31+ brain endothelial specific isolation was acquired that was gated (selection of cells indicated by light blue dots) for cell debris, single and viable cell populations confirmed by FACS (**b**). Table (**c**) indicates the gating strategy for FACS with forward scatter area (FSC-A) indicating cell size, side scatter area (SSC-A) to the granularity of the cells, viability 450 marker area (V450-A), and phycoerythrin fluorophore CD31+ marker (PE-A CD31+).

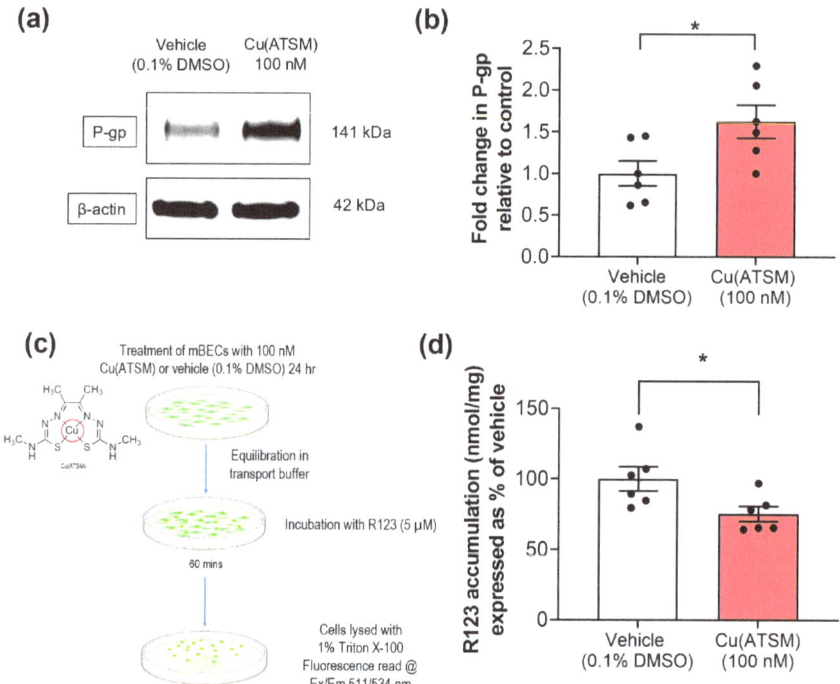

Figure 2. Primary mBECs treated with 100 nM Cu(ATSM) show increased expression (**a,b**) compared to mBECs treated with a vehicle (0.1% DMSO) for 24 h. (**a**) Representative WB of expression levels of P-gp and β-actin housekeeping protein from one gel. (**c**) P-gp function was assessed by measuring the cellular accumulation of a fluorescent substrate, R123 (nmol), normalised to total protein (mg). (**d**) Cu(ATSM) reduces accumulation of R123 in mBECs following a 24 h treatment. Data are presented as mean ± SEM (n = 6, independent cell culture isolations, 2 brains per pooled sample) expressed as fold-change for (**b**) WB or (**d**) % of vehicle (0.1% DMSO) for accumulation of R123 (nmol/mg protein) with * $p < 0.05$ when compared to the vehicle-treated group, assessed by an unpaired t-test. Illustration created with BioRender.com.

3.3. Mice Dosed with Cu(ATSM) Show Increased P-gp Abundance and Levels of Cu in Isolated Brain MEFs

To assess if the changes observed at the cellular level (mBECs) and from our previous studies (in hCMEC/D3 cells) were able to be translated into an in vivo system, mice were dosed daily with 30 mg/kg of Cu(ATSM) via oral administration over 28 days. When compared with mice dosed with vehicle (SSV), there was a visible increase in P-gp abundance in brain MEFs isolated from Cu(ATSM)-treated mice that were investigated in a pilot study to assess potential changes in expression (Figure 3a). Following this observation, all other samples were processed to assess the membrane fractions and cytosolic fractions for P-gp expression to account for functional membrane-bound P-gp. The levels of P-gp detected in the cytosolic fractions were negligible, whereas MEFs from the Cu(ATSM)-treated mice exhibited a 1.5-fold increase in membrane-bound P-gp when compared to those obtained from the vehicle (SSV)-treated group (Figure 3b,c). This correlated with an increase in the levels of Cu in the brain MEFs from the Cu(ATSM) treated mice (6.2 ± 0.88 µM/mg protein) compared to the brain MEFs isolated from vehicle-treated mice (1.7 ± 0.20 µM/mg protein) (Figure 3d).

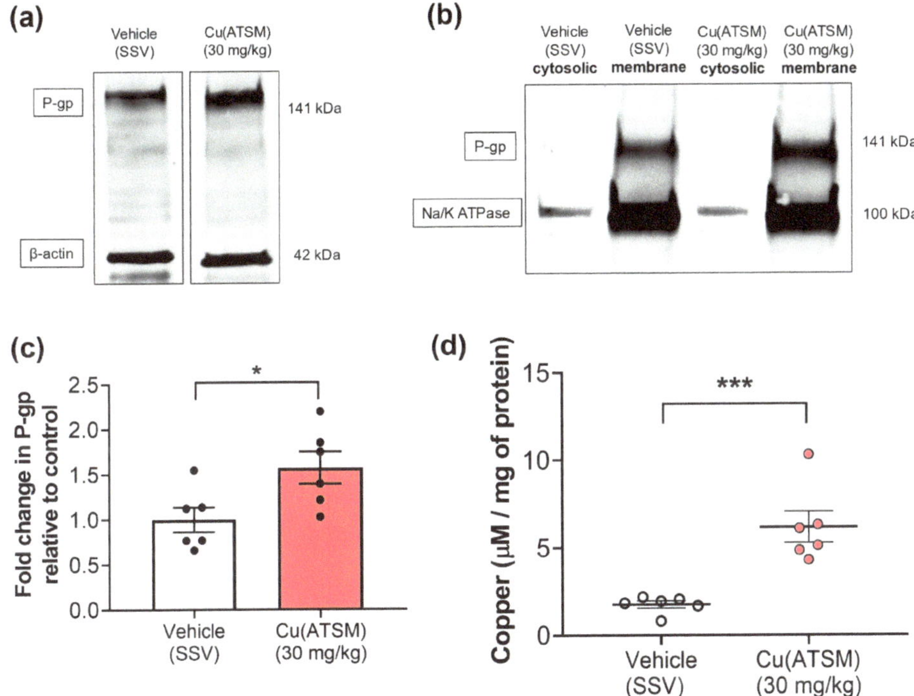

Figure 3. Cu(ATSM) upregulates P-gp expression with an increase in cytosolic copper (Cu) in MEFs isolated from mouse brain. Representative WB of expression levels of (**a**) P-gp and β-actin in MEF lysates and (**b**) P-gp and Na/K ATPase in MEF cytosolic & membrane fractions. A graphical representation of (**c**) P-gp expression as assessed by densitometry of membrane fractions and (**d**) levels of Cu in cytosolic MEFs from vehicle (SSV) or Cu(ATSM) (30 mg/kg)-treated mice. Data are presented as mean $\pm$ SEM (n = 6, 4 pooled brains per sample) expressed as fold change of vehicle (SSV) for WB or levels of Cu (µg/mg of protein) for ICP-MS with * $p < 0.05$, *** $p < 0.001$, when compared with the vehicle-treated group, assessed by an unpaired *t*-test.

3.4. Mice Dosed with Cu(ATSM) Show Increased BBB Function of P-gp

A clinically relevant substrate for P-gp, digoxin, was utilised to assess the efflux capacity of P-gp and to ascertain whether enhanced P-gp expression leads to enhanced efflux function in mice treated with Cu(ATSM). The in situ transcardiac brain perfusion studies demonstrated a 20% decrease in [3H]-digoxin uptake into the brains of mice treated with Cu(ATSM) (0.023 ± 0.00062 mL/g) compared to the vehicle-treated group (0.029 ± 0.0013 mL/g) (Figure 4). It was important to assess whether BBB integrity was maintained following the 28-day daily dosing of Cu(ATSM), and this was confirmed by assessing the transport of a paracellular diffusion marker [14C]-sucrose, which also provided a value of the microvascular volume. There was no significant difference observed in the cerebral vascular volume of [14C]-sucrose between the Cu(ATSM) (0.0077 ± 0.00025 mL/g) and vehicle-treated group (0.0078 ± 0.00027 mL/g) (Figure 4).

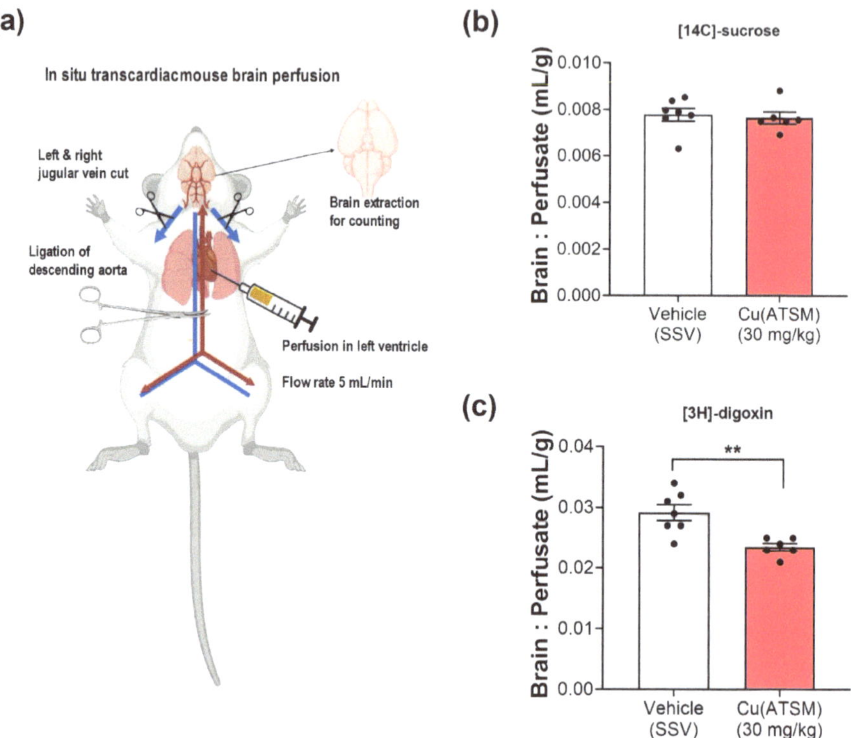

Figure 4. Cu(ATSM) enhances P-gp-mediated efflux function. In situ transcardiac brain perfusion technique (**a**) demonstrates an intact, uncompromised BBB with no difference in the brain:perfusate (B:P) ratio of [14C]-sucrose (**b**) but a decreased B:P ratio of the P-gp substrate [3H]-digoxin (**c**) in Cu(ATSM)-treated (30 mg/kg daily for 28 days) mice compared to the SSV-treated group. Data are presented as mean ± SEM (n = 6–7) expressed as brain: erfusate (mL/g) with ** $p < 0.01$ when compared with the SSV-treated group, assessed by an unpaired t-test. Illustration created with BioRender.com.

3.5. Cu(ATSM) Treatment Increases Copper and P-gp Expression in a Peripheral Organ-Specific Manner

To assess the impact of Cu(ATSM) on P-gp expression in peripheral organs and the biodistribution of Cu following the dosing regimen, WB and ICP-MS were utilised (Figure 5 & Table 2). The mice dosed with 30 mg/kg Cu(ATSM) exhibited a significantly higher Cu level in the jejunum (2.5 ± 0.26 μg/g tissue) when compared to vehicle-treated mice (1.4 ± 0.10 μg/g tissue), but this was not associated with a change in P-gp expression (Figure 5a). The biggest change in treatment-associated Cu levels was observed in the liver where Cu(ATSM)-treated mice had 60.2 ± 7.89 μg/g tissue Cu compared to 3.0 ± 0.12 μg/g tissue Cu in vehicle-treated mice (a 19.9-fold difference) (Figure 5b) and this was accompanied with a 1.3-fold increase in P-gp expression. No signs of liver toxicity or damage were observed in the tissues collected for analysis. Interestingly, there was no change in the levels of Cu nor P-gp expression in the kidneys isolated from the vehicle and Cu(ATSM)-treated mice.

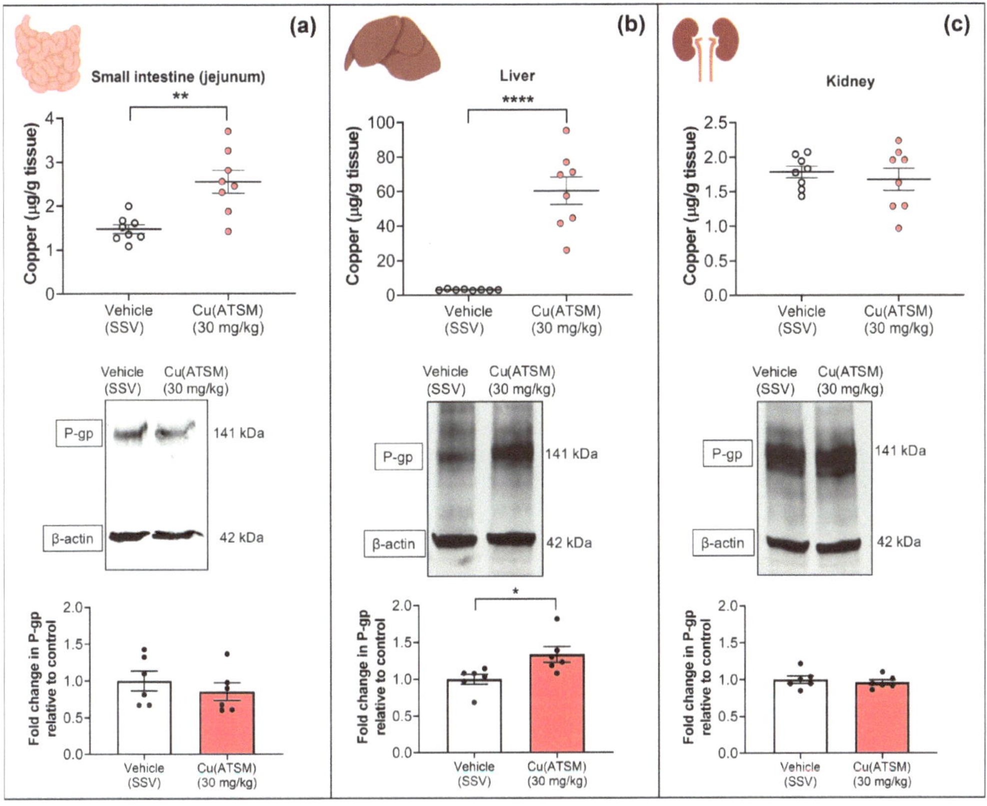

Figure 5. Cu levels detected in peripheral organs measured by ICP-MS following daily dosing of either Cu(ATSM) (30 mg/kg) or vehicle (SSV) for 28 days to C57BL/6J mice. Cu(ATSM) increased Cu in the small intestine (jejunum) (**a**) and liver (**b**) but not in the kidneys (**c**). P-gp protein expression was enhanced in the liver tissue only (**b**). Representative WBs of expression levels of P-gp and β-actin of peripheral organs and graphical representation of P-gp expression as assessed by densitometry (bottom). Data are presented as mean ± SEM (n = 6–8) as Cu levels (μg/g of tissue wet weight) or a fold change in P-gp compared to vehicle (SSV) with * $p < 0.05$, ** $p < 0.01$, **** $p < 0.0001$ when compared to the SSV-treated group as assessed by an unpaired t-test. Illustration created with BioRender.com.

Table 2. Metallomic analysis of tissue measured by ICP-MS. Concentrations of metal ions in various organs of C57BL/6J mice following a 28-day treatment with vehicle (SSV) or Cu(ATSM) (30 mg/kg/day) by oral gavage.

Tissue	Element	Concentration in Vehicle-Treated Mice (n = 6–8)	Concentration in Cu(ATSM)-Treated Mice (n = 6–8)	p Value	Fold Change to Control (If Significant)
	Sodium	237,338 ± 38,901	230,820 ± 37,462	0.9063	
	Magnesium	280.3 ± 43.39	232.3 ± 25.60	0.3634	
Microvessel	Phosphorus	20521 ± 2428	19953 ± 2564	0.8754	
Enriched	Potassium	9762 ± 1065	9590 ± 1095	0.9122	
Fraction	Calcium	1922 ± 542.1	1579 ± 273.2	0.5838	
(μg/mg protein)	Iron	183.8 ± 8.56	180.6 ± 10.35	0.8166	
	Copper	1.74 ± 0.20	6.16 ± 0.88	0.0006	3.5
	Zinc	13.08 ± 2.07	13.82 ± 2.01	0.8038	

Table 2. *Cont.*

Tissue	Element	Concentration in Vehicle-Treated Mice (n = 6–8)	Concentration in Cu(ATSM)-Treated Mice (n = 6–8)	p Value	Fold Change to Control (If Significant)
Small Intestines (µg/g tissue)	Sodium	463.6 ± 34.09	401.6 ± 17.80	0.1295	
	Magnesium	86.47 ± 4.91	92.14 ± 6.67	0.5046	
	Phosphorus	1222 ± 59.09	1082 ± 78.75	0.1746	
	Potassium	1394 ± 73.55	1163 ± 84.03	0.0578	
	Calcium	8226 ± 775.9	8970 ± 1133	0.5966	
	Iron	7.558 ± 0.57	7.558 ± 0.80	0.9997	
	Copper	1.473 ± 0.10	2.545 ± 0.26	0.0016	1.7
	Zinc	8.417 ±0.40	7.796 ± 0.56	0.3817	
Liver (µg/g tissue)	Sodium	405.7 ± 13.95	398.1 ± 17.36	0.7376	
	Magnesium	87.21 ± 2.48	77.25 ± 0.70	0.0017	0.9
	Phosphorus	1310 ± 36.79	1181 ± 19.62	0.0081	0.9
	Potassium	1547 ± 43.34	1355 ± 20.16	0.0012	0.9
	Calcium	1128 ± 97.08	1363 ± 125.6	0.1619	
	Iron	56.47 ± 2.99	46.99 ± 3.00	0.0421	0.8
	Copper	3.03 ± 0.12	60.19 ± 7.89	<0.0001	19.9
	Zinc	10.94 ± 0.31	13.77 ± 0.24	<0.0001	1.3
Kidneys (µg/g tissue)	Sodium	428.2 ± 15.71	382.6 ± 30.94	0.2103	
	Magnesium	49.20 ± 2.26	41.89 ± 3.60	0.1076	
	Phosphorus	816.1 ± 36.45	702.3 ± 64.41	0.1463	
	Potassium	868.3 ± 27.34	754.0 ± 58.65	0.0991	
	Calcium	2675 ± 203.8	2306 ± 178.0	0.1946	
	Iron	22.83 ± 1.23	20.38 ± 2.48	0.3910	
	Copper	1.781 ± 0.08	1.673 ± 0.16	0.5590	
	Zinc	4.380 ± 0.18	3.777 ± 0.32	0.1238	

Data are presented as mean ± SEM, with $p < 0.05$ deemed significantly different to the vehicle-treated group assessed by an unpaired t-test.

The analysis of the levels of other metals in the MEF and peripheral organs revealed no significant differences between the two treatment groups except within the liver, as summarised in Table 2. With regards to the liver, Cu(ATSM)-treated mice exhibited a significant reduction of between 0.8–0.9-fold in magnesium, phosphorus, potassium, and iron. However, there was a 1.3-fold increase in zinc (Zn) in the livers of Cu(ATSM)-treated mice (Table 2).

4. Discussion

Although P-gp impedes drug delivery across the BBB, it is also involved in the clearance of neurotoxic Aβ peptides from the brain, making it a crucial contributor to AD etiology [24,26,29]. The identification of potential modulators of P-gp expression and function, and understanding the regulatory mechanisms involved in maintaining the function of this gatekeeper, has numerous applications in developing neurotherapeutics. Copper influences pathways involved in P-gp regulation, and our previous findings showed modulation of key efflux transporters in hCMEC/D3 cells treated with copper-modulating compounds [36,40,50,51]. Therefore, this study investigated the impact of Cu(ATSM) in mBECs and the mouse microvasculature in vivo to determine if the observed changes in hCMEC/D3 cells translated to mice, where Cu(ATSM) has been shown to exhibit disease-reversing effects. Cu(ATSM) has a relatively stable structure compared to other bis(thiosemicarbazone) counterparts due to the presence of two additional methyl groups which increase the reduction potential and thus prevent the release of Cu [52]. However, under hypoxic conditions where the reducing capacity of the intracellular environment is increased, Cu(ATSM) is reduced, leading to an increase in bioavailable Cu. This selective property makes Cu(ATSM) a potential therapeutic agent for cells and tissues under oxygen-deprived stress in various disease states [53]. Cu(ATSM) has great potential to be utilised in diagnostics and has recently emerged as a promising pharmacotherapeutic for the treatment of various neurodegenerative disorders including amyotrophic lateral sclerosis, Parkinson's disease, ischemic stroke and neuroinflammation [45,54–57]. While Cu(ATSM) has been designed to release Cu in hypoxic environments, we demonstrated that this entity also increases Cu in hCMEC/D3 cells, an outcome associated with an increase in P-gp expression and function. This suggests that in addition to disease-modifying effects, Cu(ATSM) may have beneficial effects at the BBB, which has not been previously investigated.

This study demonstrated the successful isolation of 97.3% pure mBECs using the MACS® system and this was validated using both ICC and FACS. The purity of mBECs was comparable to other studies using other techniques for isolation such as glass bead purification, or BSA or dextran density filtration without magnetic separation, where an 87–95% purity was obtained as confirmed by FACS, with the majority of contaminating cells reported to be microglia or astrocytes [58–60]. The MACS® system proved to be a significant advance for isolating mBECs due to the commercially available complete kits that facilitated the relatively fast, reliable, and relatively simple acquisition of pure endothelial cells. The purity of cells was essential to investigate the effects of Cu(ATSM) on the expression and function of P-gp so that we could be confident that any alterations observed were indeed mBEC-specific given that P-gp is also expressed in astrocytes, pericytes and microglia [61,62]. The increase in P-gp expression and function in mBECs following Cu(ATSM) treatment was comparable to that found in hCMEC/D3 cells previously in our laboratory, where a 24 h Cu(ATSM) treatment at the same concentration (100 nM) lead to a 2-fold increase in P-gp expression and 30% increase in efflux of R123 [40]. Other studies have used alternative methods to enhance P-gp expression and function such as activation of the pregnane X receptor (PXR) as well as blocking P-gp degradation [24,63–65]. Expression and function of P-gp were significantly enhanced in all of these studies compared to controls, but direct comparison of P-gp function cannot be made as the methods to assess transport ranged from cellular transport of diverse P-gp substrates to isolated capillary transport assays, and thus no direct comparison to in situ brain perfusion studies could be made. However, taken together, these studies demonstrate that P-gp can be modulated with various interventions and similar to other approaches, Cu(ATSM) has the potential to modulate the trafficking of agents into and out of the brain.

The mechanism of P-gp upregulation following Cu(ATSM) treatment was not explored in this study but our previous studies have elucidated a potential pathway in hCMEC/D3 cells that is likely replicated in mBECs. The regulation of MDR1 transcription and subsequent P-gp expression has been associated with the activation of the mitogen-activated protein kinase (MAPK) pathway and downstream activation of ERK1/2 [66–68]. Studies, including our previous findings, have demonstrated that Cu(ATSM) activates this pathway, leading to the phosphorylation of ERK1/2 [40,69–71]. By inhibiting this pathway with U0126 (an upstream inhibitor of mitogen-activated protein kinase kinase 1/2), the beneficial effects of Cu(ATSM) are prevented [40]. As these pathways are also present in mBECs [72,73], it is likely that Cu(ATSM) activates these pathways to increase P-gp expression. Cu(ATSM)-mediated upregulation of P-gp may also be attributed to the activation of the neuroprotective molecular sensor of oxidative stress, nuclear erythroid 2–related factor 2 (Nrf2). Nrf2 was shown to promote P-gp expression and transport activity in rat brain capillaries and this potential pathway has been demonstrated to be activated by Cu(ATSM) [69,74,75]. While Nrf2 is a ligand-activated transcription factor, it has been suggested that Nrf2-mediated P-gp upregulation occurs indirectly, involving activation of NF-κB via p53 and p38 signalling [74]. Studies in murine models of traumatic brain injury have demonstrated that Nrf2 ligand administration is neuroprotective and helps to maintain BBB integrity [76]. Srivastava et al. (2016) demonstrated with both oral delivery of Cu(ATSM) in mice and in vitro that Cu(ASTM) treatment leads to a significant upregulation of Nrf2-dependent antioxidant defences that elicit the cardioprotective effects of Cu(ATSM) [69]. Furthermore, ERK1/2 activation has been shown to mediate Nrf2 phosphorylation [77], thus providing an integrated mechanism through which Cu(ATSM) may enhance Nrf2 signalling. Taken together, this suggests a potential mechanism of action for P-gp regulation in response to Cu(ATSM) suggesting that it may be mediated through the activation of, but not limited to MAPK, ERK1/2 and Nrf2 signalling. Further studies are needed to elucidate the exact mechanisms and pathways involved in the regulation of P-gp in mice following treatment with Cu(ATSM).

While our in vitro studies in mice and human brain endothelial cells were complementary, it was important to assess whether Cu(ATSM) had a similar P-gp-enhancing effect *in vivo*. Mice dosed with Cu(ATSM) over 28 days showed a 1.5-fold increase in P-gp expression in brain MEFs. The MEFs isolated from mice were separated into membrane and cytosolic fractions to probe for membrane associated-P-gp which represents the functional ATP-binding cassette efflux transporter (Figure 3b). The cytosolic fractions did not show P-gp bands after completing the WB indicating that the majority of detected P-gp was indeed the membrane-associated functional protein. The enhanced expression was accompanied by a 20% reduction in the transport of the P-gp substrate [3H]-digoxin. The brain:perfusate ratio of [3H]-digoxin was significantly reduced in Cu(ATSM) treated mice indicating enhanced P-gp mediated efflux capacity, and this ratio was comparable (~0.002–0.003 mL/g) to ratios found in other studies in our laboratory that used a 150-day old mouse model of amyotrophic lateral sclerosis [78]. This comparison suggests that P-gp-mediated [3H]-digoxin efflux is similar across different ages and strains of mice. To ensure that Cu(ATSM) did not compromise the integrity of the BBB and thus alter paracellular transport, [14C]-sucrose was used as an integrity marker. There was no significant difference in [14C]-sucrose transport between the treatment groups and we thus concluded that Cu(ATSM) does not change paracellular permeability. The B:P ratio of [14C]-sucrose was similar to that found in recent studies for vascular volume determination in transcardiac brain perfusion studies (ranging from 0.006 mL/g–0.013 mL/g) [49,78,79]. The findings of the current study demonstrate that Cu(ATSM) may be a promising drug candidate for increasing P-gp expression and function in vivo, and may represent a potential strategy to enhance clearance of Aβ across the BBB, with possible disease-modifying effects.

Having shown that Cu(ATSM) increased the BBB expression and function of P-gp, we then assessed the levels of Cu (μg/mg protein) being delivered to brain MEFs following daily oral administration of 30 mg/kg Cu(ATSM) for 28 consecutive days. To our knowledge, this is the first analysis of MEF metal levels in mice treated with Cu(ATSM) (Table 2). Using ICP-MS, a 3.5-fold increase in Cu was detected in the cytosolic MEFs from Cu(ATSM)-treated mice when compared to vehicle-treated mice. This finding was consistent with our previous in vitro studies showing a significant increase in intracellular Cu, supporting the hypothesis that Cu(ATSM) increases both bioavailable Cu and P-gp expression in vivo as well as in vitro [40]. As was observed with our studies in hCMEC/D3 cells, Cu(ATSM) treatment did not result in any detectable differences in the levels of other metal levels in the brain MEF. Interestingly, brain Cu levels have been reported to increase 1.2-fold in mice treated with clioquinol [80], an agent that we have shown increases P-gp expression in hCMEC/D3 cells [36]. It would therefore be expected that clioquinol would also increase P-gp expression in brain microvessels, however, we have shown that neither P-gp expression nor Cu levels in MEFs are significantly different between vehicle and clioquinol-treated mice [50]. This is an important distinction to make as it suggests that compounds need to modulate brain microvascular Cu levels specifically, rather than elevate Cu levels across the brain overall, in order to modulate P-gp expression, as we have demonstrated with Cu(ATSM).

To obtain a better understanding of the potential systemic effects of Cu(ATSM), we conducted a metallomic analysis of peripheral organs, investigating changes to metal biodistribution as well as P-gp expression in the small intestines, liver, and kidneys of mice treated with oral Cu(ATSM) in comparison to vehicle-treated mice (Figure 5 and Table 2). These organs abundantly express P-gp and although the blood-testis barrier and the placenta were not investigated as part of the current study, future studies could explore the impact of Cu(ATSM) on P-gp that also has a protective role in these tissues. Within the small intestines, there was a significant increase in Cu levels detected but no differences to other metals, and no changes in P-gp expression. While Cu levels are increased in the small intestine, it is possible that the suggested signalling pathways mediating P-gp regulation by Cu in the brain microvessels may not be impacted by Cu in the small intestine, or such pathways may have less of an impact on P-gp regulation in the small intestine. Remarkably, there was a 19.9-fold increase in liver levels of Cu in mice treated with Cu(ATSM) relative to vehicle-treated mice. Compared to the small intestines, the highly vascularised nature of the liver may account for the accumulation of Cu(ATSM) in hepatic tissue. This substantial increase in hepatic Cu levels was accompanied by an increase in P-gp expression. This effect of Cu(ATSM) has the potential to lead to an increase in hepatic P-gp-mediated clearance and therefore it is possible that Cu(ATSM) may affect the pharmacokinetics of other drugs that are P-gp substrates, leading to drug-drug interactions [81]. Interestingly, and only within the liver, a 1.3-fold increase in Zn with Cu(ATSM) treatment was also observed, alongside significant reductions in magnesium, phosphorus, potassium and iron. The reduction in other metals may be attributed to the higher levels of Cu in the liver which may displace these metals, as has been shown in the case of iron mobilisation [31,32]. The increase in Zn is in contrast with what has been reported previously, where an increase in Zn reduces the levels of Cu in the liver because Zn hinders the absorption of Cu and competes with it to bind to the same metal binding sites on proteins [82,83]. However, this would not be expected in this scenario given that it is unlikely that Zn would impact the passive diffusion of Cu(ATSM) into hepatocytes, the agent responsible for the increase in hepatocellular Cu. Furthermore, the kidneys did not exhibit any differences in Cu levels nor P-gp expression. A recent study also reports that following oral administration of Cu(ATSM), no significant changes were observed in levels of Cu in the kidney, but significant increases were observed in the liver and brain which aligns with our observations [84]. However, our study is the first that demonstrates increased Cu availability in the brain microvasculature and the first to associate the increase in Cu levels with an increase in P-gp expression.

The limitations of the sampling used in this investigation were recognised and acknowledged. The assays employed, namely ICP-MS and WB, were performed on homogenised whole tissue samples and not on sub-organ regions of those tissues other than the small intestines where only the jejunum was analysed. This may limit the ability to detect any region-specific or localised changes in the tissues analysed. The ICP-MS detects total Cu and cannot distinguish between bioavailable and complex-bound Cu. However, the data presented in this study provides a basis for drawing appropriate conclusions and comparisons between treatment groups. The present study paves the way for further research into the potential clinical applications of Cu(ATSM) and similar compounds that target P-gp expression and function at the BBB and offers hope for the development of more effective drug therapies for neurodegenerative diseases such as AD.

5. Conclusions

This study has successfully demonstrated the capacity of Cu(ATSM) to enhance P-gp expression and function in mBECs, as well as in vivo, given that oral administration of Cu(ATSM) resulted in increased levels of Cu and P-gp expression in brain MEFs, as well as in the liver. The implications of this research are significant, as they demonstrate the potential of Cu(ATSM) to modulate BBB transport of P-gp substrates, including many drugs as well as Aβ, the toxic peptide accumulating in the brain in AD. The outcomes of this study have provided a foundation for future research and investigation of Cu(ATSM) and its potential as a therapeutic approach for reducing Aβ levels in AD. Further studies are needed to explore the specific mechanism by which Cu(ATSM) upregulates P-gp and investigate the potential of this antioxidative, anti-inflammatory and neuroprotective compound to reduce the burden of Aβ in AD models by enhancing clearance through this key BBB transporter.

Author Contributions: J.P. conceived, planned, and carried out all experimentation, analysed experimental data, designed the figures and drafted the manuscript. H.K. synthesised and characterised the bis(thiosemicarbazone) complex. P.R. assisted with the collection of samples, animal handling and reviewing procedures. C.M. performed and processed the ICP-MS data. Y.P. trained, provided, and supervised the isolation of primary cells and animal interventions. P.S.D. and A.I.B. consulted on the study design, aided in interpreting the results and provided supervision of work. J.L.S. and J.A.N. were involved in designing and planning and supervised all aspects of the work. All authors have read and agreed to the published version of the manuscript.

Funding: Jae Pyun was supported by the Australian Government Research Training Program Scholarship. This study was also supported in part by funding from the Australian Research Council (DP200100178) and the National Health and Medical Research Council (APP2011853). The Australian Cancer Research Foundation provided support for the NMR spectrometer used in this research (Bio21). We thank the Mass Spectrometry and Proteomics Facility at Bio21 Institute, University of Melbourne. Professor Ashley I. Bush is supported by a National Health and Medical Research Council L3 Investigator Grant (1194028). Open access publishing facilitated by Monash University, as part of the MDPI—Monash University agreement via the Council of Australian University Librarians.

Institutional Review Board Statement: The study was conducted in accordance with the National Health and Medical Research Council guidelines for the care and use of animals for scientific purposes and approved by the Monash Institute of Pharmaceutical Sciences Animal Ethics Committee (protocol number, MIPS 27467, date of approval 28 February 2021) of Monash University, Faculty of Pharmacy and Pharmaceutical Sciences.

Informed Consent Statement: Not applicable.

Data Availability Statement: Data is contained within the article. The data and analysis presented in this study are available upon request.

Conflicts of Interest: Collaborative Medicinal Development LLC has licensed intellectual property related to this subject from The University of Melbourne, where Professor Paul S. Donnelly is listed as an inventor and has served as a consultant to Collaborative Medicinal Development LLC. Professor Ashley I. Bush is a paid consultant for, and has a profit share interest in, Collaborative Medicinal Development LLC and is a shareholder in Alterity Ltd., Cogstate Ltd. and Mesoblast Ltd.

References

1. Hawkins, B.T.; Davis, T.P. The blood-brain barrier/neurovascular unit in health and disease. *Pharmacol. Rev.* **2005**, *57*, 173–185. [CrossRef] [PubMed]
2. Schinkel, A.H. P-Glycoprotein, a gatekeeper in the blood–brain barrier. *Adv. Drug Deliv. Rev.* **1999**, *36*, 179–194. [CrossRef] [PubMed]
3. Ueda, K.; Clark, D.P.; Chen, C.J.; Roninson, I.B.; Gottesman, M.M.; Pastan, I. The human multidrug resistance (mdr1) gene. cDNA cloning and transcription initiation. *J. Biol. Chem.* **1987**, *262*, 505–508. [CrossRef]
4. Borst, P.; Schinkel, A.H. P-glycoprotein ABCB1: A major player in drug handling by mammals. *J. Clin. Investig.* **2013**, *123*, 4131–4133. [CrossRef] [PubMed]
5. Schinkel, A.H.; Wagenaar, E.; van Deemter, L.; A Mol, C.; Borst, P. Absence of the mdr1a P-Glycoprotein in mice affects tissue distribution and pharmacokinetics of dexamethasone, digoxin, and cyclosporin A. *J. Clin. Investig.* **1995**, *96*, 1698–1705. [CrossRef] [PubMed]
6. Schinkel, A.; Smit, J.; van Tellingen, O.; Beijnen, J.; Wagenaar, E.; van Deemter, L.; Mol, C.; van der Valk, M.; Robanus-Maandag, E.; Riele, H.T.; et al. Disruption of the mouse mdr1a P-glycoprotein gene leads to a deficiency in the blood-brain barrier and to increased sensitivity to drugs. *Cell* **1994**, *77*, 491–502. [CrossRef]
7. Schinkel, A.H.; Wagenaar, E.; Mol, C.A.; Van Deemter, L. P-glycoprotein in the blood-brain barrier of mice influences the brain penetration and pharmacological activity of many drugs. *J. Clin. Investig.* **1996**, *97*, 2517–2524. [CrossRef]
8. Callaghan, R.; Luk, F.; Bebawy, M. Inhibition of the multidrug resistance P-glycoprotein: Time for a change of strategy? *Drug Metab. Dispos.* **2014**, *42*, 623–631. [CrossRef]
9. Atadja, P.; Watanabe, T.; Xu, H.; Cohen, D. PSC-833, a frontier in modulation of P-glycoprotein mediated multidrug resistance. *Cancer Metastasis Rev.* **1998**, *17*, 163–168. [CrossRef]
10. Palmeira, A.; Sousa, E.; Vasconcelos, M.H.; Pinto, M.M. Three decades of P-gp inhibitors: Skimming through several generations and scaffolds. *Curr. Med. Chem.* **2012**, *19*, 1946–2025. [CrossRef]
11. Abbott, N.J.; Patabendige, A.A.K.; Dolman, D.E.M.; Yusof, S.R.; Begley, D.J. Structure and function of the blood–brain barrier. *Neurobiol. Dis.* **2010**, *37*, 13–25. [CrossRef] [PubMed]
12. Choi, H.; Lee, E.-H.; Han, M.; An, S.-H.; Park, J. Diminished expression of P-glycoprotein using focused ultrasound is associated with JNK-dependent signaling pathway in cerebral blood vessels. *Front. Neurosci.* **2019**, *13*, 1350. [CrossRef] [PubMed]

13. Correia, A.; Monteiro, A.; Silva, R.; Moreira, J.; Lobo, J.S.; Silva, A. Lipid nanoparticles strategies to modify pharmacokinetics of central nervous system targeting drugs: Crossing or circumventing the blood-brain barrier (BBB) to manage neurological disorders. *Adv. Drug Deliv. Rev.* **2022**, *189*, 114485. [CrossRef] [PubMed]

14. Karamanos, Y.; Gosselet, F.; Dehouck, M.-P.; Cecchelli, R. Blood–brain barrier proteomics: Towards the understanding of neurodegenerative diseases. *Arch. Med. Res.* **2014**, *45*, 730–737. [CrossRef] [PubMed]

15. Vautier, S.; Fernandez, C. ABCB1: The role in Parkinson's disease and pharmacokinetics of antiparkinsonian drugs. *Expert Opin. Drug Metab. Toxicol.* **2009**, *5*, 1349–1358. [CrossRef]

16. Qosa, H.; Lichter, J.; Sarlo, M.; Markandaiah, S.S.; McAvoy, K.; Richard, J.-P.; Jablonski, M.R.; Maragakis, N.J.; Pasinelli, P.; Trotti, D. Astrocytes drive upregulation of the multidrug resistance transporter ABCB1 (P-glycoprotein) in endothelial cells of the blood–brain barrier in mutant superoxide dismutase 1-linked amyotrophic lateral sclerosis. *Glia* **2016**, *64*, 1298–1313. [CrossRef]

17. van Assema, D.M.; Lubberink, M.; Bauer, M.; van der Flier, W.M.; Schuit, R.C.; Windhorst, A.D.; Comans, E.F.; Hoetjes, N.J.; Tolboom, N.; Langer, O.; et al. Blood–brain barrier P-glycoprotein function in Alzheimer's disease. *Brain* **2011**, *135*, 181–189. [CrossRef]

18. Sweeney, M.D.; Zhao, Z.; Montagne, A.; Nelson, A.R.; Zlokovic, B.V. Blood-brain barrier: From physiology to disease and back. *Physiol. Rev.* **2018**, *99*, 21–78. [CrossRef]

19. Zlokovic, B.V. Neurovascular mechanisms of Alzheimer's neurodegeneration. *Trends Neurosci.* **2005**, *28*, 202–208. [CrossRef]

20. Montagne, A.; Barnes, S.R.; Sweeney, M.D.; Halliday, M.R.; Sagare, A.P.; Zhao, Z.; Toga, A.W.; Jacobs, R.E.; Liu, C.Y.; Amezcua, L.; et al. Blood-brain barrier breakdown in the aging human hippocampus. *Neuron* **2015**, *85*, 296–302. [CrossRef]

21. Iturria-Medina, Y.I.; Sotero, R.C.; Toussaint, P.J.; Mateos-Pérez, J.M.; Evans, A.C. Early role of vascular dysregulation on late-onset Alzheimer's disease based on multifactorial data-driven analysis. *Nat. Commun.* **2016**, *7*, 11934. [CrossRef] [PubMed]

22. Lam, F.C.; Liu, R.; Lu, P.; Shapiro, A.B.; Renoir, J.M.; Sharom, F.J.; Reiner, P.B. β-amyloid efflux mediated by P-glycoprotein. *J. Neurochem.* **2001**, *76*, 1121–1128. [CrossRef] [PubMed]

23. Kuhnke, D.; Jedlitschky, G.; Grube, M.; Krohn, M.; Jucker, M.; Mosyagin, I.; Cascorbi, I.; Walker, L.C.; Kroemer, H.K.; Warzok, R.W.; et al. MDR1-P-glycoprotein (ABCB1) mediates transport of Alzheimer's amyloid-β peptides—Implications for the mechanisms of Aβ clearance at the blood–brain barrier. *Brain Pathol.* **2007**, *17*, 347–353. [CrossRef] [PubMed]

24. Hartz, A.M.; Miller, D.S.; Bauer, B. Restoring blood-brain barrier P-glycoprotein reduces brain amyloid-β in a mouse model of Alzheimer's disease. *Mol. Pharmacol.* **2010**, *77*, 715–723. [CrossRef] [PubMed]

25. Chai, A.B.; Hartz, A.M.S.; Gao, X.; Yang, A.; Callaghan, R.; Gelissen, I.C. New evidence for P-gp-mediated export of amyloid-β peptides in molecular, blood-brain barrier and neuronal models. *Int. J. Mol. Sci.* **2021**, *22*, 246. [CrossRef]

26. Cirrito, J.R.; Deane, R.; Fagan, A.M.; Spinner, M.L.; Parsadanian, M.; Finn, M.B.; Jiang, H.; Prior, J.L.; Sagare, A.; Bales, K.R.; et al. P-glycoprotein deficiency at the blood-brain barrier increases amyloid-β deposition in an Alzheimer disease mouse model. *J. Clin. Investig.* **2005**, *115*, 3285–3290. [CrossRef]

27. Vogelgesang, S.; Cascorbi, I.; Schroeder, E.; Pahnke, J.; Kroemer, H.K.; Siegmund, W.; Kunert-Keil, C.; Walker, L.C.; Warzok, R.W. Deposition of Alzheimer's β-amyloid is inversely correlated with P-glycoprotein expression in the brains of elderly non-demented humans. *Pharm. Genom.* **2002**, *12*, 535–541. [CrossRef]

28. Chiu, C.; Miller, M.C.; Monahan, R.; Osgood, D.P.; Stopa, E.G.; Silverberg, G.D. P-glycoprotein expression and amyloid accumulation in human aging and Alzheimer's disease: Preliminary observations. *Neurobiol. Aging* **2015**, *36*, 2475–2482. [CrossRef]

29. Deo, A.K.; Borson, S.; Link, J.M.; Domino, K.; Eary, J.F.; Ke, B.; Richards, T.L.; Mankoff, D.A.; Minoshima, S.; O'sullivan, F.; et al. Activity of P-glycoprotein, a β-amyloid transporter at the blood–brain barrier, is compromised in patients with mild Alzheimer disease. *J. Nucl. Med.* **2014**, *55*, 1106–1111. [CrossRef]

30. Jeynes, B.; Provias, J. An investigation into the role of P-glycoprotein in Alzheimer's disease lesion pathogenesis. *Neurosci. Lett.* **2011**, *487*, 389–393. [CrossRef]

31. Ayton, S.; Lei, P.; Bush, A.I. Metallostasis in Alzheimer's disease. *Free Radic. Biol. Med.* **2013**, *62*, 76–89. [CrossRef] [PubMed]

32. Lutsenko, S.; Bhattacharjee, A.; Hubbard, A.L. Copper handling machinery of the brain. *Metallomics* **2010**, *2*, 596–608. [CrossRef] [PubMed]

33. Barnham, K.J.; Bush, A.I. Metals in Alzheimer's and Parkinson's diseases. *Curr. Opin. Chem. Biol.* **2008**, *12*, 222–228. [CrossRef] [PubMed]

34. Barnham, K.J.; Bush, A.I. Biological metals and metal-targeting compounds in major neurodegenerative diseases. *Chem. Soc. Rev.* **2014**, *43*, 6727–6749. [CrossRef]

35. Agarwal, P.; Ayton, S.; Agrawal, S.; Dhana, K.; Bennett, D.A.; Barnes, L.L.; Leurgans, S.E.; Bush, A.I.; Schneider, J.A. Brain copper may protect from cognitive decline and Alzheimer's disease pathology: A community-based study. *Mol. Psychiatry* **2022**, *27*, 4307–4313. [CrossRef]

36. McInerney, M.P.; Volitakis, I.; Bush, A.I.; Banks, W.A.; Short, J.L.; Nicolazzo, J.A. Ionophore and biometal modulation of P-glycoprotein expression and function in human brain microvascular endothelial cells. *Pharm. Res.* **2018**, *35*, 83. [CrossRef]

37. Donnelly, P.S.; Caragounis, A.; Du, T.; Laughton, K.M.; Volitakis, I.; Cherny, R.A.; Sharples, R.A.; Hill, A.F.; Li, Q.-X.; Masters, C.L.; et al. Selective intracellular release of copper and zinc ions from bis (thiosemicarbazonato) complexes reduces levels of Alzheimer disease amyloid-β peptide. *J. Biol. Chem.* **2008**, *283*, 4568–4577. [CrossRef]

38. Fodero-Tavoletti, M.T.; Villemagne, V.L.; Paterson, B.M.; White, A.R.; Li, Q.X.; Camakaris, J.; O'keefe, G.; Cappai, R.; Barnham, K.J.; Donnelly, P.S. Bis (thiosemicarbazonato) Cu-64 complexes for positron emission tomography imaging of Alzheimer's disease. *J. Alzheimer's Dis.* **2010**, *20*, 49–55. [CrossRef]

39. Paterson, B.M.; Cullinane, C.; Crouch, P.; White, A.R.; Barnham, K.J.; Roselt, P.D.; Noonan, W.; Binns, D.; Hicks, R.J.; Donnelly, P.S. Modification of biodistribution and brain uptake of copper bis(thiosemicarbazonato) complexes by the incorporation of amine and polyamine functional groups. *Inorg. Chem.* **2019**, *58*, 4540–4552. [CrossRef]

40. Pyun, J.; McInnes, L.E.; Donnelly, P.S.; Mawal, C.; Bush, A.I.; Short, J.L.; Nicolazzo, J.A. Copper bis (thiosemicarbazone) complexes modulate P-glycoprotein expression and function in human brain microvascular endothelial cells. *J. Neurochem.* **2022**, *162*, 226–244. [CrossRef]

41. Syvänen, S.; Lindhe, O.; Palner, M.; Kornum, B.R.; Rahman, O.; Långström, B.; Knudsen, G.M.; Hammarlund-Udenaes, M. Species differences in blood-brain barrier transport of three positron emission tomography radioligands with emphasis on P-glycoprotein transport. *Drug Metab. Dispos.* **2009**, *37*, 635–643. [CrossRef] [PubMed]

42. Deo, A.K.; Theil, F.-P.; Nicolas, J.-M. Confounding parameters in preclinical assessment of blood–brain barrier permeation: An overview with emphasis on species differences and effect of disease states. *Mol. Pharm.* **2013**, *10*, 1581–1595. [CrossRef] [PubMed]

43. Blower, P.J.; Castle, T.C.; Cowley, A.R.; Dilworth, J.R.; Donnelly, P.S.; Labisbal, E.; Sowrey, F.E.; Teat, S.J.; Went, M.J. Structural trends in copper (II) bis (thiosemicarbazone) radiopharmaceuticals. *Dalton Trans.* **2003**, *23*, 4416–4425. [CrossRef]

44. Gingras, B.; Suprunchuk, T.; Bayley, C.H. The preparation of some thiosemicarbazones and their copper complexes: Part III. *Can. J. Chem.* **1962**, *40*, 1053–1059. [CrossRef]

45. Hung, L.W.; Villemagne, V.L.; Cheng, L.; Sherratt, N.A.; Ayton, S.; White, A.R.; Crouch, P.J.; Lim, S.; Leong, S.L.; Wilkins, S.; et al. The hypoxia imaging agent Cu^{II} (atsm) is neuroprotective and improves motor and cognitive functions in multiple animal models of Parkinson's disease. *J. Exp. Med.* **2012**, *209*, 837–854. [CrossRef]

46. Roberts, B.R.; Lim, N.K.; McAllum, E.J.; Donnelly, P.S.; Hare, D.J.; Doble, P.A.; Turner, B.J.; Price, K.A.; Lim, S.C.; Paterson, B.M.; et al. Oral treatment with Cu^{II} (atsm) increases mutant SOD1 in vivo but protects motor neurons and improves the phenotype of a transgenic mouse model of amyotrophic lateral sclerosis. *J. Neurosci.* **2014**, *34*, 8021–8031. [CrossRef]

47. Soon, C.P.; Donnelly, P.S.; Turner, B.J.; Hung, L.W.; Crouch, P.J.; Sherratt, N.A.; Tan, J.L.; Lim, N.K.H.; Lam, L.; Bica, L.; et al. Diacetylbis (N(4)-methylthiosemicarbazonato) copper (II) (Cu^{II}(atsm)) protects against peroxynitrite-induced nitrosative damage and prolongs survival in amyotrophic lateral sclerosis mouse model. *J. Biol. Chem.* **2011**, *286*, 44035–44044. [CrossRef]

48. Pan, Y.; Short, J.L.; Newman, S.A.; Choy, K.H.; Tiwari, D.; Yap, C.; Senyschyn, D.; Banks, W.A.; Nicolazzo, J.A. Cognitive benefits of lithium chloride in APP/PS1 mice are associated with enhanced brain clearance of β-amyloid. *Brain Behav. Immun.* **2018**, *70*, 36–47. [CrossRef]

49. Pan, Y.; Scanlon, M.J.; Owada, Y.; Yamamoto, Y.; Porter, C.J.H.; Nicolazzo, J.A. Fatty acid-binding protein 5 facilitates the blood–brain barrier transport of docosahexaenoic acid. *Mol. Pharm.* **2015**, *12*, 4375–4385. [CrossRef]

50. McInerney, M.P.; Pan, Y.; Volitakis, I.; Bush, A.I.; Short, J.L.; Nicolazzo, J.A. The effects of clioquinol on P-glycoprotein expression and biometal distribution in the mouse brain microvasculature. *J. Pharm. Sci.* **2019**, *108*, 2247–2255. [CrossRef]

51. Yap, C.; Short, J.L.; Nicolazzo, J.A. A combination of clioquinol, zinc and copper increases the abundance and function of breast cancer resistance protein in human brain microvascular endothelial cells. *J. Pharm. Sci.* **2020**, *110*, 338–346. [CrossRef]

52. Xiao, Z.; Donnelly, P.S.; Zimmermann, M.; Wedd, A.G. Transfer of copper between bis (thiosemicarbazone) ligands and intracellular copper-binding proteins. Insights into mechanisms of copper uptake and hypoxia selectivity. *Inorg. Chem.* **2008**, *47*, 4338–4347. [CrossRef]

53. Dearling, J.L.; Lewis, J.S.; Mullen, G.E.; Welch, M.J.; Blower, P.J. Copper bis (thiosemicarbazone) complexes as hypoxia imaging agents: Structure-activity relationships. *J. Biol. Inorg. Chem.* **2002**, *7*, 249–259. [CrossRef] [PubMed]

54. Hilton, J.B.; Kysenius, K.; Liddell, J.R.; Rautengarten, C.; Mercer, S.W.; Paul, B.; Beckman, J.S.; McLean, C.A.; White, A.R.; Donnelly, P.S.; et al. Disrupted copper availability in sporadic ALS: Implications for Cu^{II} (atsm) as a treatment option. *BioRxiv* **2020**. [CrossRef]

55. Hilton, J.B.; Mercer, S.W.; Lim, N.K.H.; Faux, N.G.; Buncic, G.; Beckman, J.S.; Roberts, B.R.; Donnelly, P.S.; White, A.R.; Crouch, P.J. Cu II (atsm) improves the neurological phenotype and survival of SOD1 G93A mice and selectively increases enzymatically active SOD1 in the spinal cord. *Sci. Rep.* **2017**, *7*, 42292. [CrossRef] [PubMed]

56. McInnes, L.E.; Noor, A.; Kysenius, K.; Cullinane, C.; Roselt, P.; McLean, C.A.; Chiu, F.C.K.; Powell, A.K.; Crouch, P.; White, J.M.; et al. Potential diagnostic imaging of Alzheimer's disease with copper-64 complexes that bind to amyloid-β plaques. *Inorg. Chem.* **2019**, *58*, 3382–3395. [CrossRef]

57. Price, K.A.; Crouch, P.J.; Lim, S.; Paterson, B.M.; Liddell, J.R.; Donnelly, P.S.; White, A.R. Subcellular localization of a fluorescent derivative of Cu II (atsm) offers insight into the neuroprotective action of Cu II (atsm). *Metallomics* **2011**, *3*, 1280–1290. [CrossRef]

58. Bernard-Patrzynski, F.; Lécuyer, M.-A.; Puscas, I.; Boukhatem, I.; Charabati, M.; Bourbonnière, L.; Ramassamy, C.; Leclair, G.; Prat, A.; Roullin, V.G. Isolation of endothelial cells, pericytes and astrocytes from mouse brain. *PLoS ONE* **2019**, *14*, e0226302. [CrossRef]

59. Paraiso, H.C.; Wang, X.; Kuo, P.-C.; Furnas, D.; Scofield, B.A.; Chang, F.-L.; Yen, J.-H.; Yu, I.-C. Isolation of mouse cerebral microvasculature for molecular and single-cell analysis. *Front. Cell. Neurosci.* **2020**, *14*, 84. [CrossRef]

60. Navone, S.E.; Marfia, G.; Invernici, G.; Cristini, S.; Nava, S.; Balbi, S.; Sangiorgi, S.; Ciusani, E.; Bosutti, A.; Alessandri, G.; et al. Isolation and expansion of human and mouse brain microvascular endothelial cells. *Nat. Protoc.* **2013**, *8*, 1680–1693. [CrossRef]

61. Lee, G.; Schlichter, L.; Bendayan, M.; Bendayan, R. Functional expression of P-glycoprotein in rat brain microglia. *J. Pharmacol. Exp. Ther.* **2001**, *299*, 204–212. [PubMed]

62. Bendayan, R.; Ronaldson, P.T.; Gingras, D.; Bendayan, M. In situ localization of P-glycoprotein (ABCB1) in human and rat brain. *J. Histochem. Cytochem.* **2006**, *54*, 1159–1167. [CrossRef]

63. Chan, G.N.Y.; Hoque, T.; Cummins, C.L.; Bendayan, R. Regulation of P-glycoprotein by orphan nuclear receptors in human brain microvessel endothelial cells. *J. Neurochem.* **2011**, *118*, 163–175. [CrossRef]

64. Ding, Y.; Zhong, Y.; Baldeshwiler, A.; Abner, E.L.; Bauer, B.; Hartz, A.M.S. Protecting P-glycoprotein at the blood–brain barrier from degradation in an Alzheimer's disease mouse model. *Fluids Barriers CNS* **2021**, *18*, 10. [CrossRef]

65. Hartz, A.M.S.; Zhong, Y.; Shen, A.N.; Abner, E.L.; Bauer, B. Preventing P-gp ubiquitination lowers Aβ brain levels in an Alzheimer's disease mouse model. *Front. Aging Neurosci.* **2018**, *10*, 186. [CrossRef] [PubMed]

66. Nwaozuzu, O.M.; Sellers, L.A.; Barrand, M.A. Signalling pathways influencing basal and H_2O_2-induced P-glycoprotein expression in endothelial cells derived from the blood–brain barrier. *J. Neurochem.* **2003**, *87*, 1043–1051. [CrossRef] [PubMed]

67. Zhou, Y.; Zhou, J.; Li, P.; Xie, Q.; Sun, B.; Li, Y.; Chen, Y.; Zhao, K.; Yang, T.; Zhu, L.; et al. Increase in P-glycoprotein levels in the blood-brain barrier of partial portal vein ligation/chronic hyperammonemia rats is medicated by ammonia/reactive oxygen species/ERK1/2 activation: In vitro and in vivo studies. *Eur. J. Pharmacol.* **2019**, *846*, 119–127. [CrossRef]

68. Shao, Y.; Wang, C.; Hong, Z.; Chen, Y. Inhibition of p38 mitogen-activated protein kinase signaling reduces multidrug transporter activity and anti-epileptic drug resistance in refractory epileptic rats. *J. Neurochem.* **2016**, *136*, 1096–1105. [CrossRef]

69. Srivastava, S.; Blower, P.J.; Aubdool, A.A.; Hider, R.C.; Mann, G.E.; Siow, R.C. Cardioprotective effects of Cu (II) ATSM in human vascular smooth muscle cells and cardiomyocytes mediated by Nrf2 and DJ-1. *Sci. Rep.* **2016**, *6*, 1–13. [CrossRef]

70. Acevedo, K.M.; Hayne, D.J.; McInnes, L.E.; Noor, A.; Duncan, C.; Moujalled, D.; Volitakis, I.; Rigopoulos, A.; Barnham, K.J.; Villemagne, V.L.; et al. Effect of structural modifications to glyoxal-bis (thiosemicarbazonato) copper (II) complexes on cellular copper uptake, copper-mediated ATP7A trafficking, and P-glycoprotein mediated efflux. *J. Med. Chem.* **2018**, *61*, 711–723. [CrossRef]

71. Crouch, P.J.; Hung, L.W.; Adlard, P.A.; Cortes, M.; Lal, V.; Filiz, G.; Perez, K.A.; Nurjono, M.; Caragounis, A.; Du, T.; et al. Increasing Cu bioavailability inhibits Aβ oligomers and tau phosphorylation. *Proc. Natl. Acad. Sci. USA* **2009**, *106*, 381–386. [CrossRef] [PubMed]

72. Lin, C.C.; Hsieh, H.L.; Shih, R.H.; Chi, P.L.; Cheng, S.E.; Yang, C.M. Up-regulation of COX-2/PGE 2 by endothelin-1 via MAPK-dependent NF-κB pathway in mouse brain microvascular endothelial cells. *Cell Commun. Signal.* **2013**, *11*, 8. [CrossRef] [PubMed]

73. Qin, L.-H.; Huang, W.; Mo, X.-A.; Chen, Y.-L.; Wu, X.-H. LPS induces occludin dysregulation in cerebral microvascular endothelial cells via MAPK signaling and augmenting MMP-2 levels. *Oxidative Med. Cell. Longev.* **2015**, *2015*, 120641. [CrossRef] [PubMed]

74. Wang, X.; Campos, C.R.; Peart, J.C.; Smith, L.K.; Boni, J.L.; Cannon, R.E.; Miller, D.S. Nrf2 upregulates ATP binding cassette transporter expression and activity at the blood–brain and blood–spinal cord barriers. *J. Neurosci.* **2014**, *34*, 8585–8593. [CrossRef]

75. Chai, A.B.; Callaghan, R.; Gelissen, I.C. Regulation of P-Glycoprotein in the brain. *Int. J. Mol. Sci.* **2022**, *23*, 14667. [CrossRef]

76. Zhao, J.; Moore, A.N.; Redell, J.B.; Dash, P.K. Enhancing expression of Nrf2-driven genes protects the blood–brain barrier after brain injury. *J. Neurosci.* **2007**, *27*, 10240–10248. [CrossRef]

77. Huang, H.-C.; Nguyen, T.; Pickett, C.B. Phosphorylation of Nrf2 at Ser-40 by protein kinase C regulates antioxidant response element-mediated transcription. *J. Biol. Chem.* **2002**, *277*, 42769–42774. [CrossRef]

78. Pan, Y.; Kagawa, Y.; Sun, J.; Turner, B.J.; Huang, C.; Shah, A.D.; Schittenhelm, R.B., Nicolazzo, J.A. Altered blood-brain barrier dynamics in the C9orf72 hexanucleotide repeat expansion mouse model of amyotrophic lateral sclerosis. *Pharmaceutics* **2022**, *14*, 2803. [CrossRef]

79. Dagenais, C.; Rousselle, C.; Pollack, G.M.; Scherrmann, J.M. Development of an in situ mouse brain perfusion model and its application to mdr1a P-glycoprotein-deficient mice. *J. Cereb. Blood Flow Metab.* **2000**, *20*, 381–386. [CrossRef]

80. Cherny, R.A.; Atwood, C.S.; E Xilinas, M.; Gray, D.N.; Jones, W.D.; A McLean, C.; Barnham, K.J.; Volitakis, I.; Fraser, F.W.; Kim, Y.-S.; et al. Treatment with a copper-zinc chelator markedly and rapidly inhibits β-amyloid accumulation in Alzheimer's disease transgenic mice. *Neuron* **2001**, *30*, 665–676. [CrossRef]

81. Elmeliegy, M.; Vourvahis, M.; Guo, C.; Wang, D.D. Effect of P-glycoprotein (P-gp) inducers on exposure of P-gp substrates: Review of clinical drug–drug interaction studies. *Clin. Pharmacokinet.* **2020**, *59*, 699–714. [CrossRef] [PubMed]

82. Stamoulis, I.; Kouraklis, G.; Theocharis, S. Zinc and the liver: An active interaction. *Dig. Dis. Sci.* **2007**, *52*, 1595–1612. [CrossRef] [PubMed]

83. Hatano, R.; Ebara, M.; Fukuda, H.; Yoshikawa, M.; Sugiura, N.; Kondo, F.; Yukawa, M.; Saisho, H. Accumulation of copper in the liver and hepatic injury in chronic hepatitis C. *J. Gastroenterol. Hepatol.* **2000**, *15*, 786–791. [CrossRef] [PubMed]

84. Nikseresht, S.; Hilton, J.B.; Liddell, J.R.; Kysenius, K.; Bush, A.I.; Ayton, S.; Koay, H.; Donnelly, P.S.; Crouch, P.J. Transdermal application of soluble Cu^{II} (atsm) increases brain and spinal cord uptake compared to gavage with an insoluble suspension. *Neuroscience* **2023**, *509*, 125–131. [CrossRef] [PubMed]

 *pharmaceutics*

Article

Altered Blood–Brain Barrier Dynamics in the C9orf72 Hexanucleotide Repeat Expansion Mouse Model of Amyotrophic Lateral Sclerosis

Yijun Pan [1,2,3,*], Yoshiteru Kagawa [1,2], Jiaqi Sun [1,3], Bradley J. Turner [3,4], Cheng Huang [5], Anup D. Shah [5,6], Ralf B. Schittenhelm [5] and Joseph A. Nicolazzo [1,*]

1. Drug Delivery, Disposition and Dynamics, Monash Institute of Pharmaceutical Sciences, Monash University, 399 Royal Parade, Parkville, VIC 3052, Australia
2. Department of Organ Anatomy, Tohoku University Graduate School of Medicine, 2-1 Seiryomachi, Aobaku, Sendai 980-0872, Miyagi, Japan
3. Florey Institute of Neuroscience and Mental Health, University of Melbourne, Parkville, VIC 3052, Australia
4. Perron Institute for Neurological and Translational Science, Queen Elizabeth Medical Centre, Nedlands, WA 6009, Australia
5. Monash Proteomics & Metabolomics Facility, Biomedicine Discovery Institute, Monash University, Clayton, VIC 3800, Australia
6. Monash Bioinformatics Platform, Biomedicine Discovery Institute, Monash University, Clayton, VIC 3800, Australia
* Correspondence: yijun.pan@monash.edu (Y.P.); joseph.nicolazzo@monash.edu (J.A.N.); Tel.: +61-3-8344-4000 (Y.P.); +61-3-9903-9605 (J.A.N.); Fax: +61-3-9903-9583 (J.A.N.)

Citation: Pan, Y.; Kagawa, Y.; Sun, J.; Turner, B.J.; Huang, C.; Shah, A.D.; Schittenhelm, R.B.; Nicolazzo, J.A. Altered Blood–Brain Barrier Dynamics in the C9orf72 Hexanucleotide Repeat Expansion Mouse Model of Amyotrophic Lateral Sclerosis. *Pharmaceutics* **2022**, *14*, 2803. https://doi.org/10.3390/pharmaceutics14122803

Academic Editors: Gert Fricker and Elena Puris

Received: 1 December 2022
Accepted: 8 December 2022
Published: 14 December 2022

Publisher's Note: MDPI stays neutral with regard to jurisdictional claims in published maps and institutional affiliations.

Abstract: For peripherally administered drugs to reach the central nervous system (CNS) and treat amyotrophic lateral sclerosis (ALS), they must cross the blood–brain barrier (BBB). As mounting evidence suggests that the ultrastructure of the BBB is altered in individuals with ALS and in animal models of ALS (e.g., SOD1^{G93A} mice), we characterized BBB transporter expression and function in transgenic C9orf72 BAC (C9-BAC) mice expressing a hexanucleotide repeat expansion, the most common genetic cause of ALS. Using an in situ transcardiac brain perfusion technique, we identified a 1.4-fold increase in ^{3}H-2-deoxy-D-glucose transport across the BBB in C9-BAC transgenic (C9) mice, relative to wild-type (WT) mice, which was associated with a 1.3-fold increase in brain microvascular glucose transporter 1 expression, while other general BBB permeability processes (passive diffusion, efflux transporter function) remained unaffected. We also performed proteomic analysis on isolated brain microvascular endothelial cells, in which we noted a mild (14.3%) reduction in zonula occludens-1 abundance in C9 relative to WT mice. Functional enrichment analysis highlighted trends in changes to various BBB transporters and cellular metabolism. To our knowledge, this is the first study to demonstrate altered BBB function in a C9orf72 repeat expansion model of ALS, which has implications on how therapeutics may access the brain in this mouse model.

Keywords: amyotrophic lateral sclerosis; blood–brain barrier; brain microvascular endothelial cells; C9orf72 gene; proteomics

1. Introduction

Amyotrophic lateral sclerosis (ALS) is a fatal neurodegenerative disease characterized by selective damage to motor neurons in the brain and spinal cord. The pathogenesis of ALS remains largely unknown, although there are a wide range of potential mechanisms related to neurodegeneration, including protein aggregation, altered RNA processing, oxidative stress, excitotoxicity and neuroinflammation [1–3]. Individuals with ALS present with progressive voluntary muscle weakness, paralysis and atrophy, followed by respiratory complications and failure [4]. ALS typically occurs between the age of 40 and 70, with an incidence rate of 2 in 100,000 people and a prevalence rate of 4–7 cases per 100,000 people [5].

Due to marked heterogeneity in its clinical phenotypes and aggressive progression of the disease, therapeutic treatments are limited, and patients usually survive for 3–5 years after symptom onset [6]. The genetic component of the pathogenesis in 70% of individuals with familial ALS has been identified; mutations in chromosome 9 open reading frame 72 (*C9orf72*) (40%), superoxide dismutase 1 (*SOD1*) (20%), fused in sarcoma (*FUS*) (5%), and TAR DNA-binding protein 43 (*TARDBP* encoding TDP-43) (5%) [2]. The repeat expansion in C9orf72 has also been identified in 5–10% of individuals with sporadic ALS [3], making C9orf72 mutations the most commonly known genetic cause of ALS. C9orf72 mutations are also a frequent cause of frontotemporal dementia (FTD) which underscores the genetic and pathological overlap with ALS [2].

Like other neurodegenerative diseases, such as Alzheimer's disease and Parkinson's disease, altered blood–brain barrier (BBB) and blood-spinal cord barrier (BSCB) dynamics have been reported in individuals with ALS and mouse models [7]. These barriers are composed of a lining of brain microvascular endothelial cells (BMECs) that serve to precisely regulate the transport of substances (including drugs) between the blood and the brain/spinal cord. In individuals with ALS and in $SOD1^{G93A}$ mice [8], reduced expression (at mRNA or protein levels) of tight junction proteins has been reported [9,10], and increased paracellular permeability has also been noted [8,11]. Increased expression/function of efflux transporters (P-glycoprotein, P-gp and breast cancer resistance protein, Bcrp) has also been reported in individuals with ALS and both $SOD1^{G93A}$ and $SOD1^{G86R}$ mice [12,13]. To date, most animal studies have been performed in SOD1-related ALS mouse models, which only account for 20% of familial ALS cases. While C9orf72 repeat expansions are the most common (40%) genetic link in familial ALS and 5–10% of sporadic ALS, BBB dynamics in a mouse model with the C9orf72 mutation has yet to be assessed.

The C9orf72 gene contains 11 exons and healthy individuals commonly have between 2 and 23 GGGGCC (G_4C_2 hexanucleotide) repeats in the first intron, while individuals with ALS usually have hundreds to thousands of repeats [14,15]. Abnormal hexanucleotide repeat expansions are proposed to cause ALS pathology via three non-exclusive mechanisms. The first mechanism is through a gain of toxicity due to bidirectional transcription of C9orf72 repeat expansions, which leads to the formation of RNA foci, resulting in sequestration of RNA binding proteins and dysregulation of RNA splicing, transport and translation [16,17]. The second mechanism involves aggregation of dipeptide repeats via repeat-associated non-ATG translation which contribute to the formation of inclusions positive for TDP-43 protein, resulting in the disruption of mRNA splicing [18]. The third mechanism is believed to be a partial loss of function of C9orf72, which has been shown to influence lysosomal trafficking and autophagy processes, the alteration of which leads to aggregation of pathological proteins including TDP-43 [19]. Overall, misfolding and mislocalization of proteins due to pathological repeat expansions in C9orf72 disrupt multiple intracellular functions such as RNA metabolism, protein homeostasis, and mitochondrial function, possibly resulting in ALS pathophysiology [20]. An ALS mouse model based on C9orf72 repeat expansion (FVB-C9orf72 BAC mouse; C9-BAC) was successfully generated in 2016, with the transgenic mice (C9) exhibiting decreased survival, hind limb paralysis, muscle denervation, and motor neuron loss that is relevant to ALS, especially in females [21]. In addition, females also revealed signs of impaired cognition and anxiety indicative of FTD [11]. While the status of the BBB in other mouse models of ALS have been reported, characterization of the BBB in the C9-BAC model is a critical knowledge gap to be bridged, and may assist to unravel approaches that could be used to improve central nervous system (CNS) drug delivery in ALS.

In this study, we characterized the dynamics of the BBB in the C9-BAC mouse model at an early symptomatic age. This age (i.e., 145–150 days old) was selected as we were interested in potential BBB alterations at an age where significant pathology is observed, albeit severe motor dysfunction has not yet occurred. We evaluated the function of the BBB by assessing the transport of common probes, e.g., ^{3}H-diazepam and ^{14}C-caffeine for passive transcellular diffusion and ^{14}C-sucrose for paracellular diffusion and determination of brain vascular volume. ^{3}H-digoxin was included as a P-gp substrate as increased P-gp activity has been reported in other rodent models of ALS [7]. ^{3}H-oleic acid was included to assess fatty acid transport processes given that oleic acid is one of the most abundant fatty acids in the brain. The BBB transport of ^{3}H-2-deoxyglucose ([3]H-2DG) was included to assess the function of glucose transporter 1 (Glut1), which is important as altered glucose metabolism has been reported in other ALS mouse models and individuals with ALS [22]. Finally, [14]C-(L)-alanine was selected as a probe to assess amino acid trafficking processes, because altered amino acid levels have been reported in the plasma and cerebrospinal fluid of individuals with ALS [23]. Proteomic analysis was also performed on brain microvascular endothelial cells (BMECs) isolated from wildtype (WT) and C9 mice to further assess any alterations to the proteome of the BBB in this model. We specifically compared the relative abundance of important tight junction and gap junction proteins, and key adenosine triphosphate (ATP)-binding cassette (ABC) transporters and solute carrier (SLC) transporters between WT and C9 mouse BMECs. Differential expression and functional enrichment analysis were also performed to explore particular proteins and cellular processes/pathways modified in the C9 BMECs. To our knowledge, this is the first study to characterize the BBB of a C9orf72 repeat expansion mouse model, both at a functional and proteomic level, providing insight into potential approaches that can be explored to improve CNS drug delivery in this mouse model, and ultimately, individuals with ALS.

2. Materials and Methods

2.1. Materials

All radioactive probes (^{3}H-diazepam, ^{14}C-caffeine, ^{14}C-sucrose, ^{3}H-digoxin, ^{3}H-oleic acid, [3]H-2DG, and [14]C-(L)-alanine) were purchased from American Radiolabeled Chemicals (Saint Louis, MO, USA). Solvable and Ultima Gold scintillation cocktail were obtained from Perkin Elmer (Waltham, MA, USA). Pierce™ BCA Protein Assay Kit, sequencing grade trypsin, TMTpro™ 16plex Label Reagent Set and other reagents used for proteomics were purchased from Thermo Fisher Scientific (Waltham, MA, USA). A mouse Glut1 ELISA kit was sourced from Wuxi Donglin Sci & Tech Development Co., Ltd. (Wuxi, China). Dulbecco's Modified Eagle Medium (DMEM), bovine serum albumin (BSA) and phosphate-buffered saline (PBS) were purchased from Sigma-Aldrich (St. Louis, MO, USA).

2.2. Animals

Animal experiments were approved by the Monash Institute of Pharmaceutical Sciences Animal Ethics Committee (MIPS.21542) and performed in accordance with the National Health and Medical Research Council guidelines for the care and use of animals for scientific purposes. Transgenic C9-BAC (FVB/NJ-Tg(C9orf72)500Lpwr/J, strain number 029099) mice were purchased from the Jackson Laboratory (Bar Harbor, ME, USA) and maintained on FVB/NJ background. Female Tg(C9orf72)500Lpwr (C9) mice and their wildtype (WT) littermates were bred at Monash Animal Research Platform (Parkville, VIC, Australia) and the genotype of the mice was confirmed using Transnetyx® Genotyping services (Cordova, TN, USA). C9 mice were generated by Liu et al. [21], who reported that hemizygous C9 mice appear normal with no overt cage behavior abnormalities up to 16 weeks of age. However, an acute, rapidly progressive disease (inactivity, labored breathing, sudden weight loss, hindlimb weakness, motor neuron degeneration/disease throughout the motor unit, paralysis and death) was reported in ~30–35% of C9 females between 20–40 weeks. As we were interested in the impact of this genetic mutation on the BBB at an early symptomatic age, female mice at 145–150 days old were used in this study. Strain matched non-transgenic littermates were used as WT controls. Therefore, WT and C9 female mice were housed with ad libitum access to standard rodent chow and water until 145–150 days of age for end point experiments.

2.3. In Situ Transcardiac Brain Perfusion to Assess BBB Transport

The transport of probe drugs across the BBB was assessed using an in situ transcardiac brain perfusion technique as previously described [24]. Surgical anesthesia of mice (n = 4–8 per genotype) was established using ketamine (133 mg/kg) and xylazine (10 mg/kg), and the mice were kept on a warm surgery pad throughout anesthesia. The thoracic cavity of the mice was cut open to expose the heart, the descending aorta was clamped to restrict blood flow to the peripheral organs, and both jugular veins were severed. Immediately thereafter, warm (37 °C) Krebs-ringer bicarbonate buffer containing the radioactive probe (0.05 µCi/mL) was injected into the left ventricle of the heart and maintained at a rate of 10 mL/min for 1 min controlled by a Harvard infusion pump (Harvard Apparatus, Holliston, MA, USA). After 1 min, the perfusion was stopped, and the brain was harvested and digested in 2 mL of Solvable™ at 70 °C for 6 h. A 200 µL aliquot of hydrogen peroxide (30% v/v) was added to neutralize the color. Ultima Gold scintillation cocktail (10 mL) was added to the brain sample and 2 mL of Ultima Gold scintillation cocktail was added to 100 µL of the radioactive perfusion fluid. Radioactivity in each sample was counted in a PerkinElmer 2800TR liquid scintillation analyzer (Waltham, MA, USA). The apparent brain distribution volume of the probe compound (brain:perfusate, mL/g) was calculated as previously described [24,25]. The vascular volume was quantified using brain:perfusate ratio of ^{14}C-sucrose in C9 and their WT littermates, which was used for vascular volume subtraction for the calculation of apparent brain distribution volume of all other probes.

2.4. Brain Microvessel Enrichment for Glut1 Quantification

Given results demonstrated that the BBB transport of [3]H-2DG was increased in C9 mice, quantification of Glut1 levels in brain microvessels from WT and C9 mice was conducted. WT and C9 mice were humanely killed via cervical dislocation under anesthesia. Brains were immediately removed and 3–4 brains were pooled for subsequent processing, as previously described [25], to obtain sufficient microvessels for ELISA analysis of Glut1. In brief, mouse brains were homogenized using a Dounce homogenizer, and the microvessel enriched fraction (MEF) was separated from the homogenate by centrifugation using 17% (w/v) BSA in DMEM. The residual BSA was removed by rinsing the MEFs with ice-cold PBS and the total protein was extracted from the MEFs via 2 freeze–thaw cycles. A total volume of 200 µL of PBS was added to the MEFs and the samples were sonicated (2 × 10 s) prior to centrifugation. The total protein count for the supernatant was assessed using a Pierce™ BCA Protein Assay Kit by comparison to a standard curve developed using known BSA standard solutions. ELISA results were presented as Glut1 levels (ng)/mg total protein.

2.5. BMEC Isolation for Proteomic Analysis

To achieve a highly purified BMEC population (compared to that achieved via brain microvessel enrichment method described above), a magnetic-activated cell sorting (MACS) technique was employed. All procedures were performed according to manufacturer's instructions (Miltenyi Biotec Inc., Bergisch Gladbach, Germany). WT and C9 mice were anesthetized using isoflurane and humanely killed via cervical dislocation. Immediately, the brain was removed, weighed, and digested using an Adult Brain Dissociation Kit (mouse and rat) (Miltenyi Biotec Inc.). From the pool of brain cells, endothelial cells were isolated using MACs mouse CD31 microbeads. The obtained BMECs were pelleted down and stored at −80 °C for proteomic analysis.

2.6. Proteomic Analysis of BMECs

BMEC samples were lysed in SDS lysis buffer (5% w/v sodium dodecyl sulphate, 100 mM HEPES, pH 8.1), heated at 95 °C for 10 min and then probe-sonicated before measuring protein concentration using a BCA kit. The lysed samples were denatured and alkylated by adding tris(2-carboxyethyl) phosphine hydrochloride and 2-chloroacetamide to a final concentration of 10 mM and 40 mM, respectively, and the mixture was incubated at 55 °C for 15 min. The proteins were trapped using S-Trap mini columns (Profiti, Farmingdale, NY, USA), and sequencing grade trypsin was added at an enzyme to protein ratio of 1:50 and incubated overnight at 37 °C. Tryptic peptides were sequentially eluted from the columns using (i) 50 mM triethylammonium bicarbonate, (ii) 0.2% v/v formic acid and (iii) 50% v/v acetonitrile, 0.2% v/v formic acid. The sequentially eluted fractions were pooled, concentrated in a vacuum concentrator and reconstituted in 40 µL of 200 mM HEPES, pH 8.5. Using a Pierce Quantitative Colorimetric Peptide Assay Kit (Thermo Fisher Scientific, Waltham, MA, USA), equal peptide amounts of each sample were labelled with the TMTpro 16plex reagent set (Thermo Fisher Scientific) according to the manufacturer's instructions, considering a labelling strategy to minimize channel leakage. Individual samples were then pooled and high-pH RP-HPLC (1260 Infinity II, Agilent) was used to fractionate each pool into 12 fractions, which were acquired individually by LC-MS/MS to maximize the number of peptide and protein identifications.

Using a Dionex UltiMate 3000 RSLCnano system equipped with a Dionex UltiMate 3000 RS autosampler, an Acclaim PepMap RSLC analytical column (75 µm × 50 cm, nanoViper, C18, 2 µm, 100 Å; Thermo Scientific) and an Acclaim PepMap 100 trap column (100 µm × 2 cm, nanoViper, C18, 5 µm, 100 Å; Thermo Fisher Scientific, the tryptic peptides were separated by increasing concentrations of 80% v/v acetonitrile/0.1% v/v formic acid at a flow of 250 nL/min for 158 min and analyzed with an Orbitrap Fusion Tribrid mass spectrometer (ThermoFisher Scientific). The instrument was operated in data-dependent acquisition mode to automatically switch between full scan ms1 (in Orbitrap), ms2 (in ion trap) and ms3 (in Orbitrap) acquisition. Each survey full scan (380–1580 m/z) was acquired with a resolution of 120,000, an automatic gain control (AGC) target of 50%, and a maximum injection time of 50 msec. Dynamic exclusion was set to 60 sec after one occurrence. Keeping the cycle time fixed at 2.5 sec, the most intense multiply charged ions ($z \geq 2$) were selected for ms2/ms3 analysis. Ms2 analysis used collision-induced dissociation (CID) fragmentation (fixed collision energy mode, 30% CID collision energy) with a maximum injection time of 150 msec, a "rapid" scan rate and an AGC target of 40%. Following the acquisition of each ms2 spectrum, an ms3 spectrum was acquired from multiple ms2 fragment ions using Synchronous Precursor Selection. The ms3 scan was acquired in the Orbitrap after higher energy collision dissociation with a resolution of 50,000 and a maximum injection time of 250 msec.

The raw data files were analyzed with Proteome Discoverer (Thermo Fisher Scientific) to obtain quantitative ms3 reporter ion intensities. Statistical analysis was performed using R statistical analysis software [26]. First, the dataset was filtered for high confident proteins. Next, the contaminant proteins were removed. The protein intensity data was converted to log2 scale and samples were grouped by conditions. The purity of the enriched BMEC samples was evaluated by assessing the abundance of protein markers for endothelial cells (von Willebrand factor, Vwf), astrocytes (Ndrg2), pericytes (platelet-derived growth factor receptor β; Pdgfr-β), and neurons (synaptophysin; Syp). It has been reported that the BBB integrity is compromised in individuals with ALS and mouse models of ALS, and therefore, we evaluated the brain vasculature proteome data for abundance changes in classical proteins implicated in BBB integrity [27], including zonula occludens-1 (Tjp1), zonula occludens-2 (Tjp2), claudin 5 (Cldn5), occludin (Ocln), cadherin 5 (Cdh5), junctional adhesion molecule 2 (Jam2), junctional adhesion molecule 3 (Jam3), F11 receptor (F11r) and endothelial cell adhesion molecule (Esam). As drug delivery is the focus of our study, the abundance of major ABC transporters and SLC transporters were compared between WT and C9 BMECs. In addition, protein-wise linear models combined with empirical Bayes statistics were used for the differential expression analyses. The limma package [28] from R Bioconductor was used to generate a list of differentially expressed proteins for each pair-wise comparison. A cutoff of the adjusted p-value of 0.05 (Benjamini-Hochberg method) along with a fold-change of 1.5 was applied to determine significantly regulated proteins in the pairwise comparison of WT vs. C9 BMECs. Functional enrichment analysis was performed using STRING online tools (https://string-db.org/, accessed on 10 July 2022).

2.7. Statistical Analysis of Data

All data are expressed as mean $\pm$ SEM. The comparisons between results obtained from WT and C9 mice were evaluated by Student's unpaired t tests and two-way ANOVA with repeated measure where appropriate. Proteomics data analysis was detailed above and specified in the Results Section. Differences with a p value of less than 0.05 were considered statistically significant unless otherwise indicated in the text.

3. Results

3.1. BBB Transport of ^{3}H-2DG Was Increased in C9 Mice Relative to WT Mice, in Association with Increased Brain Microvascular Glut1 Abundance

An in situ transcardiac brain perfusion technique was employed to characterize BBB transport processes in the C9-BAC mouse model. No significant difference in the cerebral vascular volume was noted between WT (0.025 $\pm$ 0.003 mL/g) and C9 (0.026 $\pm$ 0.003 mL/g) mice, using ^{14}C-sucrose as a marker. The BBB transport of ^{3}H-diazepam, ^{14}C-caffeine, ^{3}H-oleic acid, ^{3}H-digoxin, and ^{14}C-(L)-alanine was comparable between WT and C9 mice (Figure 1A–E), suggesting that passive diffusion, P-gp efflux, fatty acid uptake and neutral amino acid uptake was not affected in this mouse model. Interestingly, a 1.4-fold increase in the BBB transport of ^{3}H-2DG was observed in C9 mice relative to WT mice (Figure 1F). To ascertain if this increase in the BBB transport of ^{3}H-2DG was associated with increased expression of Glut1 at the BBB, brain MEFs were isolated and a 1.3-fold increase in Glut1 abundance was observed in brain MEFs isolated from C9 mice compared to WT mice (Figure 2).

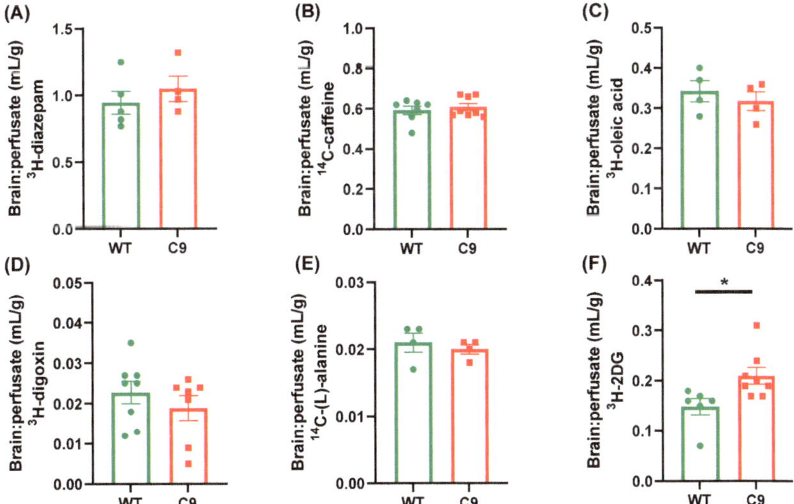

Figure 1. The brain-to-perfusate ratio of probe molecules in 145–150 day old female WT and C9 mice, with (**A**) ^{3}H-diazepam to assess lipophilic passive diffusion, (**B**) ^{14}C-caffeine to assess hydrophilic passive diffusion, (**C**) ^{3}H-oleic acid to assess fatty acid transport processes, (**D**) ^{3}H-digoxin to assess P-gp function, (**E**) ^{14}C-(L)-alanine to assess the transport of a neutral amino acid, and (**F**) ^{3}H-2DG to assess the function of Glut1. Data are presented as mean $\pm$ SEM (n = 4–8 mice per genotype), with * $p < 0.05$ using an unpaired Student's t-test.

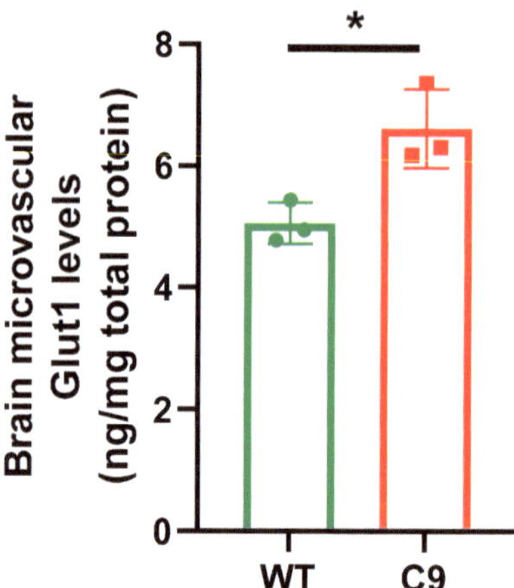

Figure 2. The abundance of Glut1 in brain MEFs from 145–150 day old female WT and C9 mice, with a significant 1.3-fold increase observed in brain MEFs from C9 mice compared to those from WT mice. Data are presented as mean $\pm$ SEM (n = 3 biological replicates per genotype; 3–4 mouse brains were pooled for each biological replicate), with * p < 0.05 using an unpaired Student's *t*-test.

3.2. The Tandem Mass Tag (TMT)-Proteomics Studies Revealed an Altered Proteome Profile of C9 BMECs Compared to WT BMECs

Based on the results demonstrating altered BBB trafficking of ^{3}H-2DG, a proteomics approach was undertaken to explore further alterations to the BBB in early symptomatic C9 mice compared to their WT littermates. Isolated BMECs from 8 biological replicates (two mouse brains were pooled for one biological replicate) were subjected to TMT-based quantitative proteomic analysis, and a total of 5267 proteins were quantified across all samples (Supplementary Materials Spreadsheet S1: Imputed normalized intensity). A relatively high abundance of the endothelial marker Vwf and low abundance of markers for astrocytes (Ndrg2), pericytes (Pdgfr-β) and neurons (Syp) were observed in the proteome profile, confirming the purity of the BMEC samples (Figure 3A). A main effect of markers ($F_{3,28}$ = 111.8, p < 0.0001), but not of genotype, was revealed by a two-way ANOVA repeated measure.

Further characterization of the BBB was undertaken by comparing the abundance of Tjp1, Tjp2, Cldn5, Ocln, Cdh5, Jam2, Jam3, F11r and Esam (Figure 3B). A two-way ANOVA with repeated measures revealed an interaction effect between genotype and proteins of interest ($F_{8,63}$ = 2.108, p = 0.048), and the post hoc Sidak's multiple comparison test demonstrated a significant (14.3 $\pm$ 9.4%, adjusted p value = 0.0003) reduction in zonula occludens-1 abundance in C9 BMECs compared to WT BMECs. In addition, a total number of 18 ABC transporters and 83 SLC transporters were identified in BMECs. Table 1 summarizes the relative intensity of the most important membrane transporters for drug and endogenous molecule disposition [29]. Of note, a significant increase in Abc9, Eaat2, Asct2, Fatp1 and a significant decrease in Bsep and Taut were observed, which are important transporters for the disposition of endogenous molecules.

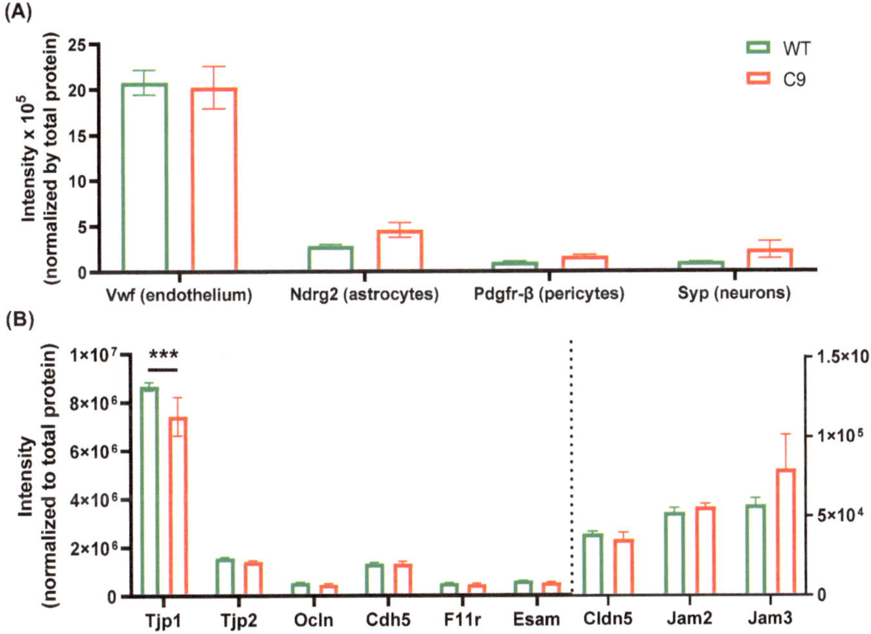

Figure 3. (**A**) The relative abundance of markers for endothelial cells (Vwf), astrocytes (Ndrg2), pericytes (Pdgfr-β) and neurons (Syp), and (**B**) microvascular integrity markers in BMECs isolated from 145–150 day old female WT and C9 mice, with the intensity of proteins on the left side of the dash line represented by the left Y axis and the intensity of proteins on the right side of the dash line represented by the right Y axis Data are presented as mean $\pm$ SEM (n = 8 mice per genotype), with *** $p < 0.001$ using a two-way ANOVA with post hoc Sidak's multiple comparison.

As per differential expression analysis, for all detected proteins, a total of 13 significantly differentially abundant proteins were identified, with 10 proteins exhibiting a >1.5 fold-change (Table 2). The volcano plot and heat map of significantly differentially abundant proteins are available in Supplementary Figures S1 and S2. STRING was used to perform functional enrichment analysis for all identified proteins irrespective of whether they were significantly regulated. Here, we provided some examples of the Biological Process (BP), Molecular Function (MF), and Cellular Component (CC) relevant to BBB transport highlighted by the analysis (Figure 4). When comparing WT with C9 BMECs, the BP was enriched with positive regulation of monovalent inorganic cation transport (GO:0015672), regulation of exocytosis (GO:0017157), and cation transmembrane transport (GO:0098655); the MF was enriched with negative regulation of organic anion transmembrane transporter activity (GO:0008514), and inorganic molecular entity transmembrane transporter activity (GO:0015077), but positive regulation of monovalent inorganic cation transmembrane transporter activity (GO:0015077); and the CC was enriched with positive regulation of mitochondrial protein complex (GO:0098798), and collagen-containing extracellular matrix (GO:0062023). We also performed KEGG analysis, where extracellular matrix organization (MMU-1474244; positive regulation), SLC-mediated transmembrane transport (MMU-425407; both positive and negative regulation), the citric acid cycle and respiratory electron transport (MMU-1428517; positive regulation), oxidative phosphorylation (MMU-00190; positive regulation), and drug metabolism—cytochrome P450 (MMU-00982; negative regulation) were highlighted. The complete results of the functional enrichment analysis are available in Supplementary Materials Spreadsheet S2: Functional enrichment analysis.

Table 1. Relative abundance of important membrane transporters in BMECs isolated from 145–150 day old female WT and C9 mice, assessed using TMT-proteomics (* $p < 0.05$).

Gene	Protein Name	WT/C9 Fold-Change	p Value
ABC transporters			
Abca2	Abc2	1.18	0.12
Abca9	Abc9	0.66 *	0.03
Abcb1a	P-gp	1.23	0.14
Abcb4	Mdr3	1.09	0.20
Abcb11	Bsep	1.35 *	0.01
Abcc1	Mrp1	1.14	0.31
Abcc9	Mrp9	0.78	0.12
Abcg2	Bcrp	1.21	0.07
SLC transporters			
Slc1a2	Eaat2	0.42 *	0.05
Slc1a5	Asct2	0.77 *	0.04
Slc6a6	Taut	1.21 *	0.05
Slc2a1	Glut1	1.22	0.23
Slc7a5	Lat1	1.32	0.08
Slc16a1	Mct1	1.32	0.13
Slc22a8	Oat3	1.15	0.33
Slc27a1	Fatp1	0.70 *	0.03
Slc29a1	Ent1	0.90	0.25
Slc38a2	Snat2	0.94	0.70
Slco1a1	Oatp1a1	3.51	0.17
Slco1a4	Oatp1a4	1.26	0.10
Slco1a6	Oatp1a6	1.33	0.08
Slco1c1	Oatp1c1	1.20	0.14
Slco2b1	Oatp2b1	0.94	0.59

Table 2. Significant differential abundant proteins between isolated from 145–150 day old female WT and C9 mice, assessed using TMT-proteomics.

Gene	Protein Name	WT/C9 Fold-Change	Adjusted p Value
Atp5mc1	ATP synthase F(0) complex subunit C1	1.63	0.0243
Bcap29	B-cell receptor-associated protein 29	1.51	0.0243
Cdc42bpa	CDC42 binding protein kinase alpha	1.27	0.0446
Gulp1	PTB domain-containing engulfment adapter protein 1	1.56	0.0258
H2ac15	histone H2A type 1-K	1.61	0.0243

Table 2. *Cont.*

Gene	Protein Name	WT/C9 Fold-Change	Adjusted *p* Value
Hmgn3	high mobility group nucleosome-binding domain-containing protein 3	1.95	0.0073
Isg15	ubiquitin-like protein ISG15	1.42	0.0243
Nckap1l	nck-associated protein 1-like	0.64	0.0292
Nono	non-POU domain-containing octamer-binding protein	1.41	0.0243
Rgs10	regulator of G-protein signaling 10	0.44	0.0243
Smco4	single-pass membrane and coiled-coil domain-containing protein 4	1.79	0.0243
Tmpo.1	lamina-associated polypeptide 2	1.55	0.0355
UPF0729	UPF0729 protein C18orf32 homolog	1.70	0.0355

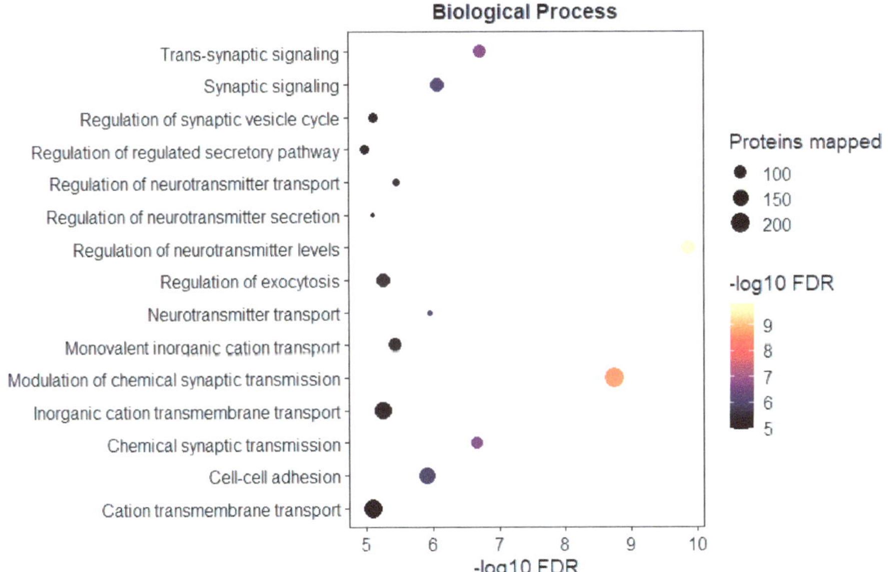

Figure 4. *Cont.*

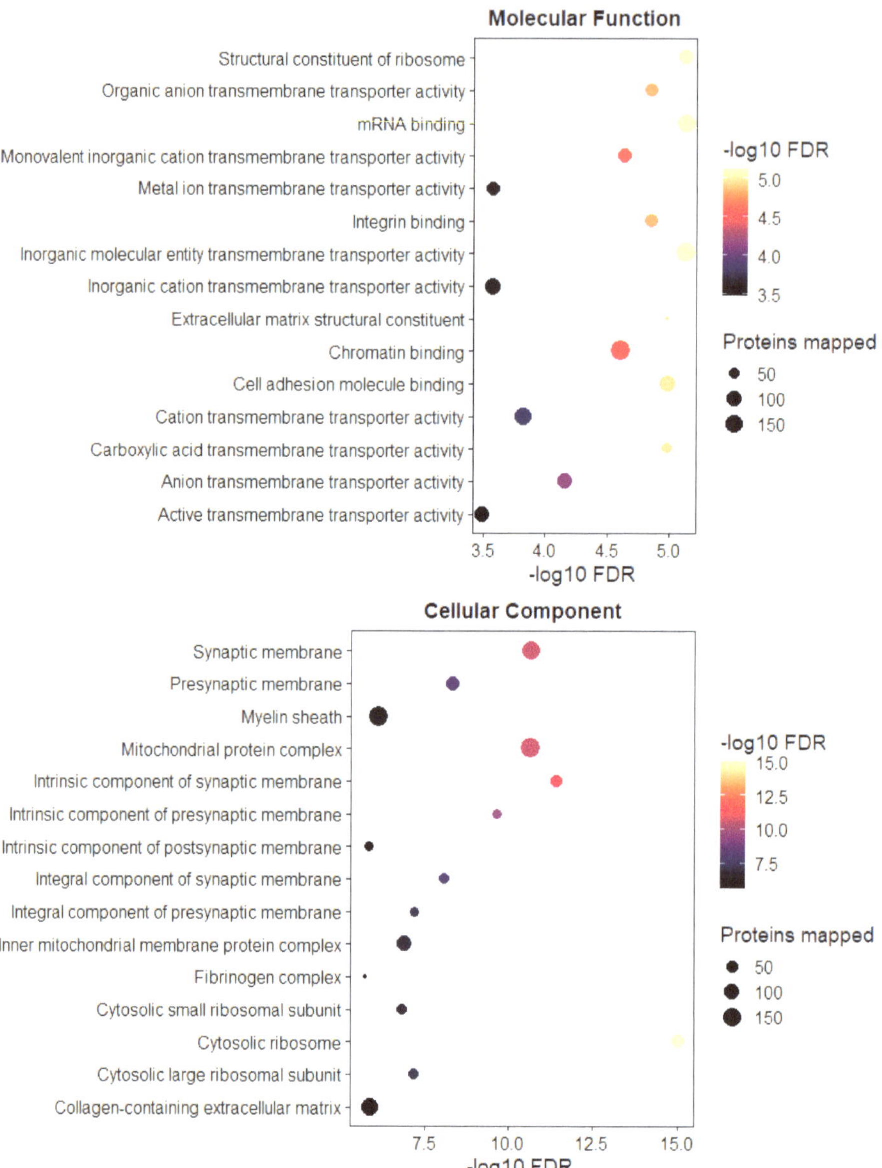

Figure 4. Representative results of functional enrichment analysis for BMECs isolated from 145–150 day old female WT and C9 mice (n = 8 per genotype).

4. Discussion

Altered BBB structure and function have been reported in SOD1-relevant rodent models of ALS and in individuals with ALS [7]. Given C9orf72 repeat expansions are the most common genetic cause of ALS, we characterized BBB transport function in the C9-BAC mouse model. The selection of probes was based on highlighting different BBB transport processes [24,30] that may be altered in this ALS model. We assessed the brain microvascular volume of these mice using ^{14}C-sucrose, and no significant difference was noted between WT and C9 mice, suggesting that paracellular permeability across the BBB was not impacted in this particular mouse model. The brain microvascular volume in WT and C9 mice was similar to that previously reported in C57BL/6 mice (~0.02 mL/g) [24,31]. ^{3}H-diazepam and ^{14}C-caffeine were used to assess transcellular BBB diffusion of both a lipophilic and hydrophilic compound, respectively, and the brain:perfusate ratios obtained in the current study, which were not different between genotype, were comparable to the reported values in rodents for diazepam [24] and caffeine [32]. Overall, these data suggested that there is no brain vascular leakiness or altered BBB passive diffusion in C9 mice, relative to WT mice, using our established in situ transcardiac brain perfusion technique [24,33].

The above-mentioned observations are not in alignment with the impaired BBB integrity and increased vascular permeability reported in individuals with ALS and SOD1^{G93A} mice [8,11]. However, it has to be noted that human samples were collected from deceased individuals with ALS and therefore the samples represent pathology at terminal disease stage, which is different to our current study where we assessed BBB trafficking at an early symptomatic stage without significant overt disease progression. For animal studies showing altered vascular integrity, this observation was in mouse spinal cord microvasculature using electron microscopy and permeability assessed using Evans Blue, as opposed to brain microvascular tissue used in our studies. Brain microvascular tissue was also more relevant in our study given C9-BAC mice model both ALS and FTD. To make direct comparison between our results in the C9-BAC model with SOD1^{G93A} mice, in situ transcardiac brain perfusions should also be performed in SOD1^{G93A} mice in future studies. Regardless, the CNS delivery of drugs that utilize passive diffusion to permeate the BBB is not expected to be affected in the C9-BAC mouse model of ALS. Given the absence of functional changes in passive transcellular and paracellular permeability, we did not proceed to histological evaluation of the cerebral microvasculature (e.g., basement membrane thickness and vascular integrity).

A selective increase in microvascular expression of P-gp during disease progression in three ALS mouse models (SOD1^{G93A}, SOD1^{G86R} and TDP-43^{A315T}) has been reported in the cerebral cortex and spinal cord [12,13]. Increased P-gp expression was identified in the brain cortex and spinal cord homogenates of symptomatic mice in all three models [12], and a 1.5-fold increase in P-gp abundance was detected in microvessels isolated from the brains of pre-symptomatic SOD1^{G86R} mice compared to age-matched WT controls [13]. We assessed P-gp function initially, rather than expression, in C9-BAC mice at an early symptomatic age, using ^{3}H-digoxin as a well established substrate of P-gp [34], and no significant difference in BBB transport of ^{3}H-digoxin was identified between WT and C9 mice. Given that there was no evidence of increased vascular leakiness as assessed using ^{14}C-sucrose (which could theoretically mask any increased function of P-gp), we concluded that P-gp activity was not modified at the BBB in the C9-BAC model. The discrepancies between our results and those in other studies could be attributed to the different ALS pathology that the C9-BAC model represents relative to the SOD1-relevant models, and the different ALS disease stage of C9 mice assessed in our studies relative to those reporting alterations in symptomatic SOD1 mice. We also assessed the BBB trafficking of ^{3}H-oleic acid and ^{14}C-alanine, given that the levels of oleic acid in the spinal cord of individuals with ALS has been shown to be increased by 10% [35], and that the levels of alanine in the cerebrospinal fluid of individuals with ALS has been shown to be increased by 21% [23]. No significant difference in the BBB trafficking of ^{3}H-oleic acid and ^{14}C-alanine was observed between WT and C9 mice, suggesting that any increases in brain levels of oleic acid or cerebrospinal fluid levels of alanine reported in ALS are less likely due to changes in their transport across the BBB.

Interestingly, a significant increase in the BBB transport of ^{3}H-2DG was noted in the C9 mice compared to their WT littermates. This observation did not appear to be in line with previous reports where glucose hypometabolism, as measured by PET, has been reported in spinal cords and several brain regions, including motor, frontal and occipital cortex of both individuals with ALS and in ALS animal models [36]. However, it has to be emphasized that our in situ transcardiac perfusion technique was performed over 1 min to evaluate the kinetics of ^{3}H-2DG trafficking across the BBB, whereas the previously reported reduction in glucose homeostasis was assessed through measuring the incorporation of ^{18}F-FDG into the brain parenchyma 40 min post-dosing via PET [36]. It has been proposed that glucose is transported across the BBB via Glut1 [37], whereas Glut3 is required to transport glucose from the brain interstitial fluid into neurons [38]. It is possible that the increased BBB transport of ^{3}H-2DG that we observed reflects an increase in the availability of glucose uptake to compensate for the decreased glucose neuronal incorporation (as has been demonstrated via PET). Associated with the increased BBB trafficking of ^{3}H-2DG was an increase in Glut1 abundance in brain MEFs isolated from C9 mice compared to WT mice. Further studies are required to explore the underlying mechanisms leading to this increased expression and function of Glut1 and to ascertain whether this may be exploited to enhance CNS delivery of therapeutics in ALS, for example by conjugating glucose-like entities to assist trafficking across the BBB [39].

We also employed proteomic analysis as an explorative approach to capture novel findings that might have been unidentified by our routine BBB functional characterization. Differentially expressed protein analysis and functional enrichment analysis were performed, which also provided insight into status of the BBB in the C9-BAC mouse model. In the current study, the purity of the BMEC samples was confirmed by the relative abundance of a marker for endothelial cells (vwf) relative to markers for astrocytes (Ndrg2), pericytes (Pdgfr-β) and neurons (syp) in the proteome profiles. While glial fibrillary acidic protein (Gfap), S100β and Ndrg2 are all recognized markers for astrocytes, Ndrg2 was selected due to its uniform expression in astrocytes (>40% of astrocytes have been found to be Gfap-negative), and its astrocyte-specific expression (S100β is also expressed in some mature oligodendrocytes, and choroid plexus epithelial cells) [40]. Pdgfr-β was selected for uniform and pericyte-specific expression [41] and Syp was employed as a neuronal marker, given it is one of the most commonly used markers of neurons [42]. A high abundance of vwf was noted in the proteomic analysis of our isolated BMECs, which was ~6 times higher than Ndrg2, and ~17 times higher than Pdgfr-β and Syp. Sharma et al. [43] have performed a proteomic analysis in different mouse brain regions, and assessed the relative abundance of these markers in brain tissues (without endothelial cell enrichment procedure). In brain homogenates, the abundance of vwf was found to be 44-, 65-, and 942-fold lower than Ndrg2, Pdgfr-β and Syp, respectively, suggesting that our samples are indeed highly enriched in BMECs.

Although we did not observe functional differences in tight junction function in C9 mice (through assessing ^{14}C-sucrose transport across the BBB), we were still interested to assess the abundance of major tight junction and gap junction proteins [44] in BMECs that were isolated from the C9 and WT mice, as altered expression of tight junction proteins in the spinal cords (though not in the brain) have been reported in individuals with ALS and in the SOD1^{G93A} mouse model [45,46]. A mild (14.3 ± 9.4%) but significant reduction in the abundance of zonula occludens-1 was identified in BMECs isolated from C9 mice compared to those isolated from WT mice. This observation is in line with a previous report where a ~50% reduction in zonula occludens-1 was observed at the protein level in the spinal cord microvasculature isolated from SOD1^{G93A} mice relative to WT mice [45]. The less robust reduction in zonula occludens-1 abundance could be due to the different techniques used (i.e., untargeted proteomics in our studies vs. Western blotting in the SOD1^{G93A} mice), or the intrinsic difference between the models and the CNS regions assessed (i.e., we assessed BMECs from the cortex and not the spinal cord). Apart from zonula occludens-1, no significant changes in other tight junction and gap junction proteins were identified in our proteomics study. Regardless, this mild reduction in zonula occludens-1 abundance, as assessed using untargeted proteomics, did not result in any functional changes to paracellular permeability in the C9 mice, overall suggesting that the paracellular route is intact in the C9-BAC mouse model of ALS.

We also specifically assessed the abundance of major ABC and SLC proteins in BMECs isolated from WT and C9 mice (Table 1). Some proteins were identified to be significantly modified in expression in C9 BMECs in this subset of proteomics data, however there are no clinically approved drugs that are known to interact with these transporters, and therefore such changes are less likely to affect CNS drug delivery in practice. Notably, no significant change to P-gp and Bcrp were identified in C9 BMECs compared to WT BMECs, while overexpression for P-gp and Bcrp has been observed in SOD1^{G93A} mice (brain cortex and spinal cord) and postmortem ALS human tissues (spinal cord) [7]. It is completely possible that the expression and function of P-gp and Bcrp are not yet altered in early symptomatic C9-BAC mice, and that such changes may occur with disease progression (as has been reported in SOD1^{G93A} mice albeit in spinal cord homogenates) [12]. In addition, different techniques used for quantitative analysis could also contribute to such discrepancies. Indeed, most other studies investigating the BBB in ALS mouse models or postmortem human tissues employed Western Blot or immunohistochemistry, rather than a high throughput proteomics technique. Despite this, we do not expect to reveal such differences in P-gp by undertaking these studies in the C9-BAC mice, given that we observed no change to P-gp function at the BBB using ^{3}H-digoxin as a P-gp substrate. While a significant increase in Glut1 was observed in the brain MEFs from C9 mice relative to WT mice using an ELISA approach, our proteomics study did not reveal such changes in BMECs. These differences could be attributed to different isolation techniques. For the brain microvessel enrichment technique, there was no enzymatic digestion step, and therefore, there may be a greater amount of astrocyte end feet connections still attached to the microvasculature; while for the BMEC isolation by MACS and subsequent enzymatic digestion, we achieved samples with much lower astrocyte contamination. It is also possible that proteomics is less sensitive to detect mild changes (<1.5-fold) than ELISA [46]. Regardless, given that Glut1 function was increased at the BBB in C9 mice, as assessed by increased BBB trafficking of ^{3}H-2DG, this transporter is still considered an exploitable target for improving CNS drug delivery in ALS.

Interestingly, alterations to the levels of some transporters, such as Eaat2 and Asct2, were identified, which are possibly important for disease pathology, albeit they may not be involved in drug transport across the BBB. Eaat2 is one of the most abundant subtypes of glutamate transporters in the CNS, playing a key role in maintaining glutamate concentration low in the extracellular space of the brain so as to avoid excitotoxicity [47]. Dysfunction of Eaat2 has been correlated with ALS [47]. It is possible that Eaat2 is upregulated at an early symptomatic stage in C9 mice as a compensatory mechanism to remove excessive glutamate, especially given that Eaat2 is expressed at the abluminal membrane of BMECs [48]. Similarly, being a glutamate transporter at the abluminal membrane of the brain microvasculature [49], the upregulation of Asct2 in C9 BMECs could also be part of the compensatory mechanism to avoid excitotoxicity from excess glutamate [50]. It was also observed that the abundance of Fatp1 (a transporter of oleic acid) [51] increased in C9 vs. WT BMECs. Though this observation is in agreement with the increased oleic acid levels observed in the postmortem spinal cord of individuals with ALS [35], it does not explain why BBB transport of ^{3}H-oleic was not increased as assessed using a transcardiac in situ brain perfusion technique. This could be due to the fact that other proteins (such as fatty acid binding protein 5) [52] are essential in oleic acid uptake, and these proteins that did not appear to be affected in this study.

In line with the differentially expressed proteins which we noted in our study comparing WT and C9 BMECs, Hmgn3 and Nono have been shown to be differentially expressed in human ALS brains [53]. While they play roles in astrocyte function and possible glucose homeostasis [54–56], their role at the BBB and any impact on trafficking of entities into the CNS remains completely unknown. On the other hand, from the functional enrichment analysis, many GO terms and KEGG pathways that are relevant to transporter function and energy metabolism were highlighted; for example, cation transmembrane transport (GO:0098655), citric acid cycle and respiratory electron transport (MMU-1428517), and oxidative phosphorylation (MMU-00190). Future studies focusing on these changes at the ALS BBB may be considered; for example, the BBB transport of quetiapine, a cationic drug that is commonly used in individuals with ALS, can be assessed given the highlighted changes in cation transmembrane transport, and the metabolism of C9 BMECs can be investigated, given the highlighted changes in citric acid cycle and oxidative phosphorylation.

5. Conclusions

Identifying novel targets to overcome the restrictive nature of the BBB has the potential to improve CNS drug delivery in ALS. Using the C9-BAC mouse model of ALS, our studies have demonstrated that the BBB in the early symptomatic phase of pathology is mildly affected with a significant increase in Glut1 expression and function, and no observed alteration to paracellular diffusion, passive diffusion or P-gp mediated efflux. Our proteomics approach has identified other possible transporters that could be further investigated to provide a more detailed characterization of the BBB, which might highlight in addition to Glut1, other transporters that could be exploited for enhancing CNS drug delivery in ALS. Building upon our proteomics data, future studies are required to evaluate the BBB transport of different classes of drugs that possibly have modified access to the brain of mouse models of ALS; this will provide invaluable insight into how to design drugs for optimum brain uptake for this disease. Knowing this in advance can also lead to tailoring the dosages of non-ALS medicines to minimize their potential for brain toxicity.

Supplementary Materials: The following supporting information can be downloaded at: https://www.mdpi.com/article/10.3390/pharmaceutics14122803/s1, Figure S1: Volcano plot of differentially expressed proteins between BMECs isolated from 145–150 day old female WT and C9 mice (n = 8 per genotype); Figure S2: Heatmap of significant differentially expressed proteins between BMECs isolated from 145–150 day old female WT and C9 mice (n = 8 per genotype); Spreadsheet S1: Imputed normalized intensity; Spreadsheet S2: Functional enrichment.

Author Contributions: Conceptualization, Y.P. and J.A.N.; methodology, Y.P., Y.K., R.B.S. and J.A.N.; validation, Y.P., J.S., C.H, A.D.S., R.B.S. and J.A.N.; formal analysis, Y.P. and A.D.S.; investigation, Y.P., Y.K., J.S. and C.H.; resources, B.J.T. and J.A.N.; writing—original draft preparation, Y.P. and Y.K.; writing—review and editing, Y.P., Y.K., J.S., B.J.T., C.H., A.D.S., R.B.S. and J.A.N.; visualization, Y.P.; supervision, Y.P. and J.A.N.; project administration, Y.P. and J.A.N.; funding acquisition, J.A.N. All authors have read and agreed to the published version of the manuscript.

Funding: This research was funded by an IMPACT grant awarded by FightMND (01_IMPACT_2019_Nicolazzo). Y.P. is supported by an Investigator Grant (GNT2007912) awarded by the National Health and Medical Research Council (NHMRC) Australia. B.J.T. is supported by a NHMRC Dementia Research Leadership Fellowship (1137024) and Stafford Fox Medical Research Foundation Grant.

Institutional Review Board Statement: The study was conducted in accordance with the National Health and Medical Research Council guidelines for the care and use of animals for scientific purposes, and approved by the Monash Institute of Pharmaceutical Sciences Animal Ethics Committee (MIPS.21542).

Informed Consent Statement: Not applicable.

Data Availability Statement: Data is contained within the article or Supplementary Materials.

Conflicts of Interest: The authors declare no conflict of interest. The funders had no role in the design of the study; in the collection, analyses, or interpretation of data; in the writing of the manuscript; or in the decision to publish the results.

References

1. Morgan, S.; Orrell, R.W. Pathogenesis of amyotrophic lateral sclerosis. *Br. Med. Bull.* **2016**, *119*, 87–98. [CrossRef]
2. Renton, A.E.; Chio, A.; Traynor, B.J. State of play in amyotrophic lateral sclerosis genetics. *Nat. Neurosci.* **2013**, *17*, 17–23. [CrossRef]

3. Byrne, S.; Elamin, M.; Bede, P.; Shatunov, A.; Walsh, C.; Corr, B.; Heverin, M.; Jordan, N.; Kenna, K.; Lynch, C.; et al. Cognitive and clinical characteristics of patients with amyotrophic lateral sclerosis carrying a C9orf72 repeat expansion: A population-based cohort study. *Lancet Neurol.* **2012**, *11*, 232–240. [CrossRef]

4. Van Es, M.A.; Hardiman, O.; Chio, A.; Al-Chalabi, A.; Pasterkamp, R.J.; Veldink, J.H.; van den Berg, L.H. Amyotrophic lateral sclerosis. *Lancet* **2017**, *390*, 2084–2098. [CrossRef]

5. Chiò, A.; Logroscino, G.; Traynor, B.; Collins, J.; Simeone, J.; Goldstein, L.; White, L. Global epidemiology of amyotrophic lateral sclerosis: A systematic review of the published literature. *Neuroepidemiology* **2013**, *41*, 118–130. [CrossRef]

6. Masrori, P.; Van Damme, P. Amyotrophic lateral sclerosis: A clinical review. *Eur. J. Neurol.* **2020**, *27*, 1918–1929. [CrossRef]

7. Pan, Y.; Nicolazzo, J.A. Altered blood–brain barrier and blood–spinal cord barrier dynamics in amyotrophic lateral sclerosis: Impact on medication efficacy and safety. *J. Cereb. Blood Flow Metab.* **2022**, *179*, 2577–2588. [CrossRef]

8. Garbuzova-Davis, S.; Saporta, S.; Haller, E.; Kolomey, I.; Bennett, S.P.; Potter, H.; Sanberg, P.R. Evidence of compromised blood-spinal cord barrier in early and late symptomatic SOD1 mice modeling ALS. *PLoS ONE* **2007**, *2*, e1205. [CrossRef]

9. Henkel, J.S.; Beers, D.R.; Wen, S.; Bowser, R.; Appel, S.H. Decreased mRNA expression of tight junction proteins in lumbar spinal cords of patients with ALS. *Neurology* **2009**, *72*, 1614–1616. [CrossRef]

10. Miyazaki, K.; Ohta, Y.; Nagai, M.; Morimoto, N.; Kurata, T.; Takehisa, Y.; Ikeda, Y.; Matsuura, T.; Abe, K. Disruption of neurovascular unit prior to motor neuron degeneration in amyotrophic lateral sclerosis. *J. Neurosci. Res.* **2011**, *89*, 718–728. [CrossRef]

11. Garbuzova-Davis, S.; Hernandez-Ontiveros, D.G.; Rodrigues, M.C.; Haller, E.; Frisina-Deyo, A.; Mirtyl, S.; Sallot, S.; Saporta, S.; Borlongan, C.; Sanberg, P.R. Impaired blood–brain/spinal cord barrier in ALS patients. *Brain Res.* **2012**, *1469*, 114–128. [CrossRef]

12. Jablonski, M.R.; Jacob, D.A.; Campos, C.; Miller, D.S.; Maragakis, N.J.; Pasinelli, P.; Trotti, D. Selective increase of two ABC drug efflux transporters at the blood–spinal cord barrier suggests induced pharmacoresistance in ALS. *Neurobiol. Dis.* **2012**, *47*, 194–200. [CrossRef]

13. Milane, A.; Fernandez, C.; Dupuis, L.; Buyse, M.; Loeffler, J.-P.; Farinotti, R.; Meininger, V.; Bensimon, G. P-glycoprotein expression and function are increased in an animal model of amyotrophic lateral sclerosis. *Neurosci. Lett.* **2010**, *472*, 166–170. [CrossRef]

14. DeJesus-Hernandez, M.; Mackenzie, I.R.; Boeve, B.F.; Boxer, A.L.; Baker, M.; Rutherford, N.J.; Nicholson, A.M.; Finch, N.A.; Flynn, H.; Adamson, J.; et al. Expanded GGGGCC hexanucleotide repeat in noncoding region of C9ORF72 causes chromosome 9p-linked FTD and ALS. *Neuron* **2011**, *72*, 245–256. [CrossRef]

15. Renton, A.E.; Majounie, E.; Waite, A.; Simon-Saánchez, J.; Rollinson, S.; Gibbs, J.R.; Schymick, J.C.; Laaksovirta, H.; van Swieten, J.C.; Myllykangas, L.; et al. A hexanucleotide repeat expansion in C9ORF72 is the cause of chromosome 9p21-Linked ALS-FTD. *Neuron* **2011**, *72*, 257–268. [CrossRef]

16. Baker, M.; Mackenzie, I.R.; Pickering-Brown, S.M.; Gass, J.; Rademakers, R.; Lindholm, C.; Snowden, J.; Adamson, J.; Sadovnick, A.D.; Rollinson, S.; et al. Mutations in progranulin cause tau-negative frontotemporal dementia linked to chromosome 17. *Nature* **2006**, *442*, 916–919. [CrossRef]

17. Kumar, V.; Hasan, G.M.; Hassan, I. Unraveling the role of RNA mediated toxicity of C9orf72 repeats in C9-FTD/ALS. *Front. Neurosci.* **2017**, *11*, 711. [CrossRef]

18. Xu, Z.; Yang, C. TDP-43—The key to understanding amyotrophic lateral sclerosis. *Rare Dis.* **2014**, *2*, e944443. [CrossRef]

19. Ormeño, F.; Hormazabal, J.; Moreno, J.; Riquelme, F.; Rios, J.; Criollo, A.; Albornoz, A.; Alfaro, I.E.; Budini, M. Chaperone mediated autophagy degrades TDP-43 protein and is affected by TDP-43 aggregation. *Front. Mol. Neurosci.* **2020**, *13*, 19. [CrossRef]

20. Mejzini, R.; Flynn, L.L.; Pitout, I.L.; Fletcher, S.; Wilton, S.D.; Akkari, P.A. ALS genetics, mechanisms, and therapeutics: Where are we now? *Front. Neurosci.* **2019**, *13*, 1310. [CrossRef]

21. Liu, Y.; Pattamatta, A.; Zu, T.; Reid, T.; Bardhi, O.; Borchelt, D.R.; Yachnis, A.T.; Ranum, L.P.W. C9orf72 BAC mouse model with motor deficits and neurodegenerative features of ALS/FTD. *Neuron* **2016**, *90*, 521–534. [CrossRef]

22. Tefera, T.W.; Steyn, F.J.; Ngo, S.T.; Borges, K. CNS glucose metabolism in amyotrophic lateral sclerosis: A therapeutic target? *Cell Biosci.* **2021**, *11*, 14. [CrossRef]

23. Perry, T.L.; Krieger, C.; Ba, S.H.; Eisen, A. Amyotrophic lateral sclerosis: Amino acid levels in plasma and cerebrospinal fluid. *Ann. Neurol.* **1990**, *28*, 12–17. [CrossRef]

24. Pan, Y.; Scanlon, M.J.; Owada, Y.; Yamamoto, Y.; Porter, C.J.H.; Nicolazzo, J.A. Fatty acid-binding protein 5 facilitates the blood–brain barrier transport of docosahexaenoic acid. *Mol. Pharm.* **2015**, *12*, 4375–4385. [CrossRef]

25. Pan, Y.; Choy, K.H.C.; Marriott, P.J.; Chai, S.Y.; Scanlon, M.J.; Porter, C.J.H.; Short, J.L.; Nicolazzo, J.A. Reduced blood-brain barrier expression of fatty acid-binding protein 5 is associated with increased vulnerability of APP/PS1 mice to cognitive deficits from low omega-3 fatty acid diets. *J. Neurochem.* **2017**, *144*, 81–92. [CrossRef]

26. R Core Team. *R: A Language and Environment for Statistical Computing*; R Foundation for Statistical Computing: Vienna, Austria, 2018; Available online: https://www.R-project.org (accessed on 19 January 2022).

27. Al-Majdoub, Z.M.; Al Feteisi, H.; Achour, B.; Warwood, S.; Neuhoff, S.; Rostami-Hodjegan, A.; Barber, J. Proteomic quantification of human blood–brain barrier SLC and ABC transporters in healthy individuals and dementia patients. *Mol. Pharm.* **2019**, *16*, 1220–1233. [CrossRef]

28. Ritchie, M.E.; Belinda, P.; Wu, D.; Hu, Y.; Law, C.W.; Shi, W.; Smyth, G.K. limma powers differential expression analyses for RNA-sequencing and microarray studies. *Nucleic Acids Res.* **2015**, *43*, e47. [CrossRef]

29. Uchida, Y.; Ohtsuki, S.; Katsukura, Y.; Ikeda, C.; Suzuki, T.; Kamiie, J.; Terasaki, T. Quantitative targeted absolute proteomics of human blood-brain barrier transporters and receptors. *J. Neurochem.* **2011**, *117*, 333–345. [CrossRef]

30. Smith, Q.R.; Allen, D.D. *The Blood-Brain Barrier: Biology and Research Protocols*; Sukriti, N., Ed.; Humana Press: Totowa, NJ, USA, 2003; pp. 209–218.

31. Cattelotte, J.; André, P.; Ouellet, M.; Bourasset, F.; Scherrmann, J.-M.; Cisternino, S. In situ mouse carotid perfusion model: Glucose and cholesterol transport in the eye and brain. *J. Cereb. Blood Flow Metab.* **2008**, *28*, 1449–1459. [CrossRef]

32. Tanaka, H.; Mizojiri, K. Drug-protein binding and blood-brain barrier permeability. *J. Pharmacol. Exp. Ther.* **1999**, *288*, 912–918.

33. Low, Y.L.; Jin, L.; Morris, E.R.; Pan, Y.; Nicolazzo, J.A. Pioglitazone increases blood-brain barrier expression of fatty acid-binding protein 5 and docosahexaenoic acid trafficking into the brain. *Mol. Pharm.* **2020**, *17*, 873–884. [CrossRef]

34. Jin, L.; Pan, Y.; Tran, N.L.L.; Polychronopoulos, L.N.; Warrier, A.; Brouwer, K.L.R.; Nicolazzo, J.A. Intestinal permeability and oral absorption of selected drugs are reduced in a mouse model of familial Alzheimer's disease. *Mol. Pharm.* **2020**, *17*, 1527–1537. [CrossRef]

35. Ilieva, E.V.; Ayala, V.; Jové, M.; Dalfó, E.; Cacabelos, D.; Povedano, M.; Bellmunt, M.J.; Ferrer, I.; Pamplona, R.; Portero-Otín, M. Oxidative and endoplasmic reticulum stress interplay in sporadic amyotrophic lateral sclerosis. *Brain* **2007**, *130*, 3111–3123. [CrossRef]

36. Tefera, T.W.; Borges, K. Metabolic dysfunctions in amyotrophic lateral sclerosis pathogenesis and potential metabolic treatments. *Front. Neurosci.* **2017**, *10*, 611. [CrossRef]

37. Devraj, K.; Klinger, M.E.; Myers, R.L.; Mokashi, A.; Hawkins, R.A.; Simpson, I.A. GLUT-1 glucose transporters in the blood-brain barrier: Differential phosphorylation. *J. Neurosci. Res.* **2011**, *89*, 1913–1925. [CrossRef]

38. Simpson, I.A.; Dwyer, D.; Malide, D.; Moley, K.H.; Travis, A.; Vannucci, S.J. The facilitative glucose transporter GLUT3: 20 years of distinction. *Am. J. Physiol. Metab.* **2008**, *295*, E242–E253. [CrossRef]

39. Patching, S.G. Glucose Transporters at the blood-brain barrier: Function, regulation and gateways for drug delivery. *Mol. Neurobiol.* **2016**, *54*, 1046–1077. [CrossRef]

40. Zhang, Z.; Ma, Z.; Zou, W.; Guo, H.; Liu, M.; Ma, Y.; Zhang, L. The appropriate marker for astrocytes: Comparing the distribution and expression of three astrocytic markers in different mouse cerebral regions. *BioMed Res. Int.* **2019**, *2019*, 9605265. [CrossRef]

41. Smyth, L.C.; Rustenhoven, J.; Scotter, E.L.; Schweder, P.; Faull, R.L.; Park, T.I.; Dragunow, M. Markers for human brain pericytes and smooth muscle cells. *J. Chem. Neuroanat.* **2018**, *92*, 48–60. [CrossRef]

42. Perry, A.; Brat, D.J. *Practical Surgical Neuropathology: A Diagnostic Approach*, 2nd ed.; Perry, A., Brat, D.J., Eds.; Elsevier: Amsterdam, The Netherlands, 2018; pp. 1–17.

43. Sharma, K.; Schmitt, S.; Bergner, C.G.; Tyanova, S.; Kannaiyan, N.; Manrique-Hoyos, N.; Kongi, K.; Cantuti, L.; Hanisch, U.-K.; Philips, M.-A.; et al. Cell type– and brain region–resolved mouse brain proteome. *Nat. Neurosci.* **2015**, *18*, 1819–1831. [CrossRef]

44. Pan, Y.; Nicolazzo, J.A. Impact of aging, Alzheimer's disease and Parkinson's disease on the blood-brain barrier transport of therapeutics. *Adv. Drug Deliv. Rev.* **2018**, *135*, 62–74. [CrossRef]

45. Zhong, Z.; Deane, R.; Ali, Z.; Parisi, M.; Shapovalov, Y.; O'Banion, M.K.; Stojanovic, K.; Sagare, A.; Boillee, S.; Cleveland, D.W.; et al. ALS-causing SOD1 mutants generate vascular changes prior to motor neuron degeneration. *Nat. Neurosci.* **2008**, *11*, 420–422. [CrossRef]

46. Mann, M.; Kelleher, N.L. Precision proteomics: The case for high resolution and high mass accuracy. *Proc. Natl. Acad. Sci. USA* **2008**, *105*, 18132–18138. [CrossRef]

47. Rosenblum, L.T.; Trotti, D. EAAT2 and the molecular signature of amyotrophic lateral sclerosis. *Adv. Neurobiol.* **2017**, *16*, 117–136. [CrossRef]

48. O'Kane, R.L.; Martínez-López, I.; DeJoseph, M.R.; Viña, J.R.; Hawkins, R.A. Na+-dependent glutamate transporters (EAAT1, EAAT2, and EAAT3) of the blood-brain barrier. *J. Biol. Chem.* **1999**, *274*, 31891–31895. [CrossRef]

49. Gliddon, C.M.; Shao, Z.; LeMaistre, J.L.; Anderson, C.M. Cellular distribution of the neutral amino acid transporter subtype ASCT2 in mouse brain. *J. Neurochem.* **2009**, *108*, 372–383. [CrossRef]

50. Scalise, M.; Pochini, L.; Console, L.; Losso, M.A.; Indiveri, C. The human SLC1A5 (ASCT2) amino acid transporter: From function to structure and role in cell biology. *Front. Cell Dev. Biol.* **2018**, *6*, 96. [CrossRef]

51. Mitchell, R.W.; Edmundson, C.L.; Miller, D.W.; Hatch, G.M. On the mechanism of oleate transport across human brain microvessel endothelial cells. *J. Neurochem.* **2009**, *110*, 1049–1057. [CrossRef]

52. Lee, G.S.; Pan, Y.; Scanlon, M.J.; Porter, C.J.; Nicolazzo, J.A. Fatty acid–binding protein 5 mediates the uptake of fatty acids, but not drugs, into human brain endothelial cells. *J. Pharm. Sci.* **2018**, *107*, 1185–1193. [CrossRef]

53. E Umoh, M.; Dammer, E.; Dai, J.; Duong, D.; Lah, J.J.; I Levey, A.; Gearing, M.; Glass, J.D.; Seyfried, N.T. A proteomic network approach across the ALS - FTD disease spectrum resolves clinical phenotypes and genetic vulnerability in human brain. *EMBO Mol. Med.* **2017**, *10*, 48–62. [CrossRef]

54. González-Romero, R.; Eirín-López, J.M.; Ausió, J. Evolution of high mobility group nucleosome-binding proteins and its implications for vertebrate chromatin specialization. *Mol. Biol. Evol.* **2014**, *32*, 121–131. [CrossRef]

55. Ueda, T.; Furusawa, T.; Kurahashi, T.; Tessarollo, L.; Bustin, M. The nucleosome binding protein HMGN3 modulates the transcription profile of pancreatic β cells and affects insulin secretion. *Mol. Cell. Biol.* **2009**, *29*, 5264–5276. [CrossRef]

56. Shelkovnikova, T.; Robinson, H.K.; Troakes, C.; Ninkina, N.; Buchman, V.L. Compromised paraspeckle formation as a pathogenic factor in FUSopathies. *Hum. Mol. Genet.* **2013**, *23*, 2298–2312. [CrossRef]

Article

Permeability of Metformin across an In Vitro Blood–Brain Barrier Model during Normoxia and Oxygen-Glucose Deprivation Conditions: Role of Organic Cation Transporters (Octs)

Sejal Sharma [1,2], Yong Zhang [1,2], Khondker Ayesha Akter [1,2], Saeideh Nozohouri [1,2], Sabrina Rahman Archie [1,2], Dhavalkumar Patel [1,2], Heidi Villalba [1,2] and Thomas Abbruscato [1,2,*]

1 Department of Pharmaceutical Sciences, Jerry H. Hodge School of Pharmacy, Texas Tech University Health Sciences Center, Amarillo, TX 79106, USA
2 Center for Blood-Brain Barrier Research, Jerry H. Hodge School of Pharmacy, Texas Tech University Health Sciences Center, Amarillo, TX 79106, USA
* Correspondence: thomas.abbruscato@ttuhsc.edu

Abstract: Our lab previously established that metformin, a first-line type two diabetes treatment, activates the Nrf2 pathway and improves post-stroke recovery. Metformin's brain permeability value and potential interaction with blood–brain barrier (BBB) uptake and efflux transporters are currently unknown. Metformin has been shown to be a substrate of organic cationic transporters (Octs) in the liver and kidneys. Brain endothelial cells at the BBB have been shown to express Octs; thus, we hypothesize that metformin uses Octs for its transport across the BBB. We used a co-culture model of brain endothelial cells and primary astrocytes as an in vitro BBB model to conduct permeability studies during normoxia and hypoxia using oxygen–glucose deprivation (OGD) conditions. Metformin was quantified using a highly sensitive LC-MS/MS method. We further checked Octs protein expression using Western blot analysis. Lastly, we completed a plasma glycoprotein (P-GP) efflux assay. Our results showed that metformin is a highly permeable molecule, uses Oct1 for its transport, and does not interact with P-GP. During OGD, we found alterations in Oct1 expression and increased permeability for metformin. Additionally, we showed that selective transport is a key determinant of metformin's permeability during OGD, thus, providing a novel target for improving ischemic drug delivery.

Keywords: metformin; ischemic stroke; repurposing; transporters; in vitro; BBB; co-culture; P-GP; Octs; permeability

Citation: Sharma, S.; Zhang, Y.; Akter, K.A.; Nozohouri, S.; Archie, S.R.; Patel, D.; Villalba, H.; Abbruscato, T. Permeability of Metformin across an In Vitro Blood–Brain Barrier Model during Normoxia and Oxygen-Glucose Deprivation Conditions: Role of Organic Cation Transporters (Octs). *Pharmaceutics* 2023, 15, 1357. https://doi.org/10.3390/pharmaceutics15051357

Academic Editors: Gert Fricker and Elena Puris

Received: 15 March 2023
Revised: 19 April 2023
Accepted: 26 April 2023
Published: 28 April 2023

1. Introduction

Metformin (MF) is a commonly prescribed, first-line therapy for treating type two diabetes mellitus [1]. The United States Food and Drug Administration (US-FDA) approved it almost thirty years ago and it has reported no major adverse effects [2]. MF reduces blood glucose levels by inhibiting peripheral and hepatic glucose production without affecting insulin sensitivity, therefore not causing hypoglycemia [3]. Outside of its application in treating diabetes, increasing numbers of pre-clinical and clinical studies show that pre- and post-treatment of the drug has a protective effect against various neurological disorders and ischemic stroke [4–9]. Stroke is the fifth leading cause of death in the US and about 87% of all strokes are ischemic strokes [10]. The most common cause of ischemic stroke is a partial or complete blockage of blood flow to the brain due to occlusion of a blood vessel from a clot, resulting in neurological loss or even death [11].

There is a clear unmet clinical need for new and better treatments for CNS diseases and ischemic stroke. Tissue plasminogen activator (tPA) is the only US-FDA-approved drug treatment for ischemic stroke, whose efficacy is compromised if not administered within 3–4 h of stroke onset [12]. Another approved intervention, mechanical or endovascular thrombectomy, is used to treat large vessel occlusion ischemia; however, the procedure is contraindicated in patients with a high risk of intracranial or systemic hemorrhage [13]. Over the last two to three decades, only a few potential CNS drugs have had the probability of getting beyond phase three clinical trials for ischemic stroke [14]. Drug repurposing or repositioning is a process that involves identifying a new indication for an approved or investigational drug that would be outside the original medical indication, with the purpose of accelerating the development of an effective therapy indication [15]. Therefore, repurposing a safe, effective, and established therapeutic such as MF could provide a much faster bench-to-patient pharmacological transition.

One underlying mechanism of MF as a neurotherapeutic is that it activates AMP-activated protein kinase (AMPK), causing downstream activation of signaling pathways such as NF-κB and mTOR that lead to reduced inflammatory and oxidative-stress responses in the brain [16–18]. Previously published data from our group also strongly suggests that MF activates nuclear factor erythroid 2-related factor (Nrf2) in the brain of stroke animals [19]. We showed that MF counteracts the cerebrovascular toxicity caused by tobacco smoking and electronic cigarette vaping (known risk factors for ischemic stroke) by protecting the blood–brain barrier's (BBB) integrity in stroke-subjected animals [8]. Another essential aspect for determining MF's neuroprotection is understanding its blood-to-brain permeability and potential interaction with uptake and efflux transporters present at the BBB during normal physiological conditions or normoxia and diseased states, such as ischemic stroke.

MF is a small hydrophilic molecule (logD −6.13) with a molecular weight of 129 Dalton that is positively charged (pKa 12.4) at physiological pH [20,21]. These physicochemical properties suggest that MF permeation across biological membranes does not occur through rapid passive diffusion unless its transport is driven either by membrane potential (carrier-mediated transport) or simply by paracellular diffusion (between the cellular gaps). The BBB mainly comprises brain endothelial cells that closely interact with supporting cells such as astrocytes, microglia, pericytes, and neurons to induce and maintain integrity and function [22–24]. Additionally, tight junctional proteins, such as occludin, claudin-5, and ZO-1, expressed by endothelial cells, are crucial for forming a paracellular seal to restrict the movement of molecules or drugs across the BBB [25,26]. This restriction by the BBB stops at least 95% of investigational molecules in the process of drug development [27]. The endothelial cells of the BBB express uptake transporters, solute carriers (SLCs) subtypes of organic cation transporters (Octs) such as Oct1 (SLC22A1), Oct2 (SLC22A2), and Oct3 (SLC22A3) for the transport of hydrophilic cationic CNS drugs into the brain [14]. Octs have been shown to interact with endogenous substrates such as monoamine neurotransmitters and with CNS drugs such as antivirals, tricyclic anti-depressants, and the anti-diabetic MF [28–30]. MF has been well known to interact with Octs in organs such as the liver and kidney [31–33].

Our lab has previously explored the contribution of organic anionic transporters (OATPs) in the transport of a potent opioid receptor agonist, biphalin, during hypoxia and further established an initial time window for brain entry [34]. Another study showed that the functional expression OATP1A2 (an isoform of OATPs) is crucial for the therapeutic efficacy of the statins during stroke recovery [35]. More recently, it was reported that Oct1 and Oct2 transporter expression variations during in vivo stroke conditions lead to enhanced permeability of memantine [36]. Furthermore, it was shown that the transporter-mediated permeation of the drug was increased by two fold and was critical to achieving therapeutic efficacy. Another study showed that expression of Octs in an in vitro human BBB is required to transport carnitine for neuronal homeostasis [37]. Thus, it can be implied that endogenous transporters at the BBB play a vital role and provide a novel approach for the development of CNS drug delivery during ischemic stroke. These reports further warrant evaluation of the permeability of a drug candidate, such as MF, which is expected to use carrier-mediated transport (as it has been shown to be a substrate of Octs in other organs) to gain access to the brain during normal and pathophysiological situations.

Besides uptake transporters, efflux transporters such as plasma glycoproteins (P-GP) are one of the major efflux transporters at the BBB that restrict brain entry [38–40]. P-GP restricts solutes or toxins from entering the brain parenchyma from blood circulation and their interaction with MF at the BBB is currently unknown. In the present study, for the first time, we examined the permeability of MF during normal physiological conditions or normoxia and oxygen glucose-deprived (OGD) conditions. We determined that MF uses a saturable mechanism for its transport across an in vitro co-culture model of BBB using a co-culture of bEnd.3 cells (immortalized mouse brain endothelial cells) and mouse primary astrocytes. In addition, we show an alteration in the expression of Oct1 in bEnd.3 cells during OGD time points and enhanced MF permeation into the brain, during which carrier-mediated transport plays a crucial role in the drug's transport. Lastly, we found that MF does not interact with one of the major efflux proteins, P-GP, in P-GP overexpressing cells.

2. Materials and Methods

2.1. Cell Culture and bEnd.3/Astrocyte Co-Culture

bEnd.3 cells passages 21–24 (ATCC, Manassas, VA, USA) were cultured in Dulbecco's modified Eagle's medium (Sigma, St. Louis, MO, USA) supplemented by 10% FBS (Atlanta Biologicals, Minneapolis, MN, USA) and 1% each of non-essential amino acid and penicillin/streptomycin (PS) solution (Sigma Aldrich). Cells were then maintained in a humidified cell culture incubator at 37 °C and with 5% CO_2/95% air. Mouse primary astrocytes were obtained from the cerebral cortices of one-day-old CD-1 mouse pups (Charles Rivers Laboratory) according to the previously published methods [41]. After isolating the brain, cerebral cortices were isolated, meninges were removed, and cortices free from meninges were placed in Hanks' balanced salt solution (HBSS) without calcium and magnesium, supplemented with gentamycin (10 µg/mL). Then, cortices were digested with 0.25% trypsin for 10–15 min at 37 °C, followed by neutralizing with FBS containing Dulbecco's modified Eagle's medium containing 10% FBS and 1% PS solution. The cells were then seeded into a cell culture T75 flask and the medium was refreshed every 3 days until reaching confluency.

For bEnd.3 and astrocyte co-culture, the transwell inserts (0.4–1 µm pore size, 12-well; Corning, Lowell, MA, USA) were inverted and astrocytes at a density of 150,000 cells per insert were seeded onto the basolateral side of the insert membrane and were allowed to adhere for 4 h. The transwell inserts were then inverted back and astrocytes were allowed to grow for 2 more days in the astrocyte medium. Then, bEnd.3 cells with a density of 50,000 cells per insert were seeded onto the apical side of the inserts. The co-culture of primary astrocytes and bEnd.3 cells was grown for 8 more days by changing the media for both cells every other day. Permeability experiments were completed on days 8–10 from day 1 of the co-culture establishment.

2.2. Barrier Integrity Measurements

The integrity of the co-culture setup was carried out using two techniques: (1) Through measurement of trans-endothelial cell resistance (TEER) of the transwell inserts and (2) by permeability assessment (apical to basolateral) of sodium fluorescein (NaF) (Sigma Aldrich), a low molecular weight BBB marker, according to previously published method [42]. All permeability experiments were conducted with the co-cultured transwell inserts that had TEER value measurements of >70 $\Omega \cdot cm^2$. The TEER was measured by EVOM resistance meter (World Precision Instruments, Sarasota, FL, USA) using the STX-2 electrodes. For NaF permeability experiments, the media was removed from both of the compartments (apical and basolateral) and rinsed with HBSS buffer and incubated at 37 °C for 30 min. Then, 500 µL of 10 µg/mL of NaF in HBSS was added to the apical chamber of the inserts and 100 µL was collected from the basolateral compartment of the wells in duplicates for concentration determination. The collected volumes were replaced with 200 μL of HBSS buffer to avoid the back diffusion of NaF. The collection was performed at time points 30, 60, and 120 min. The amount of NaF was measured at an absorbance using a fluorescent microplate reader (PerkinElmer, Waltham, MA, USA) with excitation and emission wavelengths of 485 and 520 nm. Then, the permeability coefficient (PC, in cm/min) was determined using the following equation:

$$PC = \frac{dQ}{dT} \times \frac{1}{C_0 \times A} \tag{1}$$

where dQ/dt is the diffusion rate of NaF across the membrane, A is the area of the transwell insert, and C_0 is the initial concentration of the NaF added in the donor compartment. The blank transwell insert (without cells) was added to the experimental study group to negate its PC value because blank inserts themselves provide resistance to the buffer and compound:

$$\frac{1}{PC_{cells}} = \frac{1}{PC_{total}} - \frac{1}{PC_{blank}} \tag{2}$$

The PC value of NaF (MW 376 Da) was compared with another low molecular weight BBB marker, [14C] sucrose (MW 342 Da), used previously in our lab. We found no statistically significant difference between them (Figure S1).

2.3. [14C] MF Permeability a Using Self-Inhibition Study

Permeability experiments for [14C] MF (Moravek Biochemicals, Brea, CA, USA) were conducted similarly to as explained above for NaF. 0.15 μCi/mL or 10 μM of [14C] MF was chosen as a pharmacologically relevant concentration because of the observed steady-state concentration of 1.8 μg/mL or 13 μM in human subjects. Then, MF was diluted in HEPES (in mM 120 NaCl, 1 $CaCl_2$, 25 HEPES, 1 KH_2PO_4, 2 KCl, 1 $MgSO_4$, 10 D-glucose) buffer and added to the apical chamber of the transwell inserts containing pre-incubated concentrations of unlabeled MF (Sigma Aldrich) at 1 mM, 10 mM, and 20 mM. 100 μL of samples of [14C] MF were collected from the basolateral chamber at time points 5, 15, 30, 60, and 120 min. The radioactivity of the collected samples was evaluated using a liquid scintillation counter (Beckman Coulter, Brea, CA, USA), and the permeability value for [14C] MF in the presence and absence of self-inhibitory concentrations was determined using Equations (1) and (2).

2.4. MF Permeability Using Transporter-Specific Inhibitors

Mitoxantrone (Sigma Aldrich), a known Oct1 inhibitor at 25 μM, and corticosterone (Sigma Aldrich), a known Oct1, 2, and 3 inhibitors at 150 μM, were preincubated for 30 min at the apical and basolateral chambers of transwell inserts prior to and during the transport experiments for 120 min. The transport experiment was initiated by adding 500 μL of 10 μM MF at the apical chambers and samples were collected from each group at 5, 15, 30, 60, and 120 min time points. Quantification was conducted through a highly sensitive LC-MS/MS method. After quantification, the permeability value for MF in the presence and absence of transporter-specific inhibitors was determined.

2.5. MTS Cell Viability Assay

Cell viability was measured by MTS (dimethylthiazol carboxymethoxyphenyl sulfophenyl tetrazolium) assay using the CellTiter 96 kit Aqueous assay (Promega) according to the provided manufacturer's instructions. The assay was conducted to ensure that the concentration of MF and inhibitors used in permeability studies is not toxic to the cells. bEnd.3 cells and primary astrocytes were seeded in separate 96 wells plates. After confluency, plates were incubated with different concentrations of unlabeled metformin (100 μM, 1 mM, 10 mM and 20 mM) and Octs inhibitors for 3 h. After incubation, the absorbance wavelength of formazan was read at 490 nm using fluorescent microplate reader. The values were calculated as a percentage of the control value.

2.6. LC-MS/MS Method Development, Sample Preparation and Analysis

HBSS buffer was spiked with MF to form the stock solution to achieve a calibration curve range of 3.9 ng/mL to 2000 ng/mL. In addition, two separate blank control solutions, one of them spiked with internal standard, were considered for the study. Fifty µL of the unknown samples obtained from transporter-specific inhibition, oxygen–glucose deprivation (explained below in Section 2.6), and p-GP efflux (explained below in Section 2.8) studies were spiked with 20 µL of 500 ng/mL of internal standard Metformin-D6 hydrochloride (Cayman Chemical Company). Then, the mixture was vortexed for 5 min, after which 930 µL of 2 mM of Ammonium Acetate (Sigma Aldrich) solution was added to the mixture. The solution was further vortexed for 5 min and was centrifuged at $10,000 \times g$ rpm for 5 min at 4 °C. Then, 500 µL of the supernatant was inserted in deep well plates (Waters Corporation, Milford, MA, USA) and analyzed using the LC-MS/MS method. The ultra-high performance liquid chromatography (UHPLC) system (Shimazdu Corporation, Columbia, MD, USA) equipped with AB SCIEX QTRAP 5500 triple quadrupole mass spectrometer (Foster City, CA, USA) was used. The data acquisition and analysis were performed using Analyst software 1.7.0.

Chromatographic separations were performed using an XBridge BEH C18 sorbent (50×3.0 mm, 3.5 µm, Waters Corporation, Milford, MA, USA). The mobile phase A was used as 2 mM ammonium acetate in 5% Acetonitrile (Fisher Scientific, Waltham, MA, USA) in LCMS water and the mobile phase B was used as 100% Acetonitrile. LCMS grade water (Fisher Scientific) was used as rinsing solution. The gradients separation was set up as 100% of A for 0 to 1 min, 90% of B at 1.2 min, and held until 2.5 min. At 2.6 min, the percentage of gradient was brought back to initial concentration of 100% of A and held until 4 min of total runtime. The flow rate was set at 0.35 mL/min, the column temperature was maintained at 45 °C, and injection volume was set at 2 µL. The retention time of MF and its internal standard were ~0.9 min. Multiple reaction monitoring was conducted in negative mode and the transitions for MF and internal standard were $130.1 \rightarrow 60.1$ m/z and $136.1 \rightarrow 60.1$ m/z, respectively. The collision energy for MF and IS was 20 and 20 volts, respectively.

2.7. In Vitro Hypoxia Modelling: Oxygen Glucose Deprivation (OGD)

The cells were exposed to in vitro hypoxia model by OGD, according to previously explained protocol [13,14]. In brief, the co-cultured transwell inserts were exposed to OGD for 2, 4, and 6 h for permeability experiments. In another set of study, bEnd.3 cells and primary astrocytes grown on petri dishes (Corning, Lowell, MA, USA) were exposed to OGD for 2, 4 and 6 h to check the expression of Oct1, 2, and 3. For the experiment, media was removed and replaced with media bubbled with 95% N2/5% CO_2. Dulbecco's phosphate buffered saline solution and glucose-free Earle's balanced salt solution (in mM 140 NaCl, 0.83 $MgSO_4$, 5.36 KCl, 1.02 NaH_2PO_4, 1.18 CaCl, 26.19 $NaHCO_3$, adjusted pH 7.4) to achieve an aglycemic condition. Then, the cells were transferred to custom-made hypoxic chamber (Coy Laboraties, Grasslake, MI, USA) with 95% N_2 and 5% CO_2 at 37 °C to induce hypoxia (1% oxygen).

2.8. Western Blot for Octs Expression

Caco-2 cells (ATCC, Manassas, VA, USA) was used as a positive control because they have shown to express Octs [21]. After OGD time points, the bEnd.3 cells and primary astrocytes were lysed using radioimmunoprecipitation assay buffer (RIPA) supplemented with proteinase and phosphatase inhibitor cocktails and then collected for estimation of protein concentration using a BCA assay. After solubilizing samples using Laemmli buffer, 30 µg of protein was loaded per well separated by 10% Tris–glycine polyacrylamide precast gel (Bio-Rad Laboratories, Hercules, CA, USA) and then transferred to a polyvinylidene fluoride (PVDF) membrane (Thermo Fisher, Waltham, MA, USA) for 2 h at 90 V.

Then, the membrane was incubated in blocking solution containing 5% bovine serum albumin to block nonspecific binding for 2 h at room temperature. Then, the membranes were immunoblotted with the appropriate primary antibodies overnight at 4 °C with the following dilutions: rabbit monoclonal anti-Oct-1 (1:1000) antibody (Cell Signaling Technology, Danvers, MA, USA), rabbit monoclonal anti-Oct-2 (1:1000) (Abcam) and rabbit polyclonal anti-Oct-3 (1:500) and mouse monoclonal anti-beta actin antibody (1: 10,000) (Sigma Aldrich) in TBST with 5% bovine serum albumin. After 3 times washing with TBST for 10 min each, membranes were incubated with anti-rabbit (Sigma Aldrich) or anti-mouse (Sigma Aldrich) IgG-horseradish peroxidase secondary antibody (1:10,000) in TBST with 5% bovine serum albumin for 2 h at room temperature. After 3 times of 10 min wash with TBST, the protein signals were detected using enhanced chemiluminescence substrate (Westernsure chemiluminescent substrate) and visualized in LI-COR; C-Digit blot scanner. The protein bands were quantified relative to beta-actin in Image J 1.53t software and are reported as ratios of the control group.

2.9. P-Glycoprotein Efflux Transporter Substrate Assay

For P-GP efflux assay, MDR1-MDCK 12-well transport assay (MB Biosciences LLC, Natick, MA, USA) was used. The cells were maintained in high glucose DMEM media supplied with 10% FBS, non-essential amino acid, penicillin/streptomycin, L-glutamine, and colchicine as recommended by the supplier. After the shipping medium was changed to a fresh MDR1 MDCK cell culture medium, the plate was kept in the incubator for 24 h before performing the experiment. 10 μM of MF in HBSS was added to the apical compartment of the transwell inserts in presence and absence of a potent p-GP inhibitor, Cyclosporine A (CsA) at 10 μM concentration (Sigma, St. Louis, MO, USA). The inhibitor was present on both sides of the compartments during pre-incubation of 30 min and during the transport experiment. After 120 min of incubation at 37 °C, samples were collected from the basolateral chamber and quantified by LC-MS/MS method. For groups in which MF was added in the basolateral compartment, samples were collected from the apical chamber after 120 min. The permeability coefficient (PC, cm/s) was calculated using the published literature method. The PC value from apical (A) to basolateral (B) compartment in the presence and absence of CsA was used to calculate the unidirectional flux ratio (UFR), which is defined by the following equation:

$$UFR = \frac{PC, A-to-B(+CsA)}{PC, A-to-B(-CsA)} \tag{3}$$

In a parallel study, MF was added to the basolateral compartment and samples were collected from the apical chamber to evaluate the B-to-A PC value, which allows calculating the efflux ratio (ER), defined by the following equation:

$$ER = \frac{PC, B-to-A}{PC, A-to-B} \tag{4}$$

2.10. Data Analysis

All values are presented as mean ± SD. Unpaired student's *t*-test was used to compare two groups. The comparison of three or more than three groups was conducted by one-way analysis of variance (ANOVA) followed by Tukey's post hoc multiple comparison test (Prism, version 9.0; GraphPad Software Inc., San Diego, CA, USA). *p* values of <0.05 were considered to be statistically significant.

3. Results and Discussion

3.1. MF Is Highly Permeable and Uses Saturable Transport for Its Movement across In Vitro Co-Culture Model of BBB

A desirable PC value for a compound is one of the major prerequisites in vitro ADME properties for a potential CNS drug [45,46]. According to published studies, an in vitro

PC value of $>6 \times 10^{-4}$ cm/min is considered a highly permeable drug across the BBB. Our study found MF's PC value to be $24.5 \pm 3.14 \times 10^{-4}$ cm/min, which is ~four fold greater than the high PC value threshold (Figure 1C). Additionally, we compared the PC value of MF with compound **11a**, a neurolysin activator with improved BBB permeability (11.4×10^{-4} cm/min), as shown previously by our lab and collaborators [47]. We found a two fold increase in permeability for MF (Figure 1B). Notably, the permeability experiments for MF and compound **11a** were conducted on a similar co-culture setup in our lab.

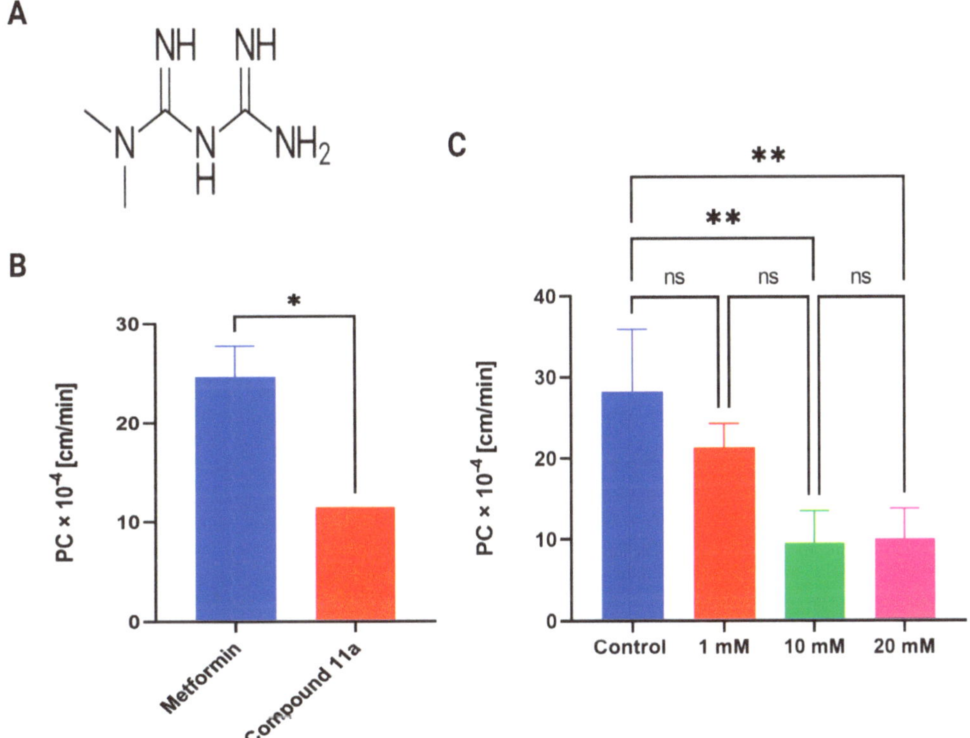

Figure 1. (**A**) Chemical structure of MF. (**B**) Permeability of MF across an in vitro co-culture setup of BBB. The permeability coefficient (PC) of MF compared to Compound **11a** [47], a neurolysin activator, experiments of which were performed under a similar setup of co-culture model of brain endothelial cells and astrocytes. Unpaired student's *t*-test (* $p < 0.05$); $n = 3$ biological replicates; mean $\pm$ SD. (**C**) Saturable transport of tracer [^{14}C] MF as shown by decrease in transport at self-inhibitory concentration of 10 mM, with no further significant decrease at 20 mM concentration. One-way ANOVA, followed by Tukey's multiple comparisons test (** $p < 0.01$); ns = non-significant; $n = 3$ biological replicates; mean $\pm$ SD.

Next, to explore whether the mechanism behind MF's transport across the in vitro BBB model is saturable, we measured the PC value of [^{14}C] MF as a tracer in the presence of an increasing concentration of cold MF at 1 mM, 10 mM, and 20 mM. We also performed MTS cell viability assay to show that these concentrations are not toxic to both bEnd.3 cells and primary astrocytes (Figure S2). As shown in Figure 1C, with a self-inhibitory concentration of 1 mM, there is no significant decrease in the permeability of [^{14}C] MF ($21.17 \pm 2.96 \times 10^{-4}$ cm/min). However, with a higher inhibitory concentration of 10 mM, there was a significant decrease in [^{14}C] MF transport ($9.4 \pm 3.98 \times 10^{-4}$ cm/min, $p < 0.01$), with no further reduction at 20 mM ($9.9 \pm 3.76 \times 10^{-4}$ cm/min), indicating the saturation of transporters. Previously, it was reported that the apparent IC_{50} value of MF for uptake of tracer [^{14}C] MF was around ~2 mM [48]. Consistent with this previously reported value, we found the saturation concentration was above 1 mM in our study. Furthermore, it can be extrapolated from Figure 1C that ~35% of MF's transport is active transporter-mediated (the inhibition component of the transport from $28.15 \pm 7.70 \times 10^{-4}$ cm/min to $9.9 \pm 3.76 \times 10^{-4}$ cm/min is active transport). In comparison, the rest, ~65%, is suggested to be due to paracellular passive diffusion of the molecule across the in vitro BBB co-culture model.

3.2. Role of Organic Cation Transporters (Octs) for MF Permeability across the BBB

Studies have shown that brain endothelial cells and brain microvessels have protein and mRNA expressions of Octs [16]. Studies reported the expression of Oct1 and Oct2 in cultured brain endothelial murine and human cell lines [49,50]. MF has been shown to be a substrate for Oct1 in the liver and Oct2 in the kidneys [51,52]. Additionally, studies have shown that MF undergoes saturable uptake in single transporter-expressing cell lines for Oct1, Oct2, and Oct3 with apparent Km values of 3.1, 0.6, and 2.6 mM, respectively [53–55]. Moreover, the 10 μm MF concentration used in our transport assays is within the reported Km values of the Oct1, Oct2, and Oct3, suggesting that these transporters are not saturated at this concentration [53]. A recently published study showed targeting Oct1 and Oct2 provided efficient pharmacotherapy for ischemic stroke treatment using the cationic drug memantine [36]. Thus, we assume exploring Octs for MF will give new insight into MF's therapeutic utility as a potential treatment during ischemic stroke.

We needed to choose highly specific inhibitors for the transporter-specific inhibition study for the respective Oct transporters. Previous studies have shown that cationic substrate affinities for Octs often have overlapping inhibition curves, with less likely cationic compounds having affinities for either of the transporters [56,57]. However, a study showed that mitoxantrone is highly selective for the inhibition of Oct1 transporter because its estimated IC_{50} value for Oct1 is 40–60 fold lower than for Oct2 and Oct3 [53]. Thus, for our study, the inhibitory concentration for mitoxantrone was chosen at 25 μM, which is at least four fold greater than its IC_{50} value for Oct1. In this concentration, the respective transporter is inhibited by >80% (Table 1). Next, at 150 μM, corticosterone was selected for overall inhibition of Oct1, 2, and 3 because its IC_{50} value for either Octs was closely valued (Table 1). We further confirmed that the concentration of inhibitors was not toxic to bEnd.3 cells (Figure S3A) and astrocytes (Figure S3B). The permeability experiment for MF involved 30 min pre-incubation and incubation throughout 120 min with the inhibitors. The TEER values were obtained to ensure intact BBB after incubation with the inhibitors to corroborate with the cell viability study (Figure S3C). With mitoxantrone, MF transport was significantly inhibited compared to the control ($14.68 \pm 0.9 \times 10^{-4}$ cm/min, $p < 0.05$), which suggests that MF uses Oct1 for its transport (Figure 2. Additionally, with corticosterone, the inhibition was also significant compared to the control ($12.86 \pm 1.27 \times 10^{-4}$ cm/min, $p < 0.01$), but not significantly different from the mitoxantrone treatment, which suggests that Oct1 is the primary transporter involved in MF's transport across the BBB co-culture of bEnd.3 cells and astrocytes.

Table 1. Cited values from a literature for IC_{50} inhibition of MF uptake by organic cationic transporters (Octs) [53].

Inhibitors/Substrates	Specificity Based on IC50 Values for 10 μM Metformin Uptake		
	Oct-1	Oct-2	Oct-3
Mitoxantrone	3.0	135.0	174.0
Corticosterone	3.2	1.3	0.2

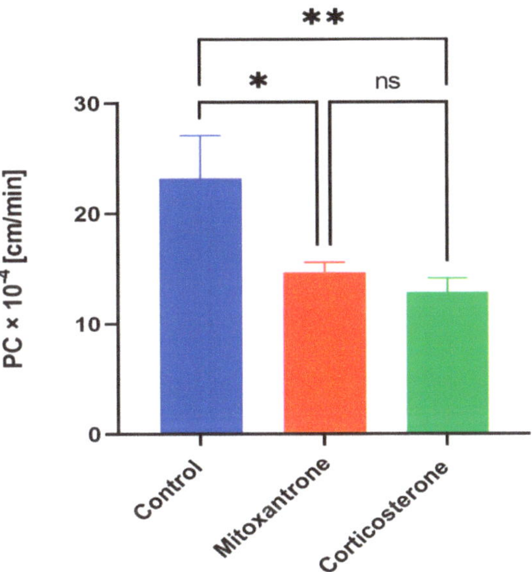

Figure 2. Permeability of MF using Octs specific transporter inhibitors. PC of MF significantly reduced with mitoxantrone (Oct1) and corticosterone (Oct1, 2 and 3) inhibitor. One-way ANOVA, followed by Tukey's multiple comparisons test (* $p < 0.05$, ** $p < 0.01$); ns = non-significant; $n = 3$ biological replicates; mean ± SD.

Cationic transporters other than Oct1, 2 and 3, such as PMAT, MATE-1, MATE-2, and SERT, which have been shown to have affinities for MF [53,58] and expressed by the brain microvascular endothelial cells [58–60], were not explored for permeability studies. This was because concentration needed for all-cationic-mediated transporter inhibition (MPP+ at 5 mM) and PMAT inhibition (desipramine at 200 μM) were toxic to the bEnd.3 cells (Figure S4). Additionally, the use of non-toxic and highly selective inhibitors (based on their IC$_{50}$ values) of PMAT, MATE-1, MATE-2, and SERT could be employed in future studies to identify their role in BBB transport of MF.

We further explore MF's transport during hypoxia using OGD conditions, as discussed below.

3.3. Permeability of MF during Oxygen Glucose Deprivation (OGD)

It is well documented that the paracellular permeability of molecules across BBB increases after 3–4 h of an ischemic stroke episode [61]. In fact, under hypoxia, brain endothelial cells have decreased protein expression of tight junctional proteins claudin-5, occludin, and ZO-1, which cause paracellular leak [62,63]. In addition to the paracellular leak, changes in transporters expression occur at the BBB following in vitro hypoxia or in vivo stroke modeling [34,36]. Such a change can drastically affect drug uptake during post-stroke recovery. To address this, we first exposed the co-culture model of BBB to 2, 4, and 6 h of OGD, examined the TEER values for the groups to have an idea of the integrity and viability of the barrier after hypoxia, and then evaluated the PC value of MF after quantification using the LC-MS/MS method.

As shown in Figure 3A, compared to normoxia, there is no decrease in TEER following 2 h of OGD. However, with increased exposure of 4 h, we found a significant reduction in the TEER of the co-cultured barrier ($p < 0.001$), suggesting compromised integrity and, thus, a paracellular leak. Additionally, with 6 h of OGD, there is a further reduction in the barrier integrity compared to 4 h of OGD exposure ($p < 0.01$). These changes suggest that the barrier integrity is compromised with longer exposure time periods of 4 h and 6 h of hypoxia to the cells. Next, we determined the PC of MF during hypoxia and compared it with the normoxic control. Similar to TEER values, we found no significant difference with 2 h of OGD exposure in MF's PC value ($21.5 \pm 2.2 \times 10^{-4}$ cm/min). However, MF's PC value significantly increased with either 4 or 6 h of OGD exposure ($32.98 \pm 2.27 \times 10^{-4}$ cm/min and $33.82 \pm 1.25 \times 10^{-4}$ cm/min, $p < 0.01$) (Figure 3B). It is also worth noting that there isn't a significant difference between 4 and 6 h of OGD exposure. However, our 6 h of TEER data showed a significant difference in barrier integrity loss compared to 4 h (Figure 3A). This made us further examine protein expression changes during the different OGD time points, as previous studies have shown differences in the regulation of transporters during such situations.

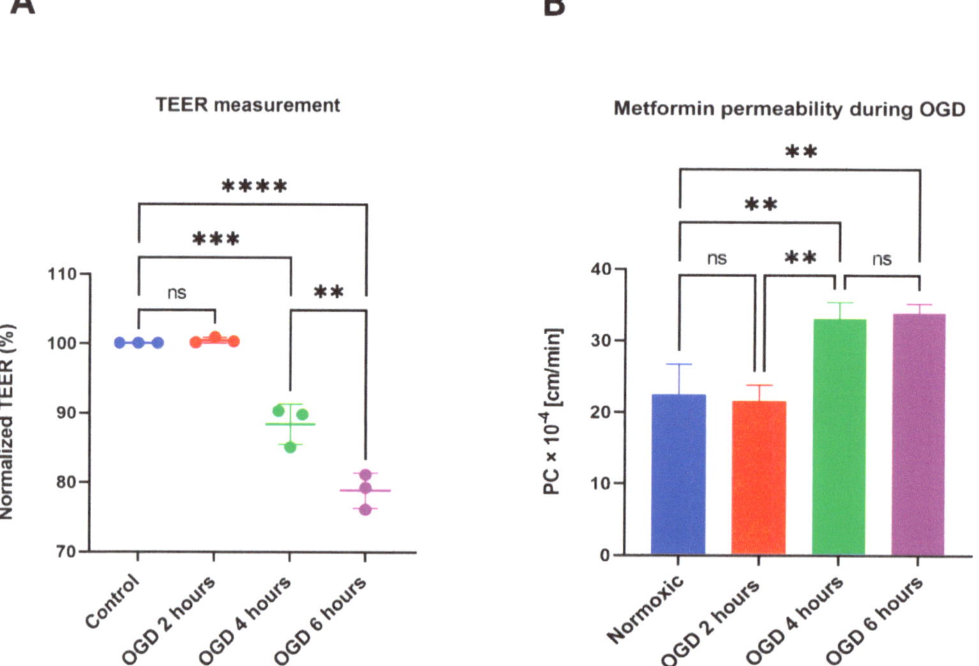

Figure 3. (**A**) OGD negatively impacts the BBB integrity, as demonstrated by TEER measurement. OGD exposure to the in vitro BBB model for 4 h significantly reduced the TEER compared to normoxia, and with 6 h exposure showed further reduction in TEER compared to 4 h exposure. One-way ANOVA, followed by Tukey's multiple comparisons test (** $p < 0.01$) (*** $p < 0.001$) (**** $p < 0.0001$); $n = 3$ biological replicates; mean $\pm$ SD. (**B**) Permeability of MF during hypoxia. The PC of MF showed no change with 2 h exposure, however, with 4 h and 6 h exposure, there was a similar increase in permeability, compared to normoxia. No significant difference between 4 h and 6 h exposure with OGD. One-way ANOVA, followed by Tukey's multiple comparisons test (** $p < 0.01$); ns = non-significant; $n = 3$ biological replicates; mean $\pm$ SD.

3.4. Protein Expression of Organic Cationic Transporters (Octs) in bEnd.3 Cells during Normoxia and Hypoxia

Previous studies have shown that participation of endogenous transporters is crucial in transport of anionic and cationic drugs across the BBB [14,34,36,64]. It is thus important to understand the functional protein expression of transporters during disease conditions. Understanding this aspect will further accelerate or maximize neuroprotection by optimizing timing and dosage during ischemic stroke. A previous study from our lab showed that hypoxia caused increased expression of the OATP1 transporter, which resulted in increased permeability of biphalin [34]. However, limited studies have examined expression of Octs during ischemia. So, to determine that, first, we checked the expression of Oct1, 2, and 3 in bEnd.3 cells during normoxia. Consistent with previously reported studies that showed luminal expression of mainly Oct1 and Oct2 in human and rat brain endothelial cells, we found that our bEnd.3 cells stably expressed Oct1 and Oct2, however, with almost undetectable expression of Oct3 (Figure 4) [50,65,66]. This is also in alignment with our permeability data that showed Oct1 inhibition with mitoxantrone for MF's permeability, thus further supporting Oct1's involvement in MF's transport. Additionally, as depicted in Figure 4, our protein expression experiment showed a stable Oct1 and Oct2 expression, but not Oct3 expression; thus, we conclude that Oct1 is the primary transporter responsible for MF's transport across the in vitro co-culture BBB model.

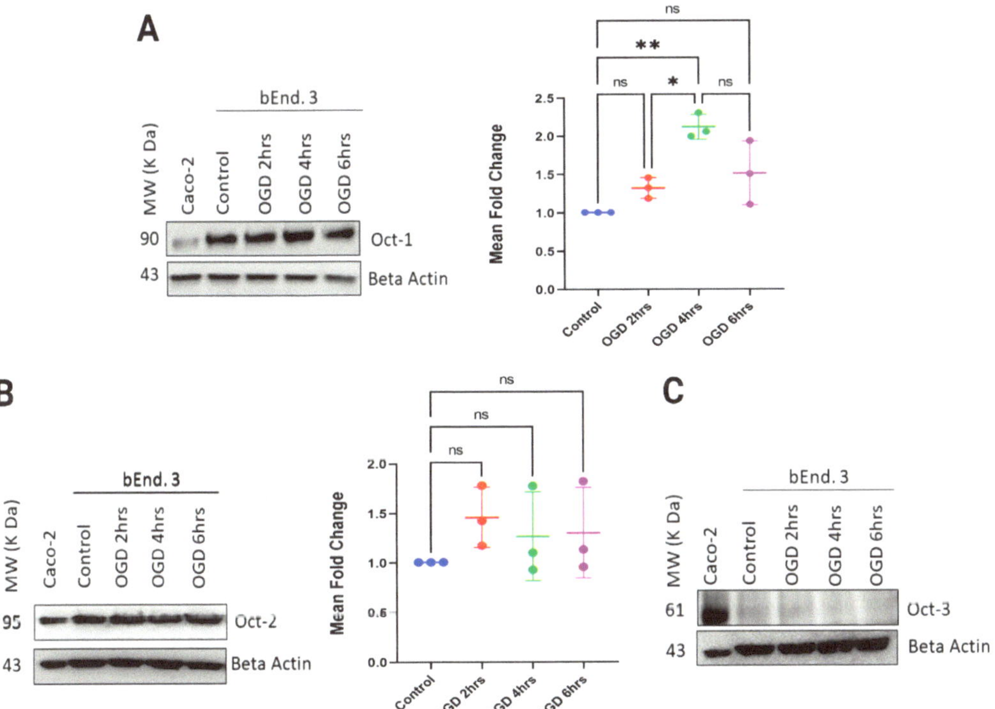

Figure 4. Densitometric analysis of organic cationic transporters (Octs) in bEnd.3 cells (immortalized mouse brain endothelial cells) during normoxia and hypoxia for 2, 4 and 6 h time points. (**A**) Expression of Oct1 in control cells (normoxia) and cells exposed for 2, 4 and, 6 h OGD (hypoxia), with evolving expression patterns seen among groups. (**B**) Stable Oct2 expression between the control and treatment groups, without significant differences in expression during OGD time-points. (**C**) Undetectable expression of Oct3 in control and treatment groups. The protein bands were quantified relative to beta-actin using Image J software and are reported as ratios of the control group. One-way ANOVA, followed by Tukey's multiple comparisons test (* $p < 0.05$, ** $p < 0.01$); ns = non-significant; $n = 3$ biological replicates; mean $\pm$ SD.

Next, we checked the expression of Octs during hypoxic time points of 2, 4, and 6 h. As depicted in Figure 4A, for Oct1 transporter, compared to the control, we did not find any change in expression with 2 h of OGD exposure. However, we found significant increase in expression at 4 h of OGD exposure compared to 2 h ($p < 0.05$). Interestingly, the expression was renormalized with 6 h of OGD exposure, similar to the control (Figure 4A).

The significantly increased expression for 4 h OGD is consistent with our increased MF permeability result, indicating that the increased permeability during hypoxic conditions could be due to the enhanced Oct1 transporter expression. Although without significantly increased Oct1 expression for 6 h of exposure (compared to control), we found that the increases in the permeability of MF during ODG 4 h and 6 h were not dissimilar (Figure 3B). This could be due to a slight increase in Oct1 expression, but not significant, as shown by densitometric analysis (Figure 4A). It is also notable that the comparison in Oct1 expression for 4 h of OGD to 6 h of OGD had no significant difference (Figure 4A). Thus, carrier-mediated transport could play an important role in enhanced permeation of MF during longer 4 and 6 h of OGD exposure. Further experiments involving Octs inhibitors during longer OGD time-points exposure to the co-cultured cells were required to prove this assumption.

Next, for Oct2 transporter, we found no change in expression among the OGD time points and normoxic control (Figure 4B). Lastly, Oct3 transporter was almost undetectable in the tested cells during the OGD time points (Figure 4C).

3.5. Oct1 Involvement in MF Permeability during different Hypoxic Time Points Using OGD

We wanted to confirm if the increased permeability of MF during 4 h of OGD exposure is due to an increase in Oct1 transporter expression. Thus, we performed permeability studies involving cells exposure to 4 h OGD in both mitoxantrone and corticosterone presence and absence. As shown in Figure 5A, the PC value of MF is reduced in the presence of inhibitors during the 4 h OGD exposure time point (from $27.58 \pm 4.53 \times 10^{-4}$ cm/min to $17.62 \pm 1.35 \times 10^{-4}$ cm/min using mitoxantrone and to $19.55 \pm 2.96 \times 10^{-4}$ cm/min using corticosterone, $p < 0.05$). This further suggests that Oct1 is a key determinant for MF transport during OGD and, thus, responsible for increased permeability of MF. It is also notable that there are no significant differences between inhibition while using mitoxantrone (Oct1 inhibitor) and corticosterone (Oct1, 2, and 3 inhibitor). It can also be extrapolated from Figure 5 that ~36% of MF transport during OGD is carrier-mediated (the inhibition component of transport during OGD is active transport), while the remaining ~64% is due to paracellular leakage. This result is almost similar compared to normoxia, where transport accounted for ~35% carrier-mediated and ~65% paracellular transport. Next, we performed permeability studies using inhibitors for 6 h of OGD exposure. As shown in Figure 5B, we found that the PC value of metformin significantly reduced in the presence of both inhibitors (from $30.71 \pm 4.65 \times 10^{-4}$ cm/min to $20.16 \pm 0.11 \times 10^{-4}$ cm/min using mitoxantrone and to $18.86 \pm 0.85 \times 10^{-4}$ cm/min using corticosterone, $p < 0.05$). This suggests that carrier-mediated transport is still a key determinant of MF transport during a longer exposure of 6 h OGD to the cells.

The altered transporter expression (nutrients, ions, or drugs) at the BBB has been shown to be either beneficial or detrimental during ischemic stroke. A previous study reported that elevated expression of nutrient transporter GLUT1, which regulates glucose levels in the brain, served a beneficial role in reducing focal ischemia [67]. However, another nutrient transporter, SGLT, responsible for cell depolarization and glucose balance, causes increased edema formation when its expression is increased during ischemia [68]. Additionally, increased expression of ion transporter Na^+-K^+-Cl-cotransporter caused brain edema formation, while decreased expression of NA^+/K^+ ATPase caused accumulation of Na^+, leading to endothelial swelling and cytotoxic edema [69,70]. Previously, our lab showed that functional expression of OATP1 is required for BBB transport of biphalin, during in vitro ischemia and reperfusion conditions [34]. Furthermore, we found that the expression of the transporter is increased, requiring consideration of the dosage and time window. Another study showed that the transport effect of Oct1 and Oct2 is critical to achieving therapeutic outcomes during in vivo ischemic stroke despite the paracellular leak [36].

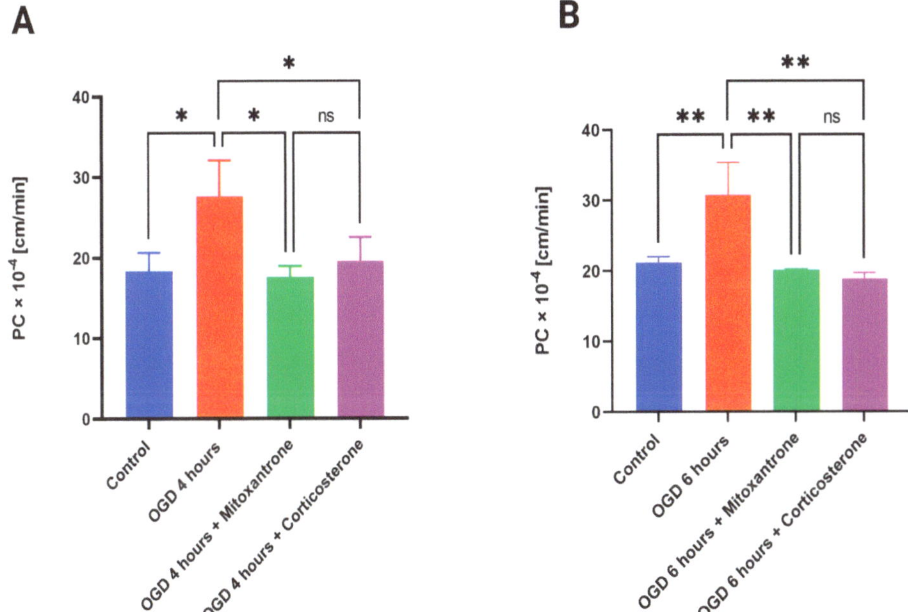

Figure 5. Permeability of MF after 4 h (**A**) and 6 h (**B**) of OGD exposure using mitoxantrone and corticosterone. The PC value of MF significantly reduces both with mitoxantrone and corticosterone in comparison to 4 and 6 h OGD exposure in absence of inhibitors. There is no significant difference in PC between both groups involving inhibitors after 4 and 6 h of OGD exposure. One-way ANOVA, followed by Tukey's multiple comparisons test (* $p < 0.05$, ** $p < 0.01$); ns = non significant; $n = 3$ biological replicates; mean ± SD.

Few scientific studies have examined the regulation of Octs. Hepatic nuclear factor 4α (HNF-4α) has been shown to regulate the gene expression of Octs, where they bind to the gene promotor regions of the transporters to modulate the mRNA expression [71]. Enhancing the cellular expression of HNF-4α has been shown to, in fact, increase Oct1 expression in human hepatocytes. Meanwhile, it is worth noting that HNF-4α has been shown to be identified as a transcription factor in brain tissues. Therefore, in future studies, checking the expression of HNF-4α along with Octs expression during various OGD time points will give us an idea of whether this transcription factor is responsible for the alteration in transporter expression during hypoxia in brain endothelial cells. Furthermore, our data suggest evolving expression of Oct1 transporter with a difference in ischemic exposure periods. Therefore, it is important to determine whether increased transporter expression is needed to reach MF's therapeutic potential for ischemic stroke treatment. Future studies that involve inhibition of Octs and examining alteration in activation pathways such as NF-κB and mTOR, through which MF has been shown to mediate its neurotherapeutic actions, are warranted. Moreover, since we found increased MF permeability during hypoxia and that ~36% was mediated through active transport and the rest ~64% paracellular (Figure 5A), thus, it is warranted to check if this increase is critical to the activation of certain pathways involved in post-stroke recovery. In vivo behavioral ischemic stroke recovery studies routinely utilized in our lab could achieve this.

3.6. Protein Expression of Octs in Primary Astrocytes during Normoxia and Hypoxia

Astrocytes play a key role in the maintenance and inductance of the BBB. Our lab previously showed that co-cultured in vitro BBB model of bEnd.3 cells and mouse primary astrocytes promoted barrier tightness compared to a monolayer BBB model of bEnD.3

cells [43]. Besides supporting barrier integrity, astrocytes have been shown to express and regulate various ions and drug transporters in the brain [72]. A previous study has reported immunoreactivity to Oct2, Oct3, and PMAT in some regions of mouse and rat brains [73]. Another study reported expression of Oct3 in primary human astrocytes [74]. Thus, we further wanted to check protein expression of Octs in mouse primary astrocytes (situated in our co-cultured model) during normoxia and different hypoxic time points. As seen in Figure 6, we found stable expression of Oct2 in the primary astrocytes without any statistically significant changes between normoxia and hypoxic OGD time points. We found no detectable expressions for Oct1 and Oct3 in our cultured mouse primary astrocytes, during both normoxia and hypoxia.

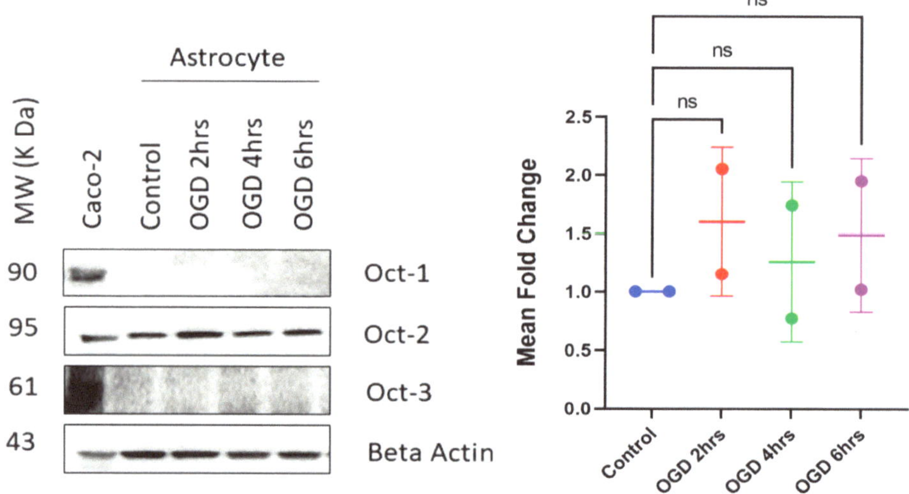

Figure 6. Densitometric analysis of organic cationic transporters (Octs) in mouse primary astrocytes during normoxia and hypoxia for 2, 4, and 6 h time points. Stable expression of Oct2 in control cells (normoxia) and cells exposed for 2, 4, and 6 h OGD (hypoxia), with no changes in expression patterns seen among groups. No detectable expression of Oct1 and Oct3 in control and treatment groups. The protein bands were quantified relative to beta-actin using Image J software and are reported as ratios of the control group. One-way ANOVA, followed by Tukey's multiple comparisons test; ns = non-significant; n = 2 biological replicates; mean $\pm$ SD.

In addition to the role of Oct1 transporter in bEnd.3 cells, the expression of Oct2 in astrocytes suggests that Oct2 might play a possible significant role in the transport/uptake of MF. Future studies involving MF uptake studies by primary astrocytes could be implemented to understand any significant uptake by the astrocytes. Additionally, uptake studies could be performed in the presence and absence of Oct inhibitors to explore their role in MF uptake in primary astrocytes. This will also help in determining if our reported high permeability value for MF is underestimated due to the amount of drug possibly being taken up by astrocytes, while getting transported from the apical to basolateral chamber of the in vitro co-culture BBB model.

3.7. Interaction of Metformin with Plasma Glycoprotein (P-GP) Using P-GP Overexpressing Cell Line

P-GP is an ATP dependent efflux protein expressed in the apical compartment of the BBB, which restricts the movement of relatively large molecules (>400 Da) into the brain, by actively transporting the molecules back to the blood [75]. The assessment of UFR and ER using transwell assays with P-GP overexpressing MDR1-MDCK cell lines is commonly used to assess bidirectional movement of molecules through the cell monolayer [76]. This assay gives an idea of whether drug molecules have limitations in gaining access to the brain. Furthermore, studies have shown that permeability studies in MDR1-MDCK cell lines accurately reflect in vivo BBB transport [77,78]. For our study, we sought to investigate whether MF interacts with the P-GP transporter and thus predict the level of its P-GP transport potential. The unidirectional PC (from A > B and B > A) of MF, along with its UFR and ER are summarized in Table 1. We found the UFR and ER values for metformin to be 1.24 ± 0.08 and 0.63 ± 0.04, respectively (Table 2). The UFR of a compound increases depending upon the potential of being a P-GP substrate (compounds with UFR < 2 are considered as less potential substrates) and compounds that have an ER of >1.99 are considered as strong P-GP substrates [79–81]. Thus, with our findings, we conclude that MF does not interact with the efflux transporter P-GP in a P-GP overexpressing cell line. This result further bolsters our conclusion from transport experiments that MF should have access to the brain as it is highly permeable across the BBB during normoxia and hypoxia.

Table 2. MF's PC in a P-GP overexpressing MDR1-MDCK cell line in presence and absence of a potent P-GP inhibitor, Cyclosporin A (CsA) at 10 μm to calculate the unidirectional flux ratio (UFR) and efflux ratio (ER). Mean ± SD of 4–5 replicates from 2 independent biological samples.

Metformin PC, A > B; 1×10^{-7} cm/s		Metformin PC, B > A; 1×10^{-7} cm/s	UFR (Unidirectional Flux Ratio)	ER (Efflux Ratio)
−CsA	+CsA			
2.58 ± 0.18	3.28 ± 0.06	1.63 ± 0.06	1.24 ± 0.08	0.63 ± 0.04

4. Conclusions and Future Studies

We conclude that MF is a highly permeable molecule across an in vitro co-culture BBB model. A key portion of its brain entry utilizes carrier-mediated active transport through Octs, while the remaining part involves passive paracellular diffusion. We found evolving expression patterns for Oct1 transport between hypoxic time periods and increased permeability for MF. Future studies should be conducted to decipher the exact mechanism causing the altered pattern of Oct1 transporter expression. In the future, MF brain pharmacokinetic studies will be completed for MF using an in vivo ischemic stroke model in our lab to determine the drug's brain uptake clearance value (Kin) and correlate with our in vitro findings from this study. Octs inhibitors could be further used in vivo to decipher whether carrier-mediated transport is necessary to achieve therapeutic efficacy during ischemic stroke.

Supplementary Materials: The following supporting information can be downloaded at: https://www.mdpi.com/article/10.3390/pharmaceutics15051357/s1. Figure S1. Permeability Coefficient (PC) of low molecular weight BBB markers, NaF (MW 376 Da), and [14C] sucrose (MW 342 Da), across an *in vitro* co-culture setup of BBB. The PC of NaF compared to [14C] sucrose was statistically non-significant. The experiments were performed under a similar setup of a co-culture model of brain endothelial cells and astrocytes. Unpaired student's *t*-test (* $p < 0.05$); $n = 3$ biological replicates; mean ± SD. Figure S2. MTS cell viability assay in bEnd.3 cells (A) and mouse primary astrocytes (B). The values are expressed as a percentage of the control value. No statistically significant differences were found among the groups. One-way ANOVA, followed by Tukey's multiple comparis ons tests; $n = 3$ biological replicates; mean ± SD. Figure S3. MTS cell viability assay in bEnd.3 cells (A) and mouse primary astrocytes (B), and TEER measurement of co-cultured cells after incubation with inhibitors (C). The values are expressed as a percentage of the control value. No statistically significant differences were found among the groups. One-way ANOVA, followed by Tukey's multiple comparisons tests; $n = 2–3$ biological replicates; mean ± SD. Figure S4. MTS cell viability assay in bEnd.3 cells using during 4 h of incubation with substrates/ inhibitors. The values are expressed as a percentage of the control value. The cell viability was significantly affected with MPP+ at 5 mM concentration, and desipramine at 100 μm and 200 μm concentrations respectively. One-way ANOVA, followed by Tukey's multiple comparisons tests; $n = 3$; mean ± SD.

Author Contributions: S.S. and T.A. conceived the study and prepared the draft of the manuscript. S.S., Y.Z., K.A.A., S.N., S.R.A., D.P. and H.V. performed the experiments. T.A. edited and revised the manuscript, assisted with drafting the manuscript, oversaw the entire project, and provided funding support. All authors reviewed the manuscript. All authors have read and agreed to the published version of the manuscript.

Funding: This work was supported by the National Institutes of Health/National Institute of Neurological Disorders and Stroke grant R01 NS#117906 to T.J.A.

Institutional Review Board Statement: Not applicable.

Informed Consent Statement: Not applicable.

Data Availability Statement: Not applicable.

Conflicts of Interest: The authors declare no conflict of interest.

Abbreviations

MF	metformin
tPA	tissue plasminogen activator
US-FDA	The United States food and drug administration
AMPK	AMP-activated protein kinase
Nrf2	Nuclear factor erythroid 2-related factor
BBB	blood-brain barrier
SLCs	solute carriers
Octs	organic cation transporters
OATPs	organic anionic transporters
P-GP	plasma glycoproteins
OGD	oxygen-glucose deprivation
UHPLC	ultra-high performance liquid chromatography
RIPA	radioimmunoprecipitation assay buffer
CsA	Cyclosporine A
PC	permeability coefficient
UFR	unidirectional flux ratio
ER	efflux ratio
ANOVA	one-way analysis of variance
HNF-4α	hepatic nuclear factor 4α

References

1. Rojas, L.B.A.; Gomes, M.B. Metformin: An old but still the best treatment for type 2 diabetes. *Diabetol. Metab. Syndr.* **2013**, *5*, 6–15. [PubMed]
2. Wróbel, M.P.; Marek, B.; Kajdaniuk, D.; Rokicka, D.; Szymborska-Kajanek, A.; Strojek, K. Metformin—A new old drug. *Endokrynol. Pol.* **2017**, *68*, 482–496. [CrossRef] [PubMed]
3. Lv, Z.; Guo, Y. Metformin and its benefits for various diseases. *Front. Endocrinol.* **2020**, *11*, 191.
4. Rotermund, C.; Machetanz, G.; Fitzgerald, J.C. The therapeutic potential of metformin in neurodegenerative diseases. *Front. Endocrinol.* **2018**, *9*, 400.
5. Markowicz-Piasecka, M.; Sikora, J.; Szydłowska, A.; Skupień, A.; Mikiciuk-Olasik, E.; Huttunen, K.M. Metformin–A future therapy for neurodegenerative diseases: Theme: Drug discovery, development and delivery in Alzheimer's disease Guest Editor: Davide Brambilla. *Pharm. Res.* **2017**, *34*, 2614–2627.
6. Saewanee, N.; Praputpittaya, T.; Malaiwong, N.; Chalorak, P.; Meemon, K. Neuroprotective effect of metformin on dopaminergic neurodegeneration and α-synuclein aggregation in *C. elegans* model of Parkinson's disease. *Neurosci. Res.* **2021**, *162*, 13–21. [CrossRef]
7. Sharma, S.; Nozohouri, S.; Vaidya, B.; Abbruscato, T. Repurposing metformin to treat age-related neurodegenerative disorders and ischemic stroke. *Life Sci.* **2021**, *274*, 119343. [CrossRef]
8. Kaisar, M.A.; Villalba, H.; Prasad, S.; Liles, T.; Sifat, A.E.; Sajja, R.K.; Abbruscato, T.J.; Cucullo, L. Offsetting the impact of smoking and e-cigarette vaping on the cerebrovascular system and stroke injury: Is Metformin a viable countermeasure? *Redox Biol.* **2017**, *13*, 353–362. [CrossRef]
9. Kadry, H.; Noorani, B.; Bickel, U.; Abbruscato, T.J.; Cucullo, L. Comparative assessment of in vitro BBB tight junction integrity following exposure to cigarette smoke and e-cigarette vapor: A quantitative evaluation of the protective effects of metformin using small-molecular-weight paracellular markers. *Fluids Barriers CNS* **2021**, *18*, 28. [CrossRef]

10. Nozohouri, S.; Sifat, A.E.; Vaidya, B.; Abbruscato, T.J. Novel approaches for the delivery of therapeutics in ischemic stroke. *Drug Discov. Today* **2020**, *25*, 535–551.

11. Sifat, A.E.; Vaidya, B.; Villalba, H.; AlBekairi, T.H.; Abbruscato, T.J. Neurovascular unit transport responses to ischemia and common coexisting conditions: Smoking and diabetes. *Am. J. Physiol. Physiol.* **2019**, *316*, C2–C15. [CrossRef] [PubMed]

12. Johnston, K.; Chapman, S.; Mehndiratta, P.; Johansen, M.C.; Southerland, A.; McMurry, T. Current perspectives on the use of intravenous recombinant tissue plasminogen activator (tPA) for treatment of acute ischemic stroke. *Vasc. Health Risk Manag.* **2014**, *10*, 75–87. [CrossRef] [PubMed]

13. Mistry, E.A.; Mistry, A.M.; Fusco, M.R. Response by mistry et al to letter regarding article, "mechanical thrombectomy outcomes with and without intravenous thrombolysis in stroke patients: A meta-analysis". *Stroke* **2017**, *48*, e334. [CrossRef]

14. Nilles, K.L.; Williams, E.I.; Betterton, R.D.; Davis, T.P.; Ronaldson, P.T. Blood–Brain Barrier Transporters: Opportunities for Therapeutic Development in Ischemic Stroke. *Int. J. Mol. Sci.* **2022**, *23*, 1898.

15. Shukla, R.; Henkel, N.D.; Alganem, K.; Hamoud, A.-R.; Reigle, J.; Alnafisah, R.S.; Eby, H.M.; Imami, A.S.; Creeden, J.; Miruzzi, S.A.; et al. Signature-based approaches for informed drug repurposing: Targeting CNS disorders. *Neuropsychopharmacology* **2020**, *46*, 116–130. [CrossRef] [PubMed]

16. Bourget, C.; Adams, K.V.; Morshead, C.M. Reduced microglia activation following metformin administration or microglia ablation is sufficient to prevent functional deficits in a mouse model of neonatal stroke. *J. Neuroinflamm.* **2022**, *19*, 146. [CrossRef] [PubMed]

17. Zemgulyte, G.; Tanaka, S.; Hide, I.; Sakai, N.; Pampuscenko, K.; Borutaite, V.; Rastenyte, D. Evaluation of the effectiveness of post-stroke metformin treatment using permanent middle cerebral artery occlusion in rats. *Pharmaceuticals* **2021**, *14*, 312. [CrossRef]

18. Hou, K.; Xu, D.; Li, F.; Chen, S.; Li, Y. The progress of neuronal autophagy in cerebral ischemia stroke: Mechanisms, roles and research methods. *J. Neurol. Sci.* **2019**, *400*, 72–82. [CrossRef]

19. Prasad, S.; Sajja, R.K.; Kaisar, M.A.; Park, J.H.; Villalba, H.; Liles, T.; Abbruscato, T.; Cucullo, L. Role of Nrf2 and protective effects of Metformin against tobacco smoke-induced cerebrovascular toxicity. *Redox Biol.* **2017**, *12*, 58–69. [CrossRef]

20. Saitoh, R.; Sugano, K.; Takata, N.; Tachibana, T.; Higashida, A.; Nabuchi, Y.; Aso, Y. Correction of permeability with pore radius of tight junctions in Caco-2 monolayers improves the prediction of the dose fraction of hydrophilic drugs absorbed by humans. *Pharm. Res.* **2004**, *21*, 749–755. [CrossRef]

21. Horie, A.; Sakata, J.; Nishimura, M.; Ishida, K.; Taguchi, M.; Hashimoto, Y. Mechanisms for membrane transport of metformin in human intestinal epithelial Caco-2 cells. *Biopharm. Drug Dispos.* **2011**, *32*, 253–260. [CrossRef] [PubMed]

22. Archie, S.R.; Sharma, S.; Burks, E.; Abbruscato, T. Biological determinants impact the neurovascular toxicity of nicotine and tobacco smoke: A pharmacokinetic and pharmacodynamics perspective. *Neurotoxicology* **2022**, *89*, 140–160. [PubMed]

23. Barthels, D.; Prateeksha, P.; Nozohouri, S.; Villalba, H.; Zhang, Y.; Sharma, S.; Anderson, S.; Howlader, S.I.; Nambiar, A.; Abbruscato, T.J.; et al. Dental Pulp-Derived Stem Cells Preserve Astrocyte Health during Induced Gliosis by Modulating Mitochondrial Activity and Functions. *Cell. Mol. Neurobiol.* **2022**, 1–23. [CrossRef] [PubMed]

24. Zhang, Y.; Archie, S.R.; Ghanwatkar, Y.; Sharma, S.; Nozohouri, S.; Burks, E.; Mdzinarishvili, A.; Liu, Z.; Abbruscato, T.J. Potential role of astrocyte angiotensin converting enzyme 2 in the neural transmission of COVID-19 and a neuroinflammatory state induced by smoking and vaping. *Fluids Barriers CNS* **2022**, *19*, 46. [CrossRef] [PubMed]

25. Sifat, A.E.; Archie, S.R.; Nozohouri, S.; Villalba, H.; Zhang, Y.; Sharma, S.; Chanwatkar, Y.; Vaidya, B.; Mara, D.; Cucullo, L.; et al. Short-term exposure to JUUL electronic cigarettes can worsen ischemic stroke outcome. *Fluids Barriers CNS* **2022**, *19*, 1–15. [CrossRef] [PubMed]

26. Sharma, S.; Archie, S.R.; Kanchanwala, V.; Mimun, K.; Rahman, A.; Zhang, Y.; Abbruscato, T. Effects of Nicotine Exposure From Tobacco Products and Electronic Cigarettes on the Pathogenesis of Neurological Diseases: Impact on CNS Drug Delivery. *Front. Drug Deliv.* **2022**, *2*, 886099. [CrossRef]

27. Shi, L.; Rocha, M.; Leak, R.K.; Zhao, J.; Bhatia, T.; Mu, H.; Wei, Z.; Yu, F.; Weiner, S.L.; Ma, F.; et al. A new era for stroke therapy: Integrating neurovascular protection with optimal reperfusion. *J. Cereb. Blood Flow Metab.* **2018**, *38*, 2073–2091. [CrossRef] [PubMed]

28. Hacker, K.; Maas, R.; Kornhuber, J.; Fromm, M.F.; Zolk, O. Substrate-dependent inhibition of the human organic cation transporter OCT2: A comparison of metformin with experimental substrates. *PLoS ONE* **2015**, *10*, e0136451. [CrossRef]

29. Akanuma, S.-I.; Han, M.; Murayama, Y.; Kubo, Y.; Hosoya, K.-I. Differences in Cerebral Distribution between Imipramine and Paroxetine via Membrane Transporters at the Rat Blood-Brain Barrier. *Pharm. Res.* **2022**, *39*, 223–237. [CrossRef]

30. Gebauer, L.; Rafehi, M.; Brockmöller, J. Stereoselectivity in the Membrane Transport of Phenylethylamine Derivatives by Human Monoamine Transporters and Organic Cation Transporters 1, 2, and 3. *Biomolecules* **2022**, *12*, 1507. [CrossRef]

31. Khanppnavar, B.; Maier, J.; Herborg, F.; Gradisch, R.; Lazzarin, E.; Luethi, D.; Yang, J.-W.; Qi, C.; Holy, M.; Jäntsch, K.; et al. Structural basis of organic cation transporter-3 inhibition. *Nat. Commun.* **2022**, *13*, 6714. [CrossRef]

32. Green, R.M.; Lo, K.; Sterritt, C.; Beier, D.R. Cloning and functional expression of a mouse liver organic cation transporter. *Hepatology* **1999**, *29*, 1556–1562. [CrossRef] [PubMed]

33. Zaïr, Z.M.; Eloranta, J.J.; Stieger, B.; Kullak-Ublick, G.A. *Pharmacogenetics of OATP (SLC21/SLCO), OAT and OCT (SLC22) and PEPT (SLC15) Transporters in the Intestine, Liver and Kidney*; Future Medicine Ltd.: London, UK, 2008.
34. Albekairi, T.H.; Vaidya, B.; Patel, R.; Nozohouri, S.; Villalba, H.; Zhang, Y.; Lee, Y.S.; Al-Ahmad, A.; Abbruscato, T.J. Brain delivery of a potent opioid receptor agonist, biphalin during ischemic stroke: Role of organic anion transporting polypeptide (OATP). *Pharmaceutics* **2019**, *11*, 467. [CrossRef] [PubMed]
35. Ronaldson, P.T.; Brzica, H.; Abdullahi, W.; Reilly, B.G.; Davis, T.P. Transport Properties of Statins by OATP1A2 and Regulation by Transforming Growth Factor-β (TGF-β) Signaling in Human Endothelial Cells. *J. Pharmacol. Exp. Ther.* **2020**. [CrossRef]
36. Stanton, J.A.; Williams, E.I.; Betterton, R.D.; Davis, T.P.; Ronaldson, P.T. Targeting organic cation transporters at the blood-brain barrier to treat ischemic stroke in rats. *Exp. Neurol.* **2022**, *357*, 114181. [CrossRef] [PubMed]
37. Inano, A.; Sai, Y.; Nikaido, H.; Hasimoto, N.; Asano, M.; Tsuji, A.; Tamai, I. Acetyl-L-carnitine permeability across the blood-brain barrier and involvement of carnitine transporter OCTN2. *Biopharm. Drug Dispos.* **2003**, *24*, 357–365. [CrossRef] [PubMed]
38. Di, L.; Kerns, E.H.; Carter, G.T. Strategies to assess blood–brain barrier penetration. *Expert Opin. Drug Discov.* **2008**, *3*, 677–687. [CrossRef]
39. Van Assema, D.M.; Lubberink, M.; Boellaard, R.; Schuit, R.C.; Windhorst, A.D.; Scheltens, P.; Lammertsma, A.A.; Van Berckel, B.N. P-glycoprotein function at the blood–brain barrier: Effects of age and gender. *Mol. Imaging Biol.* **2012**, *14*, 771–776. [CrossRef]
40. Begley, D.J. ABC Transporters and the blood-brain barrier. *Curr. Pharm. Des.* **2004**, *10*, 1295–1312. [CrossRef]
41. Du, F.; Qian, Z.M.; Zhu, L.; Wu, X.M.; Qian, C.; Chan, R.; Ke, Y. Purity, cell viability, expression of GFAP and bystin in astrocytes cultured by different procedures. *J. Cell. Biochem.* **2009**, *109*, 30–37. [CrossRef]
42. Nozohouri, S.; Noorani, B.; Al-Ahmad, A.; Abbruscato, T.J. Estimating Brain permeability using in vitro blood-brain barrier models. In *Permeability Barrier: Methods and Protocols*; Methods in Molecular Biology; 2020; pp. 47–72. Available online: https://link.springer.com/protocol/10.1007/7651_2020_311 (accessed on 25 April 2023).
43. Nozohouri, S.; Zhang, Y.; Albekairi, T.H.; Vaidya, B.; Abbruscato, T.J. Glutamate buffering capacity and blood-brain barrier protection of opioid receptor agonists biphalin and nociceptin. *J. Pharmacol. Exp. Ther.* **2021**, *379*, 260–269. [CrossRef]
44. Yang, L.; Islam, M.R.; Karamyan, V.T.; Abbruscato, T.J. In vitro and in vivo efficacy of a potent opioid receptor agonist, biphalin, compared to subtype-selective opioid receptor agonists for stroke treatment. *Brain Res.* **2015**, *1609*, 1–11. [CrossRef] [PubMed]
45. Nia, S.S.; Hossain, M.A.; Ji, G.; Jonnalagadda, S.K.; Obeng, S.; Rahman, A.; Sifat, A.E.; Nozohouri, S.; Blackwell, C.; Patel, D.; et al. Studies on diketopiperazine and dipeptide analogs as opioid receptor ligands. *Eur. J. Med. Chem.* **2023**, *254*, 115309.
46. Wager, T.T.; Hou, X.; Verhoest, P.R.; Villalobos, A. Central nervous system multiparameter optimization desirability: Application in drug discovery. *ACS Chem. Neurosci.* **2016**, *7*, 767–775. [CrossRef]
47. Rahman, M.S.; Kumari, S.; Esfahani, S.H.; Nozohouri, S.; Jayaraman, S.; Kinarivala, N.; Kocot, J.; Baez, A.; Farris, D.; Abbruscato, T.J.; et al. Discovery of first-in-class peptidomimetic neurolysin activators possessing enhanced brain penetration and stability. *J. Med. Chem.* **2021**, *64*, 12705–12722. [CrossRef] [PubMed]
48. Kimura, N.; Okuda, M.; Inui, K.-I. Metformin transport by renal basolateral organic cation transporter hOCT2. *Pharm. Res.* **2005**, *22*, 255–259. [CrossRef]
49. Sung, J.-H.; Yu, K.-H.; Park, J.-S.; Tsuruo, T.; Kim, D.-D.; Shim, C.-K.; Chung, S.-J. Saturable distribution of tacrine into the striatal extracellular fluid of the rat: Evidence of involvement of multiple organic cation transporters in the transport. *Drug Metab. Dispos.* **2004**, *33*, 440–448. [CrossRef]
50. Sekhar, G.N.; Georgian, A.R.; Sanderson, L.; Vizcay-Barrena, G.; Brown, R.C.; Muresan, P.; Fleck, R.A.; Thomas, S.A. Organic cation transporter 1 (OCT1) is involved in pentamidine transport at the human and mouse blood-brain barrier (BBB). *PLoS ONE* **2017**, *12*, e0173474. [CrossRef] [PubMed]
51. Nies, A.T.; Koepsell, H.; Winter, S.; Burk, O.; Klein, K.; Kerb, R.; Zanger, U.M.; Keppler, D.; Schwab, M.; Schaeffeler, E. Expression of organic cation transporters OCT1 (SLC22A1) and OCT3 (SLC22A3) is affected by genetic factors and cholestasis in human liver. *Hepatology* **2009**, *50*, 1227–1240. [CrossRef]
52. Kimura, N.; Masuda, S.; Tanihara, Y.; Ueo, H.; Okuda, M.; Katsura, T.; Inui, K.-I. Metformin is a superior substrate for renal organic cation transporter OCT2 rather than hepatic OCT1. *Drug Metab. Pharmacokinet.* **2005**, *20*, 379–386. [CrossRef]
53. Han, T.K.; Proctor, W.R.; Costales, C.L.; Cai, H.; Everett, R.S.; Thakker, D.R. Four cation-selective transporters contribute to apical uptake and accumulation of metformin in Caco-2 cell monolayers. *J. Pharmacol. Exp. Ther.* **2015**, *352*, 519–528. [CrossRef] [PubMed]
54. Koepsell, H.; Lips, K.; Volk, C. Polyspecific organic cation transporters: Structure, function, physiological roles, and biopharmaceutical implications. *Pharm. Res.* **2007**, *24*, 1227–1251. [PubMed]
55. Meyer, M.J.; Tuerkova, A.; Römer, S.; Wenzel, C.; Seitz, T.; Gaedcke, J.; Oswald, S.; Brockmöller, J.; Zdrazil, B.; Tzvetkov, M.V. Differences in metformin and thiamine uptake between human and mouse organic cation transporter 1: Structural determinants and potential consequences for intrahepatic concentrations. *Drug Metab. Dispos.* **2020**, *48*, 1380–1392. [CrossRef] [PubMed]
56. Koepsell, H. Update on drug-drug interaction at organic cation transporters: Mechanisms, clinical impact, and proposal for advanced in vitro testing. *Expert Opin. Drug Metab. Toxicol.* **2021**, *17*, 635–653. [CrossRef]

57. Liu, H.C.; Goldenberg, A.; Chen, Y.; Lun, C.; Wu, W.; Bush, K.T.; Balac, N.; Rodriguez, P.; Abagyan, R.; Nigam, S.K. Molecular properties of drugs interacting with SLC22 transporters OAT1, OAT3, OCT1, and OCT2: A machine-learning approach. *J. Pharmacol. Exp. Ther.* **2016**, *359*, 215–229. [CrossRef]

58. Wakayama, K.; Ohtsuki, S.; Takanaga, H.; Hosoya, K.-I.; Terasaki, T. Localization of norepinephrine and serotonin transporter in mouse brain capillary endothelial cells. *Neurosci. Res.* **2002**, *44*, 173–180. [CrossRef]

59. Wu, K.-C.; Lu, Y.-H.; Peng, Y.-H.; Hsu, L.-C.; Lin, C.-J. Effects of lipopolysaccharide on the expression of plasma membrane monoamine transporter (PMAT) at the blood-brain barrier and its implications to the transport of neurotoxins. *J. Neurochem.* **2015**, *135*, 1178–1188. [CrossRef]

60. Sekhar, G.N.; Fleckney, A.L.; Boyanova, S.T.; Rupawala, H.; Lo, R.; Wang, H.; Farag, D.B.; Rahman, K.M.; Broadstock, M.; Reeves, S.; et al. Region-specific blood–brain barrier transporter changes leads to increased sensitivity to amisulpride in Alzheimer's disease. *Fluids Barriers CNS* **2019**, *16*, 1–19. [CrossRef]

61. Sifat, A.E.; Vaidya, B.; Abbruscato, T.J. Blood-brain barrier protection as a therapeutic strategy for acute ischemic stroke. *AAPS J.* **2017**, *19*, 957–972.

62. Zhu, H.; Wang, Z.; Xing, Y.; Gao, Y.; Ma, T.; Lou, L.; Lou, J.; Gao, Y.; Wang, S.; Wang, Y. Baicalin reduces the permeability of the blood–brain barrier during hypoxia in vitro by increasing the expression of tight junction proteins in brain microvascular endothelial cells. *J. Ethnopharmacol.* **2012**, *141*, 714–720. [CrossRef]

63. Koto, T.; Takubo, K.; Ishida, S.; Shinoda, H.; Inoue, M.; Tsubota, K.; Okada, Y.; Ikeda, E. Hypoxia disrupts the barrier function of neural blood vessels through changes in the expression of claudin-5 in endothelial cells. *Am. J. Pathol.* **2007**, *170*, 1389–1397. [CrossRef] [PubMed]

64. Williams, E.I.; Betterton, R.D.; Davis, T.P.; Ronaldson, P.T. Transporter-mediated delivery of small molecule drugs to the brain: A critical mechanism that can advance therapeutic development for ischemic stroke. *Pharmaceutics* **2020**, *12*, 154. [PubMed]

65. Kurosawa, T.; Higuchi, K.; Okura, T.; Kobayashi, K.; Kusuhara, H.; Deguchi, Y. Involvement of proton-coupled organic cation antiporter in varenicline transport at blood-brain barrier of rats and in human brain capillary endothelial cells. *J. Pharm. Sci.* **2017**, *106*, 2576–2582.

66. Lin, C.J.; Tai, Y.; Huang, M.T.; Tsai, Y.F.; Hsu, H.J.; Tzen, K.Y.; Liou, H.H. Cellular localization of the organic cation transporters, OCT1 and OCT2, in brain microvessel endothelial cells and its implication for MPTP transport across the blood-brain barrier and MPTP-induced dopaminergic toxicity in rodents. *J. Neurochem.* **2010**, *114*, 717–727. [CrossRef]

67. Li, X.; Han, H.; Hou, R.; Wei, L.; Wang, G.; Li, C.; Li, D. Progesterone treatment before experimental hypoxia-ischemia enhances the expression of glucose transporter proteins GLUT1 and GLUT3 in neonatal rats. *Neurosci. Bull.* **2013**, *29*, 287–294. [CrossRef]

68. Nian, K.; Harding, I.C.; Herman, I.M.; Ebong, E.E. Blood-brain barrier damage in ischemic stroke and its regulation by endothelial mechanotransduction. *Front. Physiol.* **2020**, *11*, 605398. [CrossRef] [PubMed]

69. Hladky, S.B.; Barrand, M.A. Fluid and ion transfer across the blood–brain and blood–cerebrospinal fluid barriers; a comparative account of mechanisms and roles. *Fluids Barriers CNS* **2016**, *13*, 1–69. [CrossRef]

70. Jayakumar, A.R.; Norenberg, M.D. The Na–K–Cl co-transporter in astrocyte swelling. *Metab. Brain Dis.* **2010**, *25*, 31–38. [CrossRef]

71. Maher, J.M.; Slitt, A.L.; Callaghan, T.N.; Cheng, X.; Cheung, C.; Gonzalez, F.J.; Klaassen, C.D. Alterations in transporter expression in liver, kidney, and duodenum after targeted disruption of the transcription factor HNF1α. *Biochem. Pharmacol.* **2006**, *72*, 512–522. [CrossRef]

72. McNeill, J.; Rudyk, C.; Hildebrand, M.E.; Salmaso, N. Ion channels and electrophysiological properties of astrocytes. Implications for emergent stimulation technologies. *Front. Cell. Neurosci.* **2021**, *15*, 644126.

73. Koepsell, H. General overview of organic cation transporters in brain. In *Organic Cation Transporters in the Central Nervous System*; Handbook of Experimental Pharmacology; 2021; pp. 1–39. Available online: https://link.springer.com/chapter/10.1007/164_2021_449 (accessed on 14 March 2023).

74. Yoshikawa, T.; Naganuma, F.; Iida, T.; Nakamura, T.; Harada, R.; Mohsen, A.S.; Kasajima, A.; Sasano, H.; Yanai, K. Molecular mechanism of histamine clearance by primary human astrocytes. *Glia* **2013**, *61*, 905–916. [CrossRef]

75. Schinkel, A.H. P-Glycoprotein, a gatekeeper in the blood–brain barrier. *Adv. Drug Deliv. Rev.* **1999**, *36*, 179–194. [CrossRef]

76. Brouwer, K.L.R.; Keppler, D.; Hoffmaster, K.A.; Bow, D.A.J.; Cheng, Y.; Lai, Y.; Palm, J.E.; Stieger, B.; Evers, R. In vitro methods to support transporter evaluation in drug discovery and development. *Clin. Pharmacol. Ther.* **2013**, *94*, 95–112. [CrossRef] [PubMed]

77. Doan, K.M.M.; Humphreys, J.E.; Webster, L.O.; Wring, S.A.; Shampine, L.J.; Serabjit-Singh, C.J.; Adkison, K.K.; Polli, J. Passive permeability and p-glycoprotein-mediated efflux differentiate central nervous system (CNS) and non-CNS marketed drugs. *J. Pharmacol. Exp. Ther.* **2002**, *303*, 1029–1037. [CrossRef] [PubMed]

78. Wang, Q.; Rager, J.D.; Weinstein, K.; Kardos, P.S.; Dobson, G.L.; Li, J.; Hidalgo, I.J. Evaluation of the MDR-MDCK cell line as a permeability screen for the blood–brain barrier. *Int. J. Pharm.* **2005**, *288*, 349–359. [CrossRef]

79. Summerfield, S.G.; Stevens, A.J.; Cutler, L.; Osuna, M.D.C.; Hammond, B.; Tang, S.-P.; Hersey, A.; Spalding, D.J.; Jeffrey, P. Improving the in vitro prediction of in vivo central nervous system penetration: Integrating permeability, p-glycoprotein efflux, and free fractions in blood and brain. *J. Pharmacol. Exp. Ther.* **2005**, *316*, 1282–1290. [CrossRef]

80. Ohashi, R.; Watanabe, R.; Esaki, T.; Taniguchi, T.; Torimoto-Katori, N.; Watanabe, T.; Ogasawara, Y.; Takahashi, T.; Tsukimoto, M.; Mizuguchi, K. Development of simplified in vitro P-glycoprotein substrate assay and in silico prediction models to evaluate transport potential of P-glycoprotein. *Mol. Pharm.* **2019**, *16*, 1851–1863. [CrossRef] [PubMed]
81. Nozohouri, S.; Esfahani, S.H.; Noorani, B.; Patel, D.; Villalba, H.; Ghanwatkar, Y.; Rahman, S.; Zhang, Y.; Bickel, U.; Trippier, P.C.; et al. In-Vivo and ex-vivo brain uptake studies of peptidomimetic neurolysin activators in healthy and stroke animals. *Pharm. Res.* **2022**, *39*, 1587–1598. [CrossRef] [PubMed]

 *pharmaceutics*

Article

The Active Glucuronide Metabolite of the Brain Protectant IMM-H004 with Poor Blood–Brain Barrier Permeability Demonstrates a High Partition in the Rat Brain via Multiple Mechanisms

Jianwei Jiang [1,2,3,†], Lijun Luo [1,2,3,†], Ziqian Zhang [1,2,3,*], Xiao Liu [1,2,3], Naihong Chen [1,4], Yan Li [1,2,3] and Li Sheng [1,2,3,*]

1   State Key Laboratory of Bioactive Substances and Functions of Natural Medicines, Institute of Materia Medica, Chinese Academy of Medical Sciences and Peking Union Medical College, Beijing 100050, China; jwjiang98@zju.edu.cn (J.J.); luolijun@imm.ac.cn (L.L.)
2   Department of Drug Metabolism, Institute of Materia Medica, Chinese Academy of Medical Sciences and Peking Union Medical College, Beijing 100050, China
3   Beijing Key Laboratory of Non-Clinical Drug Metabolism and PK/PD Study, Institute of Materia Medica, Chinese Academy of Medical Sciences and Peking Union Medical College, Beijing 100050, China
4   Department of Pharmacology, Institute of Materia Medica, Chinese Academy of Medical Sciences and Peking Union Medical College, Beijing 100050, China
*   Correspondence: zhangziqian@imm.ac.cn (Z.Z.); shengli@imm.ac.cn (L.S.); Tel.: +86-010-63165191 (Z.Z.); +86-010-63165185 (L.S.)
†   These authors contributed equally to this work.

Citation: Jiang, J.; Luo, L.; Zhang, Z.; Liu, X.; Chen, N.; Li, Y.; Sheng, L. The Active Glucuronide Metabolite of the Brain Protectant IMM-H004 with Poor Blood–Brain Barrier Permeability Demonstrates a High Partition in the Rat Brain via Multiple Mechanisms. *Pharmaceutics* **2024**, *16*, 330. https://doi.org/10.3390/pharmaceutics16030330

Academic Editor: Yoshiyuki Kubo

Received: 24 January 2024
Revised: 21 February 2024
Accepted: 23 February 2024
Published: 27 February 2024

Abstract: Background: Glucuronidation is an essential metabolic pathway for a variety of drugs. IMM-H004 is a novel neuroprotective agent against ischemic stroke, and its glucuronide metabolite IMM-H004G exhibits similar pharmacological activity. Despite possessing a higher molecular weight and polarity, brain exposure of IMM-H004G is much higher than that of IMM-H004. This study aimed to investigate the brain metabolism and transport mechanisms of IMM-H004 and IMM-H004G. Methods: First, the possibility of IMM-H004 glucuronidation in the brain was evaluated in several human brain cell lines and rat homogenate. Subsequently, the blood–brain barrier carrier-mediated transport mechanism of IMM-H004 and IMM-H004G was studied using overexpression cell models. In addition, intracerebroventricular injection, in situ brain perfusion model, and microdialysis/microinjection techniques were performed to study the distribution profiles of IMM-H004 and IMM-H004G. Results: IMM-H004 could be metabolized to IMM-H004G in both rat brain and HEB cells mediated by UGT1A7. However, IMM-H004G could not be hydrolyzed back into IMM-H004. Furthermore, the entry and efflux of IMM-H004 in the brain were mediated by the pyrilamine-sensitive H^+/OC antiporter and P-gp, respectively, while the transport of IMM-H004G from the blood to the brain was facilitated by OATP1A2 and OATP2B1. Ultimately, stronger concentration gradients and OATP-mediated uptake played a critical role in promoting greater brain exposure of IMM-H004G. Conclusions: The active glucuronide metabolite of the brain protectant IMM-H004 with poor blood–brain barrier permeability demonstrates a high partition in the rat brain via multiple mechanisms, and our findings deepen the understanding of the mechanisms underlying the blood–brain barrier metabolism and transport of active glucuronide conjugates.

Keywords: blood–brain barrier; glucuronide; conjugate metabolite; uridine diphosphate-glucuronosyltransferase; organic anion-transporting polypeptide

1. Introduction

According to the World Stroke Organization, the incidence of stroke currently ranks second among the leading causes of death and third in terms of combined mortality and disability worldwide in 2022 [1]. Thrombolytic tissue plasminogen activator (tPA) has been the only agent approved by the Food and Drug Administration (FDA) for the treatment of ischemic stroke. However, the narrow time window for thrombolytic treatment (4.5 h) and the risk of hemorrhagic transformation limit the number of stroke patients who can benefit from tPA [2]. Therefore, there is an urgent need for neuroprotective agents that have extended treatment time windows and fewer side effects.

IMM-H004, a novel 3-piperazinyl coumarin derivative, is currently under development for the treatment of cerebral ischemia. Previous studies have shown that administering IMM-H004 6 h after cerebral ischemia can prevent brain damage by activating the anti-inflammatory pathway of CKLF1/CCR4 [3,4]. IMM-H004 has also been found to protect against global cerebral ischemia in rats by inhibiting apoptosis and maintaining the integrity of synaptic structures [5]. Additionally, it significantly reduces cerebral ischemia–reperfusion injury and subsequent inflammation in spontaneously hypertensive rats [6]. These findings suggest that IMM-H004 holds promise as a treatment for cerebral ischemia.

Despite its effectiveness, IMM-H004 has a very short elimination half-life (<1 h) in rat plasma and brain tissues after intravenous dosing [7]. Subsequent studies have identified four metabolites of IMM-H004 in rats, namely two demethylated metabolites, a glucuronide conjugate (IMM-H004G), and a sulfated conjugate. Among these metabolites, IMM-H004G is found to be the major metabolite in both rats and cultured human hepatocytes (Figure 1a). Multiple UDP glucuronosyltransferases (UGTs) are involved in the formation of IMM-H004G, with the highest contribution of UGT1A7, followed by UGT1A9, 1A8, and 1A1, whereas UGT1A3, 1A10, 2B15 and 1A6 exhibited less activity toward IMM-H004G [7]. Interestingly, brain exposure of IMM-H004G in rats was found to be six times higher than that of the parent drug [8]. Furthermore, pharmacological studies have shown that IMM-H004G exhibits neuroprotective activity similar to that of the parent drug in both oxygen–glucose-deprived PC12 cells and rats with transient middle cerebral artery occlusion–reperfusion (MCAO/R) injury. Hence, after administering IMM-H004, the active glucuronide metabolite IMM-H004G may prolong the duration of drug action and contribute to the efficacy of anticerebral ischemia treatment.

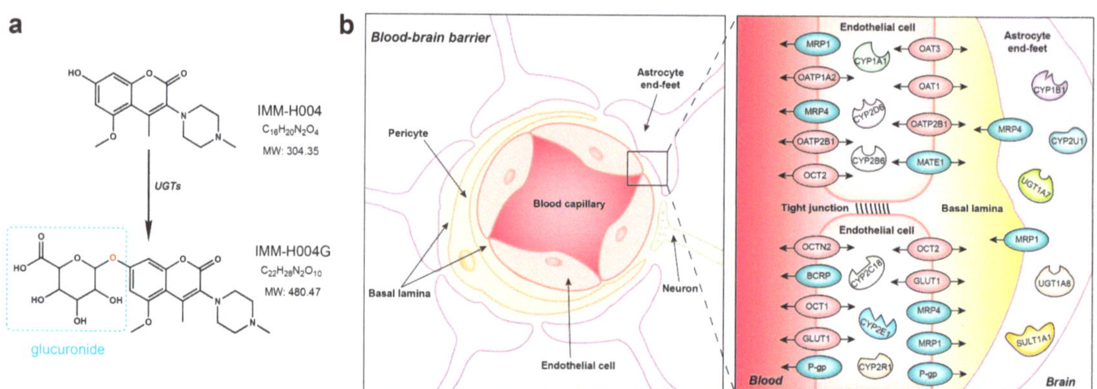

Figure 1. Structures of IMM-H004/IMM-H004G and the blood–brain barrier: (**a**) partial chemical information of IMM-H004 and IMM-H004G; (**b**) schematic diagram of the structure of the blood–brain barrier. Abbreviations: BCRP, breast cancer resistance protein; CYP, cytochrome P450; GLUT, glucose transporter; MATE, multidrug and toxin extrusion transporter; MRP, multidrug resistance protein; OAT, organic anion transporter; OATP, organic anion-transporting polypeptide; OCT, organic cation transporter; P-gp, P-glycoprotein; SULT, sulfotransferase; UGT, uridine diphosphate glucuronosyltransferase.

Generally, molecules with high lipophilicity and small molecular size can easily diffuse across the blood–brain barrier (BBB) following the concentration gradient [9]. Compared to IMM-H004, IMM-H004G is a highly polar and hydrophilic compound with a large molecular weight. However, brain exposure of IMM-H004G is significantly higher than that of IMM-H004. Due to its physicochemical properties, IMM-H004G is not favored for passive diffusion across the BBB, suggesting that it may depend on a transport system to reach high concentrations in the brain (Figure 1b). Additionally, it is widely recognized that during the transportation across the BBB, drugs have the potential to undergo metabolic transformations facilitated by drug-metabolizing enzymes that are present within the BBB [10]. Therefore, does the high brain exposure of IMM-H004G result from the in situ metabolism of IMM-H004, or is it transported into the brain through the BBB?

In this study, the impact of drug-metabolizing enzymes and transporters on the brain distribution of IMM-H004 and IMM-H004G was evaluated. This work strengthens our understanding of the metabolic and transport mechanisms of active glucuronide metabolites in the brain, which will be fundamental to the investigation of the toxicity and efficacy of novel neuroprotective agents while also promoting further research into active glucuronide compounds.

2. Materials and Methods

2.1. Chemicals and Materials

IMM-H004 and IMM-H004 citrates (purity > 99%) were synthesized by the Laboratory of Chemical Synthesis, and IMM-H004G (purity > 99%) was provided by the Laboratory of Biosynthesis of Natural Products, Institute of Materia Medica, Chinese Academy of Medical Science, as described previously [7,8]. Propranolol hydrochloride (internal standard, IS), digoxin, mitoxantrone, methotrexate, tetraethylammonium chloride (TEA), and metformin hydrochloride were purchased from Sigma-Aldrich (St. Louis, MO, USA). Diphenhydramine (DPH), cimetidine, quinidine, verapamil hydrochloride, L-carnitine, and benzbromarone were purchased from J&K Scientific, Ltd. (Beijing, China). Atenolol, pyrimethamine, rosuvastatin calcium (RSV), estradiol 17-(β-D-glucuronide) (E17βG), rifampicin (RFP), PSC833, MK571, and Ko143 were purchased from MedChemExpress (Monmouth Junction, NJ, USA). Estrone 3-sulfate sodium (E3S) was obtained from Krre Inc. (Beijing, China). Rifamycin SV was purchased from China National Institutes for Food and Drug Control. Type I collagen from rat tail was purchased from Thermo Fisher Scientific (Waltham, MA, USA). The 0.25% trypsin–EDTA solution was purchased from Gibco (Carlsbad, CA, USA). The transwell insert (0.4 μm) was obtained from Millipore (Billerica, MA, USA). The SH-SY5Y cell line was obtained from the Cell Resource Center, Peking Union Medical College (Beijing, China). The normal human astrocyte cell line HEB was purchased from the Experimental Animal Center of Sun Yat-Sen University of Medical Sciences (Guangzhou, China). Immortalized human brain microvascular endothelial cells hCMEC/D3 and related culture media were purchased from Millipore (Billerica, MA, USA). Madin–Darby canine kidney II (MDCKII) cells were purchased from the American Type Culture Collection (Manassas, VA, USA). Wild-type Chinese hamster ovary (CHO) cells were a gift from Prof. Naihong Chen (Institute of Materia Medica, Chinese Academy of Medical Sciences and Peking Union Medical College, Beijing, China). The passage number of the cell lines was between 5 and 20.

2.2. Animals

All animal studies were approved by the Animal Care and Welfare Committee of Peking Union Medical College (approval code: 00007839) and strictly adhered to the guidelines for the use and care of laboratory animals issued by the Institute Animal Care and Welfare Committee. Male Sprague Dawley rats (260–280 g) were purchased from Vital River Experimental Animal Co., Ltd. (Beijing, China). The animals were acclimated for one week prior to the treatment and provided with free access to water and food.

2.3. Metabolism in the Brain

2.3.1. In Vitro Metabolism in Human Brain Cells and Rat Brain Homogenate

SH-SY5Y, HEB, and hCMEC/D3 cells were utilized to investigate the in vitro metabolism of IMM-H004 and IMM-H004G in the brain. SH-SY5Y and HEB cells were cultured routinely in Dulbecco's modified Eagle's medium (DMEM) containing 10% fetal bovine serum and a 100 U/mL penicillin–100 µg/mL streptomycin solution. hCMEC/D3 cells were seeded on collagen type I coated 24-well plates and cultured in EndoGRO™-MV complete medium following the manufacturer's guidelines. The cells were incubated at 37 °C, with 5% CO_2, and were maintained in a humidified environment. Once the confluence reached 70–80%, the cells were digested using a 0.25% trypsin–EDTA solution, centrifuged at $200\times g$ for 5 min, and then resuspended in a fresh culture medium. For each experiment, after seeding and 24 h incubation, the cells were washed twice with a prewarmed culture medium, and then IMM-H004 or IMM-H004G (1–100 µmol/L) dissolved in the culture medium was added. Incubation was carried out at 37 °C for 5 h or 24 h. After incubation, the supernatants were collected. Subsequently, the cells were washed with cold HBSS three times and lysed in a RIPA lysis buffer (Beyotime Inc., Shanghai, China). The concentrations of IMM-H004 and IMM-H004G were determined by liquid chromatography with tandem mass spectrometry (LC-MS/MS).

Rat brains were homogenized in chilled saline (1:3, w/v) on ice to obtain the brain homogenate. The incubation mixtures composed of the brain homogenate (2 mg protein/mL), alamethicin (50 µg/mg protein), IMM-H004 (10 µmol/L), and UDPGA (5 mmol/L) in a final volume of 0.2 mL of Tris-HCl buffer (50 mmol/L, pH 7.4) containing 5 mmol/L of $MgCl_2$. Before initiating the reaction, the brain homogenate was preincubated with alamethicin on ice for 15 min. Subsequently, IMM-H004 and UDPGA were added to initiate the reaction. After incubating at 37 °C for 60 min, the reaction was quenched by adding two volumes of acetonitrile containing IS (0.1 µg/mL). On the other hand, IMM-H004G (1 µmol/L) was incubated with the brain homogenate (2 mg protein/mL) for up to 4 h. The concentrations of IMM-H004 and IMM-H004G were determined by LC-MS/MS.

2.3.2. Quantification of UGT Expression in SH-SY5Y and HEB Cells by qRT-PCR

The expression levels of UGT1A1, UGT1A3, UGT1A6, UGT1A7, UGT1A8, UGT1A9, UGT1A10, and UGT2B15 in SH-SY5Y and HEB cells were compared using qRT-PCR. Total RNA was isolated using the RNeasy micro kit (Qiagen, Hilden, Germany) following the manufacturer's instructions. The concentrations and purity of the RNA samples were accessed using a NanoPhotometer (Implen Gmbh, Munich, Germany). cDNA synthesis was performed using High-Capacity cDNA Reverse Transcription Kits (Applied Biosystems, Wakefield, RI, USA). For each reaction, 1 µg of cDNA was combined with a universal master mix, sense and antisense primers (0.4 µmol/L each), and oligonucleotide probes (0.2 µmol/L). Primer and probe for UGTs and GAPDH were purchased from Thermo Fisher Scientific (assay ID: Hs02511055_s1, Hs04194492_g1, Hs01592477_m1, Hs02517015_s1, Hs01592482_m1, Hs02516855_sH, Hs02516990_s1, and Hs00870076_s1 for 8 UGTs; Hs02786624_g1 for GAPDH). The GAPDH gene was used as a reference gene. qRT-PCR was conducted on the Step-One Plus Real-Time PCR System (Applied Biosystems). The PCR reaction procedure was as follows: 95 °C for 20 s, followed by 45 cycles of 95 °C for 3 s, and 62 °C for 30 s. Expression data were normalized to GAPDH expression using the comparative Ct method to determine the relative mRNA expression of the eight UGTs.

2.3.3. In Vivo Metabolism of IMM-H004 in Rat Brain after Intracerebroventricular Injection

The metabolism of IMM-H004 in the brain was investigated through the intracerebroventricular (icv) injection of IMM-H004 in rats, and the drug concentration in each brain region was analyzed. The rats were fixed on a benchmark stereotactic instrument (Leica Biosystems, Deer Park, IL, USA) under isoflurane anesthesia. Using a microinjector connected to a syringe pump, a 5 µL solution of IMM-H004 was dissolved in artificial cerebrospinal fluid (ACSF) and injected (3.8 µg/kg) into the lateral ventricle. The injection coordinates were as follows: medial/lateral = 1.5 mm; anterior/posterior = −0.8 mm; dorsal/ventral = −4 mm. Animals were sacrificed at 6, 15, 30, 45, and 60 min after drug administration. The brains were collected and dissected on ice into the following regions: the cerebellum, brainstem, hypothalamus, hippocampus, midbrain, striatum, and cortex. The subsequent procedure for processing tissue samples was the same as the procedure described in Section 2.3.1.

2.4. Transport Studies of IMM-H004 and IMM-H004G

2.4.1. Studies in hCMEC/D3 Cells

Kinetic Studies

Uptake assays into hCMEC/D3 cells were conducted in 24-well plates. All incubations were carried out at 37 °C in triplicate unless otherwise noted. In time-dependent uptake studies, hCMEC/D3 cells were incubated with IMM-H004 (10 μmol/L) and IMM-H004G (50 μmol/L) dissolved in HBSS under gentle shaking for up to 30 min. To investigate the impact of incubation temperature on the uptake of IMM-H004 and IMM-H004G by hCMEC/D3 cells, incubation was performed at 4 or 37 °C for 10 min. To determine the kinetic parameters, the cells were incubated with IMM-H004 and IMM-H004G (2–400 μmol/L) for 10 min. The uptake data were fitted to the following equation using nonlinear least-square regression analysis in GraphPad Prism 9 (GraphPad, San Diego, CA, USA):

$$V = \frac{V_{max} \times S}{K_m + S} + K_{ns} \times S \tag{1}$$

where V is the uptake velocity (nmol/mg protein/min), and S is the initial concentration (μmol/L). V_{max}, K_m, and K_{ns} represent the maximum uptake velocity (nmol/mg protein/min), the Michaelis–Menten constant (μmol/L), and nonsaturable uptake clearance (μL/mg protein/min), respectively. The saturable uptake of both IMM-H004 and IMM-H004G was plotted with the Eadie–Hofstee method by subtracting passive uptake from the total uptake.

Identification of Transporters Using Inhibitors

The toxicity of all compounds, including inhibitors and substrates, toward hCMEC/D3 cells, was assessed using the 3-(4,5-Dimethylthiazol-2-yl)-2,5-diphenyltetrazolium bromide (MTT) reduction assay as described previously [11]. To assess the effects of pyrilamine-sensitive proton-coupled organic cation (H$^+$/OC) antiporter, organic anion transporter (OAT), organic anion-transporting polypeptides (OATPs), organic cation transporter (OCT), and novel organic cation–carnitine transporter (OCTN) inhibitors on the uptake process, IMM-H004 or IMM-H004G (10 μmol/L) combined with the inhibitors were added to the medium after preincubating the cells with the inhibitors for 30 min. For evaluating the impact of the multidrug and toxin extrusion transporter (MATE), P-glycoprotein (P-gp), multidrug resistance protein (MRP), and breast cancer resistance protein (BCRP) on the uptake, IMM-H004 or IMM-H004G (10 μmol/L) was preincubated with the cells for 2 h before the addition of the inhibitors. After removing the supernatant, the cells were washed twice with prewarmed HBSS and incubated with fresh HBSS for 20 min. At the end of incubation, the supernatants were collected. The cells were washed with cold HBSS and lysed using RIPA buffer. The protein content of the lysed cells was determined using a bicinchoninic acid (BCA) assay kit (Beyotime Inc., Shanghai, China). The results were expressed as a percentage of the vehicle control. The substrates and inhibitors used herein are summarized in Table 1.

Table 1. Concentrations of the substrates and inhibitors for transporter identification assays in hCMEC/D3 cells.

Transporter	Substrate	Concentration (µmol/L)	Inhibitor	Concentration (µmol/L)	Reference
Pyrilamine-sensitive H⁺/OC antiporter	Diphenhydramine	1	Quinidine, Verapamil	100	[12–14]
OAT	Methotrexate	10	Benzbromarone	10	[15,16]
OCT	Metformin	10	Cimetidine	100	[17,18]
OCTN	Tetraethylammonium	20	L-Carnitine	5	[19–21]
OATP	Estrone 3-sulfate	5	Rifampicin	100	[22,23]
MATE	Metformin	10	Pyrimethamine	1	[24]
P-gp	Digoxin	5	PSC833	5	[25,26]
BCRP	Mitoxantrone	5	Ko143	5	[27,28]
MRP	Methotrexate	10	MK571	10	[29,30]

2.4.2. Identification of Transporters Using Overexpressing Cells

CHO cells and MDCKII cells were stably transfected with lentiviruses carrying human SCLO1A2 (encoding OATP1A2) and SCLO2B1 (encoding OATP2B1), respectively. Cell transfected with lentiviruses carrying empty vectors served as control. Lentiviral plasmid packages were obtained from OBiO Technology (Shanghai, China), and procedures followed the instructions of the virus manual. Stably transfected cells were routinely grown in a 100 mm dish in DMEM with 10% fetal bovine serum. Prior to the experiments, the seeding density of the overexpressing cells in 24-well plates was 4×10^5 cells per well, and the medium was supplemented with 2 mM sodium butyrate for 24 h. The cultured cells were washed twice with prewarmed HBSS after removing the medium and then treated with 10 µmol/L IMM-H004G and 100 µmol/L rifampicin (or an equal amount of solvent dimethyl sulfoxide) for 10 min. After incubation, the cells were washed four times with cold HBSS. Follow-up sample processing procedures were as described in Section 2.3.1. E17βG (5 µmol/L) and rosuvastatin (10 µmol/L) were used as positive substrates in CHO-OATP1A2 and MDCKII-OATP2B1 cells, respectively. The MDCKII-MDR1 cells were previously established by our laboratory [11].

To further validate the effect of P-gp on the transport of IMM-H004G or IMM-H004, MDCKII-MDR1 cells were seeded onto polycarbonate membrane transwell inserts in 24-well plates and used for transport studies after 5 days or when the effective transepithelial electric resistance (TEER) values exceeded 300 $\Omega \cdot cm^2$. The monolayers were initially rinsed twice with HBSS. Subsequently, IMM-H004G or IMM-H004 (10 µmol/L) was added to either the apical or basolateral side of the monolayer. Aliquots of 50 µL were collected from the opposite compartment at various time points, covering a maximum duration of 120 min. Digoxin (5 µmol/L) was used as a positive substrate, while PSC833 (5 µmol/L) was used as an inhibitor in MDCKII-MDR1 cells. The apparent permeability value (P_{app}) was calculated using the following equation:

$$P_{app} = \frac{dQ/dt}{AC_0} \tag{2}$$

where dQ/dt is the slope of the cumulative transport amount over time during the study period. A is the base area of the insert, and C_0 is the initial addition concentration.

The efflux ratio (ER) was calculated as follows:

$$ER = P_{app(B-A)}/P_{app(A-B)} \tag{3}$$

2.4.3. In Situ Brain Perfusion

The rat brain was perfused following a previously described method [31]. Briefly, the common carotid arteries on both sides were cannulated and connected to the perfusion system, while the external carotid artery was ligated. At the beginning of perfusion, the jugular veins on both sides were quickly incised. The perfusion procedure consisted of a preperfusion wash with saline for 1 min, followed by drug infusion (2 µmol/L IMM-H004 or IMM-H004G) for 2, 5, and 10 min, and concluded with a postperfusion wash (saline for 5 min) at a rate of 4.0 mL/min. To examine the impact of P-gp (PSC833, 5 µmol/L) and OATP (rifamycin SV, 100 µmol/L) on BBB transport, the perfusate containing the inhibitor was perfused for 1 min, and then the mixture of IMM-H004 or IMM-H004G (10 µmol/L) and the inhibitor was perfused for 10 min. Following the perfusion, the rats were decapitated, and their brains were homogenized (1:3, *w/v*) in chilled saline. The concentrations of IMM-H004 and IMM-H004G were then determined by LC-MS/MS.

The unidirectional transfer constant K_{in} (mL/min/g) was calculated using the following equation:

$$K_{in} = \frac{Q_{br}}{C_{pf} \times T} \tag{4}$$

where Q_{br}/C_{pf} is the apparent brain distribution volume, Q_{br} is the measured amount of compound in brain tissue, C_{pf} is the concentration of compound in the perfusate, and T is the perfusion time.

2.5. Pharmacokinetic Study in Rats

2.5.1. Plasma Protein Binding

The rat plasma protein binding of IMM-H004 and IMM-H004G (0.5–10 µg/mL) was investigated using a rapid equilibrium dialysis (RED) assay. The RED plate inserts utilized in this assay contained a dialysis membrane with a molecular weight cut-off of 8 kDa (Thermo Fisher Scientific, Waltham, MA, USA). For the assay, aliquots of plasma (0.2 mL) and phosphate-buffered saline (0.35 mL) were added to the sample chamber and buffer chamber, respectively. The plate was then sealed and incubated at 37 °C, 200 rpm for 4 h. Following the dialysis period, the concentrations of IMM-H004 and IMM-H004G were determined using LC-MS/MS. The ratio of the drug concentration in the buffer to that in the plasma chamber was used to calculate the unbound fraction (fu) of the drugs.

2.5.2. Pharmacokinetics

Rats were anesthetized with isoflurane. Microdialysis intracerebral guide cannulas and CMA 12 Elite Probe (PAES membrane with a MWCO of 20 kDa, 2 mm membrane length (CMA/Microdialysis AB, Stockholm, Sweden)) were implanted into the cerebral cortex at the coordinates relative to bregma: medial/lateral = 3.0 mm; anterior/posterior = 0.5 mm; dorsal/ventral = −3.0 mm. After a 24 h recovery period, the intracerebral probe was connected to the microdialysis cannula, and Ringer's solution was perfused at a flow rate of 2.0 µL/min. Then, IMM-H004 citrate or IMM-H004G (10 mg/kg) was injected intravenously (iv). Dialysate samples were collected every 5 to 60 min for a duration of 7 h. The probe recovery in vitro was determined by dialyzing a standard mixture of IMM-H004 (50 ng/mL) and IMM-H004G (300 ng/mL) infused with Ringer's solution.

Blood samples were collected from the orbital plexus into heparinized tubes and subsequently centrifuged to separate the plasma. The concentrations of IMM-H004 and IMM-H004G in both the dialysate and plasma samples were determined using LC-MS/MS analysis. Subsequently, the free concentrations in the brain and plasma were corrected with probe recovery and fu for further analysis, respectively. The pharmacokinetic parameters were calculated via noncompartmental analysis (NCA) using Phoenix WinNonlin 6.3 (Pharsight Corporation, Sunnyvale, CA, USA). The brain–plasma partition coefficient (Kp) was determined as the ratio of the area under the concentration–time curve for the brain (AUC_{brain}) to that of the plasma (AUC_{plasma}).

2.6. Determination of IMM-H004 and IMM-H004G Using LC-MS/MS

All samples were mixed with two volumes of ice-cold acetonitrile containing IS (propranolol, 0.1 µg/mL) and centrifugation at 18,800× *g* for 5 min. The supernatant containing IMM-H004 and IMM-H004G was analyzed using LC-MS/MS according to the previously reported methods [8,32]. In brief, the LC-MS/MS system consisted of an API 4000 mass spectrometer (AB SCIEX, Framingham, MA, USA) coupled with a Shimadzu UPLC LC-30A (Shimadzu Corporation, Kyoto, Japan). The analytical column used was a Zorbax Eclipse Plus C18 column (2.1 × 50 mm, 3.5 μm, Agilent, Santa Clara, CA, USA). The separation was performed at a flow rate of 0.3 mL/min. The mobile phase consisted of methanol and water both containing 0.5% formic acid. Quantification was carried out using the multiple reaction monitoring transition m/z 305.1→248.1 (IMM-H004), m/z 481.3→305.1 (IMM-H004G), and m/z 260.1→183.1 (propranolol, IS).

2.7. Statistical Analysis

Data were processed using Excel 2019 (Microsoft Inc., Washington, DC, USA) and are presented as mean ± SD. The statistical analysis of the data between the two groups was performed using an unpaired t-test. Statistical significance was considered as $p < 0.05$.

3. Results

3.1. Metabolism of IMM-H004 and IMM-H004G in Brain Cells and Brain Homogenate

Brain cells consist primarily of neurons, neuroglial cells, and cerebral endothelial cells. SH-SY5Y, HEB, and hCMEC/D3 are three commonly used human-derived cell lines, employed for the study of neurons, neuroglial cells, and cerebral microvascular endothelial cells, respectively. Thus, the metabolism of IMM-H004 and IMM-H004G in the brain and the species differences were evaluated using these three human brain cell lines and the rat brain homogenate (Figure 2a).

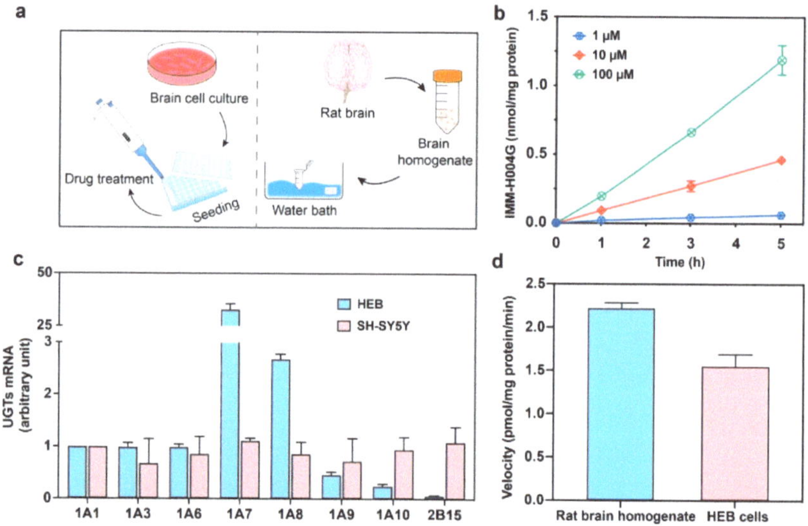

Figure 2. Metabolism of IMM-H004 and IMM-H004G in human brain cells and rat brain homogenate: (**a**) schematic diagram of the metabolism study in human brain cell lines and rat brain homogenate; (**b**) production of IMM-H004G when IMM-H004 (1–100 μmol/L) was incubated with HEB cells for up to 5 h ($n = 2$); (**c**) expression of UGTs mRNA in HEB and SH-SY5Y cells ($n = 3$); (**d**) in vitro generation rates of IMM-H004 to IMM-H004G in rat brain homogenate and HEB cells ($n = 3$).

In the SH-SY5Y and hCMEC/D3 cell systems, IMM-H004G was not detected, while in the HEB system, the production of IMM-H004G depended on both time and concentration (Figure 2b). Conversely, when IMM-H004G was incubated with these cells for 24 h, no significant amounts of IMM-H004 were detected.

It was reported that human brain microvessels did not exhibit obvious expression of any UGT genes [33]. Since IMM-H004 can form its glucuronide conjugate in HEB cells, the expression of eight UGT genes involved in the glucuronidation of IMM-H004 was investigated in both HEB and SH-SY5Y cells to determine any intercellular difference. As shown in Figure 2c, all eight UGT isoforms were detected in HEB cells, with UGT1A7 being the predominant UGT gene expressed. A smaller amount of UGT1A8 transcript was detected (approximately 10-fold less), while the expression of other UGT genes was weak (0.3–3.5%). In comparison, SH-SY5Y cells exhibited detectable levels of all eight UGT genes but at much lower levels than HEB cells. In vitro studies using cultured human brain cell lines have revealed that the interconversion of IMM-H004 and IMM-H004G is primarily facilitated by UGTs, with a predominant transformation of IMM-H004 into IMM-H004G. This metabolic conversion predominantly occurs in neuroglial cells, where UGT1A7 is believed to play a crucial role in the glucuronidation of IMM-H004.

To investigate the potential species differences in the glucuronidation of IMM-H004 between rats and human brains, IMM-H004 (10 μmol/L) was incubated with the rat brain homogenate for 1 h. The formation of IMM-H004G was observed to be time-dependent. Interestingly, both the rat brain homogenate and HEB cells exhibited a similar metabolic rate in generating IMM-H004G, as depicted in Figure 2d. These results suggest that the rat brain may possess a comparable ability to humans in catalyzing the glucuronidation of IMM-H004.

Additionally, similar to the findings observed in human brain cell lines, incubating IMM-H004G (1 μmol/L) with the rat brain homogenate (2 mg/mL protein) for 4 h did not result in a reduction in its concentration (Supplementary Information, Figure S1). This provides additional evidence that the conversion of IMM-H004G back into IMM-H004 is difficult within the brain.

3.2. Metabolism of IMM-H004 after Intracerebroventricular Injection

It is established that icv injection can provide insights into the real-time metabolic state of drugs within the brain. To assess this, IMM-H004 was injected into the lateral ventricle of rats, and the concentrations of IMM-H004 and its glucuronide metabolite IMM-H004G in different brain regions were analyzed (Figure 3). The determination of dosage was based on the initial concentration of IMM-H004 in the brain following iv injection.

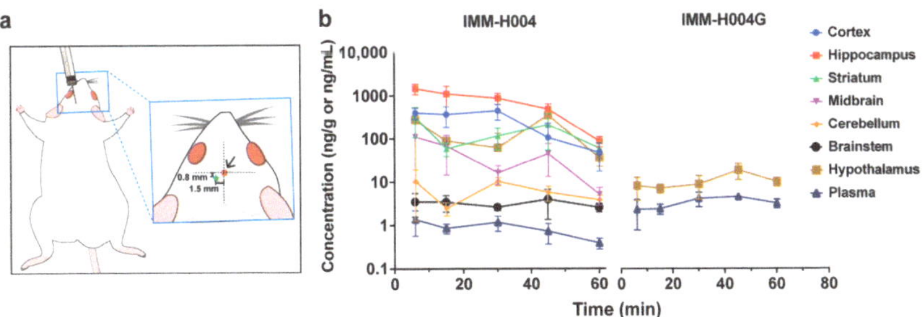

Figure 3. Metabolism of IMM-H004 after intracerebroventricular injection: (**a**) schematic diagram of intracerebroventricular injection. The red dot indicates the position of the bregma and the green pentagram represents the injection site; (**b**) concentration–time profiles of IMM-H004 and IMM-H004G in different regions of rat brain after intracerebroventricular injection of IMM-H004 at 3.8 μg/kg ($n = 4$).

The results demonstrated that IMM-H004 was detected in all brain regions, including the cerebellum, brainstem, hypothalamus, hippocampus, midbrain, striatum, and cortex, within 6–60 min after icv injection. The highest concentration of IMM-H004 was found in the hippocampus near the injection site, followed by the cortex, striatum, and hypothalamus. Lower concentrations were observed in the midbrain, cerebellum, and brainstem. In contrast, IMM-H004G was only detected at lower concentrations in the hypothalamus. Additionally, trace amounts of IMM-H004 and IMM-H004G were also detected in the plasma. The hypothalamus-to-plasma AUC ratio of IMM-H004G was 3.13, suggesting that IMM-H004G in the hypothalamus originated from in situ metabolism. As IMM-H004G was exclusively detected in the hypothalamus after icv injection and at significantly lower levels than iv injection, these findings confirmed that the glucuronidation of IMM-H004 occurs in the brain but to a limited extent.

3.3. Uptake Kinetics in hCMEC/D3 Cells

Brain microvascular endothelial cells constitute the blood–brain barrier and express various transporters to protect the brain against exogenous substances. The hCMEC/D3 cell line, derived from isolated human brain capillary endothelial cells, is widely utilized as an in vitro model for studying the transport mechanisms of drugs across the BBB. This is attributed to its expression of 144 SLC transporters and 23 ABC efflux transporters [34]. To investigate the transport characteristics of IMM-H004 and IMM-H004G across the BBB, a study was conducted on the uptake kinetics of IMM-H004 and IMM-H004G in hCMEC/D3 cells.

The uptake time profiles of IMM-H004 (10 μmol/L) and IMM-H004G (50 μmol/L) are illustrated in Figure 4a. It was found that the uptake of IMM-H004 was markedly higher than IMM-H004G. Moreover, both IMM-H004 and IMM-H004G exhibited time-dependent accumulation in hCMEC/D3 cells within 30 min. Therefore, a 10 min uptake period was selected for subsequent uptake kinetic and inhibition studies.

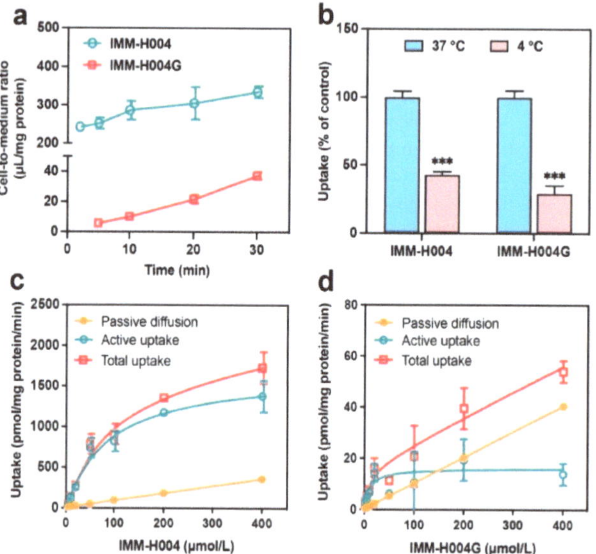

Figure 4. Uptake characteristics of IMM-H004 and IMM-H004G in hCMEC/D3 cells: (**a**) time-dependent uptake of IMM-H004 (10 μmol/L) and IMM-H004G (50 μmol/L); (**b**) the effect of temperature on the uptake of IMM-H004 and IMM-H004G (10 μmol/L, 10 min); (**c**,**d**) the concentration-dependent uptake of IMM-H004 and IMM-H004G (2–400 μmol/L, 10 min). $n = 3$. *** $p < 0.001$ vs. control group.

In general, energy consumption and active transport processes are minimal at temperatures between 0 and 4 °C. While the diffusion rate may diminish with a decreased temperature, the rate of mediated transport falls sharply. Therefore, comparing the uptake of substances at 4 °C and 37 °C can confirm the involvement of carrier-mediated transport processes in drug transport.

As shown in Figure 4b, the uptake of IMM-H004 and IMM-H004G showed a significant decrease at 4 °C, with respective reductions of 57.1% and 70.6%, compared to 37 °C ($p < 0.01$). These results suggest that the transport of these compounds is dependent on temperature and energy, implying that carrier-mediated processes contribute to their transport.

Furthermore, the concentration-dependent curves of IMM-H004 and IMM-H004G both exhibited an initial rapid increase, followed by a gentle, constant, and positive slope, indicating the occurrence of both saturable and unsaturated uptake (Figure 4c,d). The occurrence of saturable uptake of IMM-H004 and IMM-H004G further confirms the involvement of transporters in mediating the uptake of these compounds. The K_m values for IMM-H004 and IMM-H004G were determined as 89.5 and 10.7 µmol/L, respectively, based on the calculated kinetic parameters. The V_{max}/K_m ratios for the transporter-mediated uptake of IMM-H004 and IMM-H004G were 18.8 and 1.50 µL/min/mg protein, respectively. The K_{ns} values for the passive diffusion rate of IMM-H004 and IMM-H004G were 0.877 and 0.100 µL/min/mg protein, respectively. The K_{ns} values for the passive diffusion rate of IMM-H004 were 8.77 times that of IMM-H004G, indicating the greater membrane permeability of IMM-H004.

3.4. Identification of the Transporters in hCMEC/D3 Cells by Inhibitors

To investigate the transport mechanisms of IMM-H004 and IMM-H004G in more detail, we performed inhibition assays on nine transporters that exhibit the highest expression levels in the BBB and are involved in drug transport. All the positive substrates and inhibitors used in this study were tested at nontoxic concentrations, as confirmed by the MTT assay with a cell survival rate greater than 90%. DPH, metformin, TEA, methotrexate, E3S, digoxin, and mitoxantrone were employed as substrates for pyrilamine-sensitive H^+/OC antiporter, OCT and MATE, OCTN, OAT and MRP, OATP, P-gp, and BCRP, respectively.

As shown in Figure 5, the uptake or efflux of positive substrates in hCMEC/D3 cells decreased by 30–90% in the presence of corresponding inhibitors, confirming the reliability of the experiment system. Notably, quinidine and verapamil, inhibitors of the pyrilamine-sensitive H^+/OC antiporter, reduced the uptake of IMM-H004 by 57.7% and 61.5%, respectively, indicating the involvement of the pyrilamine-sensitive H^+/OC antiporter in the uptake transport of IMM-H004. However, the uptake of IMM-H004G was not affected by these inhibitors. Interestingly, the OATP inhibitor rifamycin resulted in a 39.6% reduction in the uptake of IMM-H004G. This suggests that IMM-H004G may be a substrate for the OATP uptake transporter. Moreover, the P-gp inhibitor PSC833 increased the uptake of IMM-H004 to 1.3 times that of the control, without affecting the uptake of IMM-H004G. This indicates that P-gp is selectively involved in the efflux of IMM-H004. No significant impact on the uptake of IMM-H004 and IMM-H004G was observed with inhibitors targeting other transporters, including OAT, OCT, OCTN, BCRP, and MRP. This suggests that these transporters are unlikely to be associated with the transport of IMM-H004 and IMM-H004G.

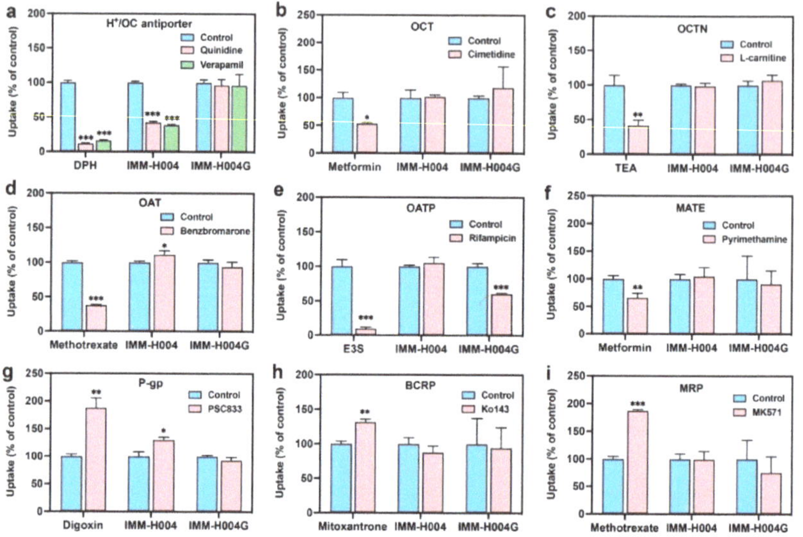

Figure 5. Effect of transporter inhibitors on the accumulation of IMM-H004 and IMM-H004G in hCMEC/D3 cells: (**a**) DPH (diphenhydramine, 1 µmol/L) is the substrate, and quinidine and verapamil (100 µmol/L) are the inhibitors of H⁺/OC antiporter; (**b**) metformin (10 µmol/L) and cimetidine (100 µmol/L) are the substrate and inhibitor of OCT, respectively; (**c**) TEA (tetraethylammonium, 20 µmol/L) and L-carnitine (5 µmol/L) are the substrate and inhibitor of OCTN, respectively; (**d**) methotrexate (10 µmol/L) and benzbromarone (10 µmol/L) are the substrate and inhibitor of OAT, respectively; (**e**) E3S (estrone 3-sulfate, 5 µmol/L) and rifamycin (100 µmol/L) are the substrate and inhibitor of OATP, respectively; (**f**) metformin (10 µmol/L) and pyrimethamine (1 µmol/L) are the substrate and inhibitor of MATE, respectively; (**g**) digoxin (5 µmol/L) and PSC833 (5 µmol/L) are the substrate and inhibitor of P-gp, respectively; (**h**) mitoxantrone (5 µmol/L) and Ko143 (5 µmol/L) are the substrate and inhibitor of BCRP, respectively; (**i**) methotrexate (10 µmol/L) and MK571 (10 µmol/L) are the substrate and inhibitor of BCRP, respectively. $n = 3$. *** $p < 0.001$, ** $p < 0.01$, * $p < 0.05$ vs. control group.

3.5. Transport in Overexpressing Cells

To further explore the potential role of OATPs in facilitating the entry of IMM-H004G into the brain, we successfully constructed CHO-OATP1A2 and MDCKII-OATP2B1 cells overexpressing these transporters using a lentiviral vector system (Figure 6a,b). Overexpression was validated by the increased uptake of positive substrates, showing a 10.8-fold and 5.6-fold increase compared to the control cells (Figure 6c,d). In OATP1A2- and OATP2B1-overexpressing cells, the IMM-H004G uptake was markedly higher, with 9.7-fold and 190-fold increases compared to the control cells, respectively (Figure 6e,f), suggesting that IMM-H004G is a substrate for OATP and exhibits stronger selectivity toward OATP2B1.

The MDCKII-MDR1 cell model is commonly used for identifying P-gp substrates. In the presence of PSC833, a P-gp inhibitor, the ER of digoxin, a known P-gp substrate, decreased from 21.14 to 1.41, confirming the reliable P-gp transport activity of the system (Figure 6g). The ER of IMM-H004 was 2.96, exceeding the threshold ER of 2, and it decreased to 1.10 in the presence of the P-gp inhibitor, indicating that IMM-H004 has the potential to be a P-gp substrate.

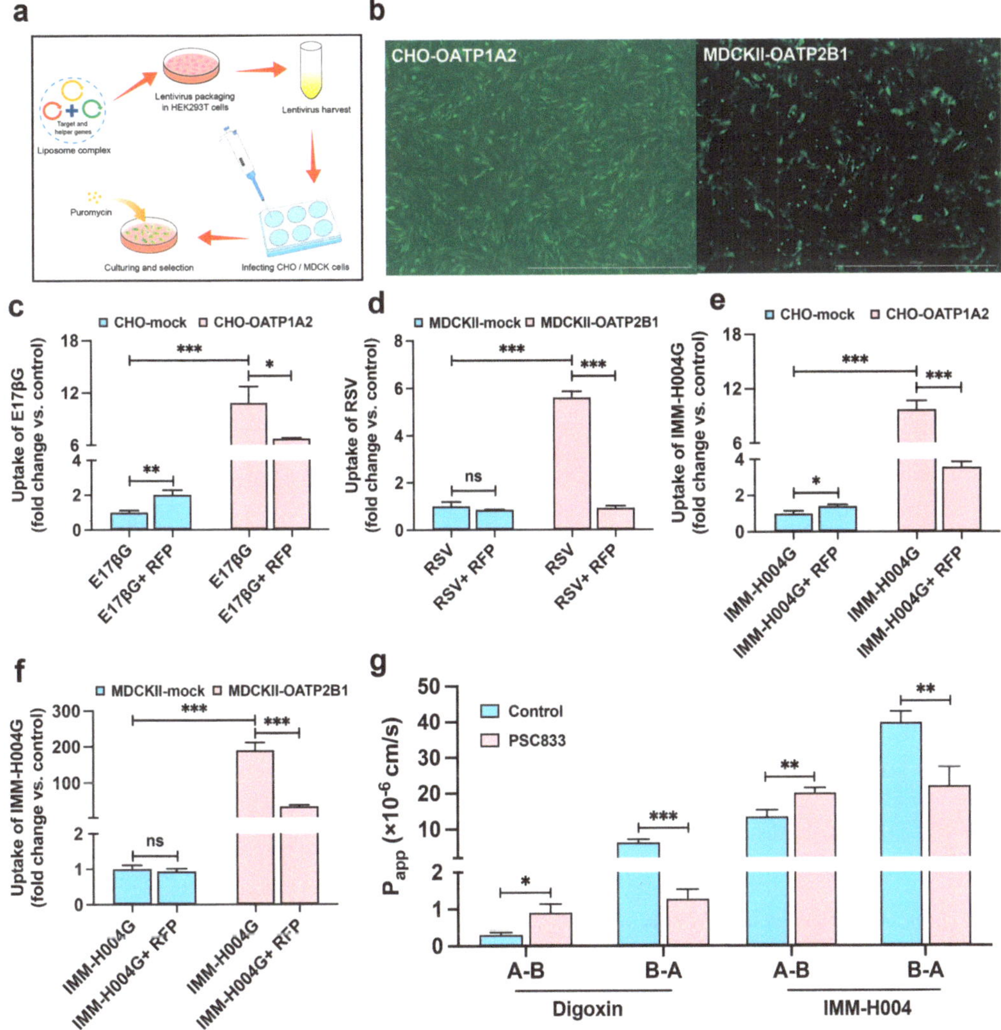

Figure 6. Transmembrane process of IMM-H004 and IMM-H004G mediated by P-gp and OATPs: (**a**) schematic diagram of the construction of transporter-overexpressing cells; (**b**) the photographs of green fluorescent protein (GFP) of CHO-OATP1A2 and MDCKII-OATP2B1 cells, respectively, scale bar = 1000 μm; (**c**) E17βG (estradiol-17β-glucuronide, 5 μmol/L) was used as a positive group in CHO cells; (**d**) RSV (rosuvastatin, 10 μmol/L) was used as a positive group in MDCKII cells; (**e**,**f**) the uptake of IMM-H004G (10 μmol/L) in OATP-overexpressing and mock cells; RFP (rifampicin, 100 μmol/L) served as a pan-OATP inhibitor; (**g**) the P_{app} values of IMM-H004 (10 μmol/L) in monolayers of MDCKII-MDR1 cells. $n = 3$. *** $p < 0.001$, ** $p < 0.01$, * $p < 0.05$ vs. control group. ns, not significant.

3.6. In Situ Brain Perfusion in Rats

In situ brain perfusion studies can comprehensively evaluate the influence of the physicochemical properties and transporter effects of IMM-H004 and IMM-H004G on their entry into the brain (Figure 7a). Both IMM-H004 and IMM-H004G exhibited time-dependent BBB crossing within 10 min (Supplementary Information Figure S2). As displayed in Figure 7b, the K_{in} value of IMM-H004 was 18 times higher than that of IMM-

H004G, indicating a faster rate of entry into the brain for IMM-H004, based on the drug's physicochemical properties and transporter effects. Furthermore, the P-gp inhibitor PSC833 led to an increase in the K_{in} value of IMM-H004 by 4.3 times compared to the control, suggesting that P-gp extremely limited the penetration of IMM-H004 into the brain. On the other hand, when an OATP inhibitor was added to the perfusion solution, there was no significant change in the K_{in} value of IMM-H004G, demonstrating that the facilitating effect of OATP on the brain uptake of IMM-H004G in rats may be limited.

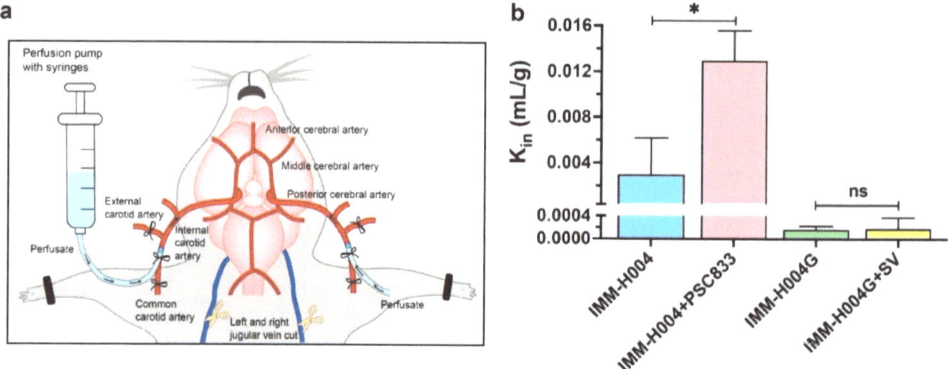

Figure 7. In situ brain perfusion of IMM-H004 and IMM-H004G in rats: (**a**) schematic diagram of in situ brain perfusion; (**b**) effect of P-gp and OATP inhibitors on the uptake characteristics of IMM-H004 and IMM-H004G using the in situ brain perfusion method. The concentration of IMM-H004 and IMM-H004G was 10 μmol/L. The concentrations of P-gp inhibitor PSC833 and the OATP inhibitor rifamycin SV were 5 μmol/L and 100 μmol/L, respectively. $n = 3$. * $p < 0.05$ vs. control group. ns, not significant.

3.7. Pharmacokinetics in Rats

To gain a deeper understanding of the origin and brain entry mechanism of IMM-H004G, separate intravenous injections of IMM-H004 and IMM-H004G were administered to rats. The microdialysis technique was employed to assess the levels of the drugs in the brain, which measures the free drug concentration (Figure 8a). The fu values of IMM-H004 and IMM-H004G in rat plasma (0.5–10 μg/mL) were determined as 0.025 ± 0.002 and 0.315 ± 0.008, respectively (Supplementary Information, Figure S3). Consequently, the total plasma concentration was converted to the free plasma concentration using the fu value, and the pharmacokinetic parameters were calculated based on the free plasma concentration.

As presented in Figure 8b, following the iv injection of IMM-H004G, both IMM-H004G and a trace amount of IMM-H004 were detected in the plasma and brain. The plasma concentrations of IMM-H004G and IMM-H004 at 2 min were 40,215 nmol/L and 39.8 nmol/L, respectively. The peak brain concentrations were 1712 nmol/L and 35.1 nmol/L, respectively. The AUC$_{brain}$ values were 1923 and 93.8 nmol/L·h, respectively. The higher concentration of IMM-H004G observed in the brain after intravenous injection indicates that IMM-H004G can penetrate the BBB and be distributed in the brain without depending on the in situ metabolism of IMM-H004.

After the iv injection of IMM-H004 citrate, The AUC$_{plasma}$ values for IMM-H004G and IMM-H004 were determined as 16,418 and 141 nmol/L·h, respectively. The AUC$_{brain}$ values were 1053 and 196 nmol/L·h, respectively. The Kp (AUC$_{brain}$/AUC$_{plasma}$) values for IMM-H004G and IMM-H004 were 0.06 and 1.39, respectively. Notably, the AUC$_{plasma}$ and AUC$_{brain}$ for IMM-H004G were 116 times and 5.4 times higher than those of IMM-H004, respectively.

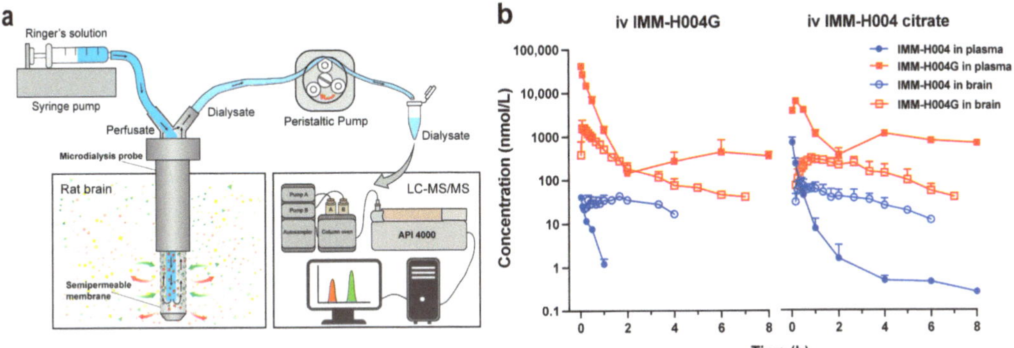

Figure 8. Pharmacokinetics of IMM-H004 and IMM-H004G in rats: (**a**) schematic diagram of the cerebral microdialysis technique; (**b**) free concentration–time profiles of IMM-H004 and IMM-H004G following intravenous injection of IMM-H004G and IMM-H004 citrate at a dose of 10 mg/kg ($n = 3$).

4. Discussion

Glucuronidation, catalyzed by UDP glucuronosyltransferases (UGTs), is the most important phase II metabolic pathway [35]. Although glucuronidation has traditionally been viewed as a drug detoxification mechanism, there is a growing recognition that certain glucuronidation products possess pharmacological effects, such as anti-inflammatory, antioxidant, and antitumor activities, serving as the basis for the efficacy of drugs [36–42]. Particularly, certain active glucuronide conjugates can also exert effects on the central nervous system (CNS). For instance, morphine-6-glucuronide (M6G) acts as a potent agonist of opioid receptors and is involved in the analgesic process of morphine [43]. Quercetin-3-O-β-D-glucuronide (Q3GA), a metabolite of orally administered quercetin in rats, can penetrate the blood–brain barrier, improve oxidative stress in the brain, and exert neuroprotective effects [44].

UGTs have been detected in the brains of humans and various animal species, including mice, rats, and macaques [45]. Some studies suggested that several drugs, such as morphine, resveratrol, and efavirenz, can generate glucuronide conjugates in the brain microsomes, tissue homogenates, or cells [46–48]. Although their activity in the brain is lower than in the liver or other extra hepatic organs, and UGTs have a minor impact on the overall glucuronidation of xenobiotics, they surely play a crucial role in maintaining the steady state of endogenous compounds [49,50]. It is speculated that UGTs can also have a major effect on the local efficacy of substances that enter the brain. Therefore, the primary purpose of this study was to explore the possibility of the in situ metabolism of IMM-H004 into IMM-H004G after entering brain tissue.

The location of UGT expression in brain tissue is controversial [3–5,32,49]. According to de Leon [51], UGTs are present in relatively low levels within brain tissue but appear to be more concentrated at brain capillaries, playing a role in the BBB. Ouzzine et al. [50] also reported that UGTs are primarily expressed in endothelial cells and astrocytes. Even in brain microvessel endothelial cells, there has been evidence of lamotrigine glucuronidation activity [52]. However, Shawahna et al. [33] did not observe any expression of UGT genes in human brain microvessels. The current consensus suggests that UGT activity is present in both neuronal and glial cells, with significantly higher activity in glial cells than in neurons. Our in vitro studies indicate the detection of UGT1A7 expression in human glial cells, with minimal expression observed in human neuronal cells. Furthermore, only the glial cells demonstrated the ability to glucuronidate IMM-H004. Similarly, UGT1A7 mRNA expression has been observed in rat astrocytes [53], and glucuronidation activity toward IMM-H004 has been detected in the rat brain, suggesting the consistent glucuronidation of IMM-H004 in both rat and human brains. To further uncover the glucuronidation activity in different brain regions, Asai et al. [54] compared the glucuronidation activity of β-estradiol, a UGT1A substrate, in microsomal proteins from various brain regions in rats. They observed that the hippocampus exhibited the highest activity, which was approximately twice that of the cerebrum. Consistently, following the icv injection of IMM-H004 in rats, IMM-H004G was only detected in the hippocampus, indicating the highest activity in the hippocampus and thus pointing to the limited glucuronidation of IMM-H004 in the brain.

Due to the tight junctions in the endothelial cells of the BBB, drugs in the vascular system cannot enter through the gaps between microvascular endothelial cells. They can only pass through the cell membrane. After undergoing glucuronidation, drugs have increased molecular weight and polarity, which theoretically poses a challenge for their entry into the brain. Based on the physicochemical properties, IMM-H004G is not favored for passive diffusion across the BBB; thus, it may require a transport system to achieve high concentrations in the brain.

Transporters presented on the BBB can be divided into two categories: ATP-binding cassette (ABC) superfamily transporters and solute carrier (SLC) superfamily transporters. ABC transporters, including P-gp, BCRP, and MRPs, mediate drug efflux transport, while SLC transporters, serving as bidirectional or uptake transporters, facilitate drug entry into the brain. The top four SLC transporters include glucose transporter 1 (GLUT1), L-type amino acid transporter 1 (LAT1), monocarboxylate transporters (MCTs), and nucleoside transporters (ENTs), which are primarily involved in the transportation of endogenous substances in the brain [55,56]. Following these, OATP2B1, OATP1A2, OCTs, and OCTNs participate in the brain transport of drugs such as nefazodone, zolmitriptan, lamotrigine, and memantine. There is limited knowledge about the brain transport of glucuronide conjugates. According to the literature, M6G is reported to be a substrate for GLUT-1 and Oatp2, while 17β-estradiol and its glucuronide metabolite, as well as edaravone glucuronide, are substrates for Oatp2 and MRP4, respectively [57–59].

Both IMM-H004 and IMM-H004G are amphoteric molecules, each containing a piperazine ring as the basic group and featuring a phenolic hydroxyl and carboxyl group as the acidic groups, respectively. The analysis using ADMET Predictor 8.5 software (Simulations Plus, Inc., Lancaster, CA, USA) showed that, under physiological conditions (pH = 7.4), IMM-H004 and IMM-H004G may mainly exist in cationic form and zwitterionic form, respectively (Figure 9). Correspondingly, inhibitor assays conducted in hCMEC/D3 cells revealed that the uptake of IMM-H004 and IMM-H004G is primarily mediated by the H$^+$/OC antiporter and OATPs, respectively. Other cationic transporters, such as OCT and OCTN, do not play a role in the transport of either compound. Additionally, P-gp is potentially involved in the efflux of IMM-H004. Although the glucose transporter GLUT has been reported to be involved in the transport of morphine-6-glucuronide [57], ourunpublished experiments using GLUT1-silenced hCMEC/D3 cells did not yield positive results.

Overexpressing cells are an effective tool for further validating the function and involvement of transporters. Regrettably, the sequence and encoding gene for the pyrilamine-sensitive H$^+$/OC antiporter protein have not been determined yet [60]. Consequently, it is currently impossible to construct an overexpression cell line to confirm its involvement in the uptake of IMM-H004. However, we validated the efflux effect of P-gp protein on IMM-H004 using the MDCKII-MDR1 cell transwell model.

OATP1A2 and OATP2B1 are considered to be the most abundant OATPs at the BBB. OATP1A2 is predominantly localized on the apical side of the brain capillary endothelial cells, while OATP2B1 is primarily distributed on the basolateral sides [61]. Although both OATP1A2 and OATP2B1 have been implicated in the transport of neuroactive steroids and various exogenous compounds including statins, antidiabetics, antihistamines, and central active drugs like triptans [61], there are still subtle differences between them. OATP1A2 exhibits broad substrate specificity for the BBB, while OATP2B1 has a relatively narrower substrate specificity [62].

Consistent with the findings from the OATP inhibitor assay in hCMEC/D3 cells, the uptake of IMM-H004G was significantly higher in cells overexpressing OATP1A2 and OATP2B1 compared to the control cells. Particularly noteworthy was the striking 190-fold increase in uptake observed in MDCKII-OATP2B1 cells compared to the control cells. Furthermore, we explored the involvement of other OATPs by using MDCKII cells overexpressing OATP1B1 and OATP1B3. Although the uptake of IMM-H004G by both cell lines was notably higher than that of the control cells, it did not exceed a tenfold increase (Supplementary Information, Figure S4). Taken together, these results indicate that IMM-H004G may have a significantly higher selectivity for OATP2B1 than other OATP subtypes.

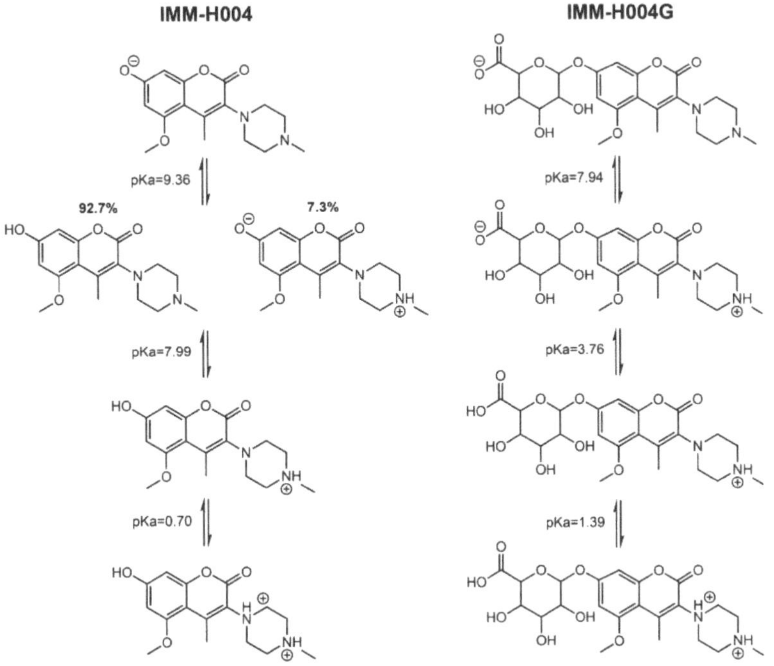

Figure 9. Software prediction of pKa values for IMM-H004 and IMM-H004G (2-column fitting image; color is not needed).

Drug permeability across the BBB is primarily governed by passive diffusion and carrier-mediated transport. Compared to cellular models, which are valuable tools for examining the impact of individual factors, animal models can offer a more comprehensive understanding of drug permeation. In situ brain perfusion is widely regarded as the gold standard method for measuring BBB permeability [63]. Therefore, we conducted in situ rat brain perfusion studies. Consistent with the uptake kinetic assay results using hCMEC/D3 cells (K_{ns} values), the K_{in} value of IMM-H004 was markedly higher than that of IMM-H004G, highlighting the higher brain entry rate of IMM-H004. The addition of the P-gp inhibitor PSC833 noticeably increased the K_{in} value of IMM-H004, underscoring the significant role of P-gp-mediated efflux in limiting the overall accumulation and extent of IMM-H004 penetration into the brain. Conversely, IMM-H004G, despite its slower entry rate, achieved a greater extent of brain exposure, which is attributed to the lesser effect of efflux transporters such as P-gp on its BBB permeability. However, the OATP inhibitor showed no substantial effect on the entry process of IMM-H004G into the brain.

In contrast to what is known for humans, very little is known about the substrate specificity and affinity of rodent orthologues of OATP2B1. Some researchers argue that species differences should be taken into consideration when studying OATP2B1 [64]. For example, Hussner et al. [65] reported that there are species differences in the recognition of OATP2B1 transporter for steroid sulfate conjugates between rats and humans. Since IMM-H004G exhibits significantly higher selectivity for OATP2B1 than for other OATPs, we speculate that species differences in OATP2B1 may be a key factor contributing to the negative result of the OATP inhibition assay in the rat brain perfusion. Humanized animal models may provide a feasible solution for addressing potential species differences.

Based on tissue cells containing β-glucuronidase activity that can hydrolyze glucuronide conjugates to produce aglycones, Terao et al. [66] proposed that glucuronide conjugates of quercetin serve as prodrugs in the vascular system, and their pharmacological effects are exerted through hydrolysis to release the parent drug. In the case of IMM-H004G, a glucuronide conjugate that exhibits higher brain exposure than its parent form with similar activity both in vitro and in vivo, the interconversion of IMM-H004 and IMM-H004G in the brain was investigated. Our findings demonstrate that IMM-H004G cannot be hydrolyzed back into IMM-H004 in either rat brain tissue homogenates or human brain cells, suggesting that IMM-H004G exerts pharmacological effects on the brain on its own.

5. Conclusions

The formation of IMM-H004G in the brain was confirmed in the present study. More importantly, although IMM-H004G had a slower entry rate, it achieved a greater extent of exposure in the brain. This is because its BBB permeability is less affected by efflux transporters such as P-gp, and it benefits from OATP uptake, enabling it to exert its independent pharmacological effects. Overall, these findings deepen our understanding of the metabolism and transport mechanisms underlying CNS drugs and provide valuable insights for the development of CNS drugs that undergo glucuronidation.

Supplementary Materials: The following supporting information can be downloaded at: https://www.mdpi.com/article/10.3390/pharmaceutics16030330/s1, Figure S1: Incubation of IMM-H004G (1 μmol/L) in rat brain homogenate (2 mg/mL protein) for up to 4 h (n = 3). Figure S2: In situ brain perfusion of IMM-H004 and IMM-H004G (2 μmol/L) in rats for up to 10 min (n = 3). Figure S3: Rat plasma protein binding and free fractions (fu values) of IMM-H004 and IMM-H004G (0.5–10 μg/mL, n = 3). Figure S4: Uptake of IMM-H004G in MDCKII-OATP1B1 and MDCKII-OATP1B3 cells.

Author Contributions: Conceptualization, J.J., L.L., Z.Z. and L.S.; methodology, J.J., L.L., Z.Z. and X.L.; software, L.L. and L.S.; validation, J.J., L.L. and Z.Z.; formal analysis, J.J., L.L., Z.Z. and L.S.; investigation, J.J., L.L., Z.Z. and X.L.; resources, L.S. and N.C.; data curation, J.J., L.L. and Z.Z.; writing—original draft preparation, L.L. and L.S.; writing—review and editing, L.S.; visualization, L.L. and L.S.; supervision, L.S. and Y.L.; project administration, L.S.; funding acquisition, L.S. and Y.L. All authors have read and agreed to the published version of the manuscript.

Funding: This research was funded by the National Natural Science Foundation of China (grant number 81603190), CAMS Innovation Fund for Medical Sciences (Grant number 2021-I2M-1-028), and the National Key R&D Program of China (Grant number 2022YFA0806400).

Institutional Review Board Statement: Animal experiments were approved by the Animal Care and Welfare Committee of Peking Union Medical College (approval code: 00007839).

Informed Consent Statement: Not applicable.

Data Availability Statement: Data will be available upon rational request.

Acknowledgments: We were very grateful to Jungui Dai (Institute of Materia Medica, Chinese Academy of Medical Sciences, and Peking Union Medical College, Beijing, China) for providing IMM-H004G powder.

Conflicts of Interest: The authors declare no conflicts of interest.

References

1. Feigin, V.L.; Brainin, M.; Norrving, B.; Martins, S.; Sacco, R.L.; Hacke, W.; Fisher, M.; Pandian, J.; Lindsay, P. World Stroke Organization (WSO): Global Stroke Fact Sheet 2022. *Int. J. Stroke* **2022**, *17*, 18–29. [CrossRef]
2. Jin, X.; Liu, J.; Liu, W. Early Ischemic Blood Brain Barrier Damage: A Potential Indicator for Hemorrhagic Transformation Following Tissue Plasminogen Activator (tPA) Thrombolysis? *Curr. Neurovascular Res.* **2014**, *11*, 254–262. [CrossRef]
3. Ai, Q.; Chen, C.; Chu, S.; Luo, Y.; Zhang, Z.; Zhang, S.; Yang, P.; Gao, Y.; Zhang, X.; Chen, N. IMM-H004 Protects against Cerebral Ischemia Injury and Cardiopulmonary Complications via CKLF1 Mediated Inflammation Pathway in Adult and Aged Rats. *Int. J. Mol. Sci.* **2019**, *20*, 1661. [CrossRef]
4. Ai, Q.D.; Chen, C.; Chu, S.; Zhang, Z.; Luo, Y.; Guan, F.; Lin, M.; Liu, D.; Wang, S.; Chen, N. IMM-H004 Therapy for Permanent Focal Ischemic Cerebral Injury via CKLF1/CCR4-Mediated NLRP3 Inflammasome Activation. *Transl. Res.* **2019**, *212*, 36–53. [CrossRef]
5. Zuo, W.; Zhang, W.; Han, N.; Chen, N.-H. Compound IMM-H004, a Novel Coumarin Derivative, Protects against CA1 Cell Loss and Spatial Learning Impairments Resulting from Transient Global Ischemia. *CNS Neurosci. Ther.* **2015**, *21*, 280–288. [CrossRef]
6. Yang, P.-F.; Song, X.-Y.; Zeng, T.; Ai, Q.-D.; Liu, D.-D.; Zuo, W.; Zhang, S.; Xia, C.-Y.; He, X.; Chen, N.-H. IMM-H004, a Coumarin Derivative, Attenuated Brain Ischemia/Reperfusion Injuries and Subsequent Inflammation in Spontaneously Hypertensive Rats through Inhibition of VCAM-1. *RSC Adv.* **2017**, *7*, 27480–27495. [CrossRef]
7. Zhang, Z.; Liu, D.; Jiang, J.; Song, X.; Zou, X.; Chu, S.; Xie, K.; Dai, J.; Chen, N.; Sheng, L.; et al. Metabolism of IMM-H004 and Its Pharmacokinetic-Pharmacodynamic Analysis in Cerebral Ischemia/Reperfusion Injured Rats. *Front. Pharmacol.* **2019**, *10*, 631. [CrossRef] [PubMed]
8. Jiang, J.; Zhang, Z.; Zou, X.; Wang, R.; Bai, J.; Zhao, S.; Fan, X.; Sheng, L.; Li, Y. Determination of IMM-H004 and Its Active Glucuronide Metabolite in Rat Plasma and Ringer's Solution by Ultra-Performance Liquid Chromatography-Tandem Mass Spectrometry. *J. Chromatogr. B Anal. Technol. Biomed. Life Sci.* **2018**, *1074–1075*, 16–24. [CrossRef] [PubMed]
9. Gosselet, F.; Loiola, R.A.; Roig, A.; Rosell, A.; Culot, M. Central Nervous System Delivery of Molecules across the Blood-Brain Barrier. *Neurochem. Int.* **2021**, *144*, 104952. [CrossRef] [PubMed]
10. Ghosh, C.; Puvenna, V.; Gonzalez-Martinez, J.; Janigro, D.; Marchi, N. Blood-Brain Barrier P450 Enzymes and Multidrug Transporters in Drug Resistance: A Synergistic Role in Neurological Diseases. *Curr. Drug Metab.* **2011**, *12*, 742–749. [CrossRef] [PubMed]
11. Li, X.; Hu, J.; Wang, B.; Sheng, L.; Liu, Z.; Yang, S.; Li, Y. Inhibitory Effects of Herbal Constituents on P-Glycoprotein in Vitro and in Vivo: Herb–Drug Interactions Mediated via P-Gp. *Toxicol. Appl. Pharmacol.* **2014**, *275*, 163–175. [CrossRef] [PubMed]
12. Higuchi, K.; Kitamura, A.; Okura, T.; Deguchi, Y. Memantine Transport by a Proton-Coupled Organic Cation Antiporter in hCMEC/D3 Cells, an in Vitro Human Blood-Brain Barrier Model. *Drug Metab. Pharmacokinet.* **2015**, *30*, 182–187. [CrossRef]
13. Shimomura, K.; Okura, T.; Kato, S.; Couraud, P.-O.; Schermann, J.-M.; Terasaki, T.; Deguchi, Y. Functional Expression of a Proton-Coupled Organic Cation (H+/OC) Antiporter in Human Brain Capillary Endothelial Cell Line hCMEC/D3, a Human Blood-Brain Barrier Model. *Fluids Barriers CNS* **2013**, *10*, 8. [CrossRef] [PubMed]
14. Wang, X.; Qi, B.; Su, H.; Li, J.; Sun, X.; He, Q.; Fu, Y.; Zhang, Z. Pyrilamine-Sensitive Proton-Coupled Organic Cation (H+/OC) Antiporter for Brain-Specific Drug Delivery. *J. Control. Release* **2017**, *254*, 34–43. [CrossRef]
15. El-Sheikh, A.A.K.; Greupink, R.; Wortelboer, H.M.; van den Heuvel, J.J.M.W.; Schreurs, M.; Koenderink, J.B.; Masereeuw, R.; Russel, F.G.M. Interaction of Immunosuppressive Drugs with Human Organic Anion Transporter (OAT) 1 and OAT3, and Multidrug Resistance-Associated Protein (MRP) 2 and MRP4. *Transl. Res.* **2013**, *162*, 398–409. [CrossRef]
16. Iwaki, M.; Shimada, H.; Irino, Y.; Take, M.; Egashira, S. Inhibition of Methotrexate Uptake via Organic Anion Transporters OAT1 and OAT3 by Glucuronides of Nonsteroidal Anti-Inflammatory Drugs. *Biol. Pharm. Bull.* **2017**, *40*, 926–931. [CrossRef] [PubMed]
17. Umehara, K.-I.; Iwatsubo, T.; Noguchi, K.; Usui, T.; Kamimura, H. Effect of Cationic Drugs on the Transporting Activity of Human and Rat OCT/Oct 1-3 in Vitro and Implications for Drug-Drug Interactions. *Xenobiotica* **2008**, *38*, 1203–1218. [CrossRef]
18. Ito, S.; Kusuhara, H.; Yokochi, M.; Toyoshima, J.; Inoue, K.; Yuasa, H.; Sugiyama, Y. Competitive Inhibition of the Luminal Efflux by Multidrug and Toxin Extrusions, but Not Basolateral Uptake by Organic Cation Transporter 2, Is the Likely Mechanism Underlying the Pharmacokinetic Drug-Drug Interactions Caused by Cimetidine in the Kidney. *J. Pharmacol. Exp. Ther.* **2012**, *340*, 393–403. [CrossRef]
19. Tamai, I.; Nakanishi, T.; Kobayashi, D.; China, K.; Kosugi, Y.; Nezu, J.; Sai, Y.; Tsuji, A. Involvement of OCTN1 (SLC22A4) in pH-Dependent Transport of Organic Cations. *Mol. Pharm.* **2004**, *1*, 57–66. [CrossRef]
20. Tamai, I.; Ohashi, R.; Nezu, J.; Yabuuchi, H.; Oku, A.; Shimane, M.; Sai, Y.; Tsuji, A. Molecular and Functional Identification of Sodium Ion-Dependent, High Affinity Human Carnitine Transporter OCTN2. *J. Biol. Chem.* **1998**, *273*, 20378–20382. [CrossRef]
21. Nishida, K.; Takeuchi, K.; Hosoda, A.; Sugano, S.; Morisaki, E.; Ohishi, A.; Nagasawa, K. Ergothioneine Ameliorates Oxaliplatin-Induced Peripheral Neuropathy in Rats. *Life Sci.* **2018**, *207*, 516–524. [CrossRef]
22. Wu, L.-X.; Guo, C.-X.; Qu, Q.; Yu, J.; Chen, W.-Q.; Wang, G.; Fan, L.; Li, Q.; Zhang, W.; Zhou, H.-H. Effects of Natural Products on the Function of Human Organic Anion Transporting Polypeptide 1B1. *Xenobiotica* **2012**, *42*, 339–348. [CrossRef]
23. Fattinger, K.; Cattori, V.; Hagenbuch, B.; Meier, P.J.; Stieger, B. Rifamycin SV and Rifampicin Exhibit Differential Inhibition of the Hepatic Rat Organic Anion Transporting Polypeptides, Oatp1 and Oatp2. *Hepatology* **2000**, *32*, 82–86. [CrossRef]

24. Ito, S.; Kusuhara, H.; Kuroiwa, Y.; Wu, C.; Moriyama, Y.; Inoue, K.; Kondo, T.; Yuasa, H.; Nakayama, H.; Horita, S.; et al. Potent and Specific Inhibition of mMate1-Mediated Efflux of Type I Organic Cations in the Liver and Kidney by Pyrimethamine. *J. Pharmacol. Exp. Ther.* **2010**, *333*, 341–350. [CrossRef]

25. Schinkel, A.H.; Wagenaar, E.; Mol, C.A.; van Deemter, L. P-Glycoprotein in the Blood-Brain Barrier of Mice Influences the Brain Penetration and Pharmacological Activity of Many Drugs. *J. Clin. Investig.* **1996**, *97*, 2517–2524. [CrossRef] [PubMed]

26. Shen, F.; Bailey, B.J.; Chu, S.; Bence, A.K.; Xue, X.; Erickson, P.; Safa, A.R.; Beck, W.T.; Erickson, L.C. Dynamic Assessment of Mitoxantrone Resistance and Modulation of Multidrug Resistance by Valspodar (PSC833) in Multidrug Resistance Human Cancer Cells. *J. Pharmacol. Exp. Ther.* **2009**, *330*, 423–429. [CrossRef] [PubMed]

27. Doyle, L.A.; Yang, W.; Abruzzo, L.V.; Krogmann, T.; Gao, Y.; Rishi, A.K.; Ross, D.D. A Multidrug Resistance Transporter from Human MCF-7 Breast Cancer Cells. *Proc. Natl. Acad. Sci. USA* **1998**, *95*, 15665–15670. [CrossRef] [PubMed]

28. Allen, J.D.; van Loevezijn, A.; Lakhai, J.M.; van der Valk, M.; van Tellingen, O.; Reid, G.; Schellens, J.H.M.; Koomen, G.-J.; Schinkel, A.H. Potent and Specific Inhibition of the Breast Cancer Resistance Protein Multidrug Transporter in Vitro and in Mouse Intestine by a Novel Analogue of Fumitremorgin C. *Mol. Cancer Ther.* **2002**, *1*, 417–425.

29. Yokooji, T. Role of ABC efflux transporters in the oral bioavailability and drug-induced intestinal toxicity. *Yakugaku Zasshi* **2013**, *133*, 815–822. [CrossRef]

30. Hara, Y.; Sassi, Y.; Guibert, C.; Gambaryan, N.; Dorfmüller, P.; Eddahibi, S.; Lompré, A.-M.; Humbert, M.; Hulot, J.-S. Inhibition of MRP4 Prevents and Reverses Pulmonary Hypertension in Mice. *J. Clin. Investig.* **2011**, *121*, 2888–2897. [CrossRef]

31. Takasato, Y.; Rapoport, S.I.; Smith, Q.R. An in Situ Brain Perfusion Technique to Study Cerebrovascular Transport in the Rat. *Am. J. Physiol.* **1984**, *247*, H484–H493. [CrossRef] [PubMed]

32. Zhang, Z.; Wu, X.; Zhao, M.; Yang, Y.; Wang, Y.; Hu, J.; Wang, B.; Sheng, L.; Li, Y. Determination of IMM-H004, a Novel Neuroprotective Agent, in Rat Plasma and Brain Tissue by Liquid Chromatography-Tandem Mass Spectrometry. *J. Chromatogr. B Anal. Technol. Biomed. Life Sci.* **2017**, *1048*, 49–55. [CrossRef] [PubMed]

33. Shawahna, R.; Uchida, Y.; Declèves, X.; Ohtsuki, S.; Yousif, S.; Dauchy, S.; Jacob, A.; Chassoux, F.; Daumas-Duport, C.; Couraud, P.-O.; et al. Transcriptomic and Quantitative Proteomic Analysis of Transporters and Drug Metabolizing Enzymes in Freshly Isolated Human Brain Microvessels. *Mol. Pharm.* **2011**, *8*, 1332–1341. [CrossRef] [PubMed]

34. Helms, H.C.; Abbott, N.J.; Burek, M.; Cecchelli, R.; Couraud, P.-O.; Deli, M.A.; Förster, C.; Galla, H.J.; Romero, I.A.; Shusta, E.V.; et al. In Vitro Models of the Blood-Brain Barrier: An Overview of Commonly Used Brain Endothelial Cell Culture Models and Guidelines for Their Use. *J. Cereb. Blood Flow Metab.* **2016**, *36*, 862–890. [CrossRef] [PubMed]

35. Jančová, P.; Šiller, M. Phase II Drug Metabolism. In *Topics on Drug Metabolism*; IntechOpen: London, UK, 2012; pp. 35–60. [CrossRef]

36. Cho, Y.-C.; Park, J.; Cho, S. Anti-Inflammatory and Anti-Oxidative Effects of Luteolin-7-O-Glucuronide in LPS-Stimulated Murine Macrophages through TAK1 Inhibition and Nrf2 Activation. *Int. J. Mol. Sci.* **2020**, *21*, 2007. [CrossRef] [PubMed]

37. Zhang, Y.; Song, T.T.; Cunnick, J.E.; Murphy, P.A.; Hendrich, S. Daidzein and Genistein Glucuronides in Vitro Are Weakly Estrogenic and Activate Human Natural Killer Cells at Nutritionally Relevant Concentrations. *J. Nutr.* **1999**, *129*, 399–405. [CrossRef] [PubMed]

38. Harada, M.; Kan, Y.; Naoki, H.; Fukui, Y.; Kageyama, N.; Nakai, M.; Miki, W.; Kiso, Y. Identification of the Major Antioxidative Metabolites in Biological Fluids of the Rat with Ingested (+)-Catechin and (-)-Epicatechin. *Biosci. Biotechnol. Biochem.* **1999**, *63*, 973–977. [CrossRef]

39. Ebner, T.; Wagner, K.; Wienen, W. Dabigatran Acylglucuronide, the Major Human Metabolite of Dabigatran: In Vitro Formation, Stability, and Pharmacological Activity. *Drug Metab. Dispos.* **2010**, *38*, 1567–1575. [CrossRef]

40. Ryder, T.F.; Calabrese, M.F.; Walker, G.S.; Cameron, K.O.; Reyes, A.R.; Borzilleri, K.A.; Delmore, J.; Miller, R.; Kurumbail, R.G.; Ward, J. Acyl Glucuronide Metabolites of 6-Chloro-5-[4-(1-Hydroxycyclobutyl) Phenyl]-1 H-Indole-3-Carboxylic Acid (PF-06409577) and Related Indole-3-Carboxylic Acid Derivatives Are Direct Activators of Adenosine Monophosphate-Activated Protein Kinase (AMPK). *J. Med. Chem.* **2018**, *61*, 7273–7288. [CrossRef]

41. Trescot, A.M.; Datta, S.; Lee, M.; Hansen, H. Opioid Pharmacology. *Pain Physician* **2008**, *11*, S133–S153. [CrossRef]

42. Yamamoto, M.; Jokura, H.; Hashizume, K.; Ominami, H.; Shibuya, Y.; Suzuki, A.; Hase, T.; Shimotoyodome, A. Hesperidin Metabolite Hesperetin-7-O-Glucuronide, but Not Hesperetin-3'-O-Glucuronide, Exerts Hypotensive, Vasodilatory, and Anti-Inflammatory Activities. *Food Funct.* **2013**, *4*, 1346–1351. [CrossRef]

43. Osborne, P.B.; Chieng, B.; Christie, M.J. Morphine-6β-Glucuronide Has a Higher Efficacy than Morphine as a Mu-Opioid Receptor Agonist in the Rat Locus Coeruleus. *Br. J. Pharmacol.* **2000**, *131*, 1422–1428. [CrossRef]

44. Moon, J.-H.; Tsushida, T.; Nakahara, K.; Terao, J. Identification of Quercetin 3-O-β-D-Glucuronide as an Antioxidative Metabolite in Rat Plasma after Oral Administration of Quercetin. *Free Radic. Biol. Med.* **2001**, *30*, 1274–1285. [CrossRef]

45. Wheeler, A.M.; Orsburn, B.C.; Bumpus, N.N. Biotransformation of Efavirenz and Proteomic Analysis of Cytochrome P450s and UDP-Glucuronosyltransferases in Mouse, Macaque, and Human Brain-Derived In Vitro Systems. *Drug Metab. Dispos.* **2023**, *51*, 521–531. [CrossRef]

46. Yamada, H.; Ishii, K.; Ishii, Y.; Ieiri, I.; Nishio, S.; Morioka, T.; Oguri, K. Formation of Highly Analgesic Morphine-6-Glucuronide Following Physiologic Concentration of Morphine in Human Brain. *J. Toxicol. Sci.* **2003**, *28*, 395–401. [CrossRef] [PubMed]

47. Sabolovic, N.; Heurtaux, T.; Humbert, A.-C.; Krisa, S.; Magdalou, J. Cis- and Trans-Resveratrol Are Glucuronidated in Rat Brain, Olfactory Mucosa and Cultured Astrocytes. *Pharmacology* **2007**, *80*, 185–192. [CrossRef] [PubMed]

48. Togna, A.R.; Antonilli, L.; Dovizio, M.; Salemme, A.; De Carolis, L.; Togna, G.I.; Patrignani, P.; Nencini, P. In Vitro Morphine Metabolism by Rat Microglia. *Neuropharmacology* **2013**, *75*, 391–398. [CrossRef]
49. Kutsuno, Y.; Hirashima, R.; Sakamoto, M.; Ushikubo, H.; Michimae, H.; Itoh, T.; Tukey, R.H.; Fujiwara, R. Expression of UDP-Glucuronosyltransferase 1 (UGT1) and Glucuronidation Activity toward Endogenous Substances in Humanized UGT1 Mouse Brain. *Drug Metab. Dispos.* **2015**, *43*, 1071–1076. [CrossRef]
50. Ouzzine, M.; Gulberti, S.; Ramalanjaona, N.; Magdalou, J.; Fournel-Gigleux, S. The UDP-Glucuronosyltransferases of the Blood-Brain Barrier: Their Role in Drug Metabolism and Detoxication. *Front. Cell Neurosci.* **2014**, *8*, 349. [CrossRef] [PubMed]
51. de Leon, J. Glucuronidation Enzymes, Genes and Psychiatry. *Int. J. Neuropsychopharmacol.* **2003**, *6*, 57–72. [CrossRef] [PubMed]
52. Ghosh, C.; Hossain, M.; Puvenna, V.; Martinez-Gonzalez, J.; Alexopolous, A.; Janigro, D.; Marchi, N. Expression and Functional Relevance of UGT1A4 in a Cohort of Human Drug-Resistant Epileptic Brains. *Epilepsia* **2013**, *54*, 1562–1570. [CrossRef]
53. Gradinaru, D.; Minn, A.-L.; Artur, Y.; Minn, A.; Heydel, J.-M. Effect of Oxidative Stress on UDP-Glucuronosyltransferases in Rat Astrocytes. *Toxicol. Lett.* **2012**, *213*, 316–324. [CrossRef]
54. Asai, Y.; Sakakibara, Y.; Onouchi, H.; Nadai, M.; Katoh, M. Characterization of β-Estradiol 3-Glucuronidation in Rat Brain. *Biol. Pharm. Bull.* **2017**, *40*, 1556–1560. [CrossRef]
55. Billington, S.; Salphati, L.; Hop, C.E.C.A.; Chu, X.; Evers, R.; Burdette, D.; Rowbottom, C.; Lai, Y.; Xiao, G.; Humphreys, W.G.; et al. Interindividual and Regional Variability in Drug Transporter Abundance at the Human Blood-Brain Barrier Measured by Quantitative Targeted Proteomics. *Clin. Pharmacol. Ther.* **2019**, *106*, 228–237. [CrossRef]
56. Bao, X.; Wu, J.; Xie, Y.; Kim, S.; Michelhaugh, S.; Jiang, J.; Mittal, S.; Sanai, N.; Li, J. Protein Expression and Functional Relevance of Efflux and Uptake Drug Transporters at the Blood-Brain Barrier of Human Brain and Glioblastoma. *Clin. Pharmacol. Ther.* **2020**, *107*, 1116–1127. [CrossRef]
57. Bourasset, F.; Cisternino, S.; Temsamani, J.; Scherrmann, J.-M. Evidence for an Active Transport of Morphine-6-Beta-d-Glucuronide but Not P-Glycoprotein-Mediated at the Blood-Brain Barrier. *J. Neurochem.* **2003**, *86*, 1564–1567. [CrossRef] [PubMed]
58. Sugiyama, D.; Kusuhara, H.; Shitara, Y.; Abe, T.; Meier, P.J.; Sekine, T.; Endou, H.; Suzuki, H.; Sugiyama, Y. Characterization of the Efflux Transport of 17β-Estradiol-d-17β-Glucuronide from the Brain across the Blood-Brain Barrier. *J. Pharmacol. Exp. Ther.* **2001**, *298*, 316–322. [PubMed]
59. Mizuno, N.; Takahashi, T.; Kusuhara, H.; Schuetz, J.D.; Niwa, T.; Sugiyama, Y. Evaluation of the Role of Breast Cancer Resistance Protein (BCRP/ABCG2) and Multidrug Resistance-Associated Protein 4 (MRP4/ABCC4) in the Urinary Excretion of Sulfate and Glucuronide Metabolites of Edaravone (MCI-186; 3-Methyl-1-Phenyl-2-Pyrazolin-5-One). *Drug Metab. Dispos.* **2007**, *35*, 2045–2052. [CrossRef] [PubMed]
60. Kurosawa, T.; Tega, Y.; Uchida, Y.; Higuchi, K.; Tabata, H.; Sumiyoshi, T.; Kubo, Y.; Terasaki, T.; Deguchi, Y. Proteomics-Based Transporter Identification by the PICK Method: Involvement of TM7SF3 and LHFPL6 in Proton-Coupled Organic Cation Antiport at the Blood-Brain Barrier. *Pharmaceutics* **2022**, *14*, 1683. [CrossRef] [PubMed]
61. Schäfer, A.M.; Meyer Zu Schwabedissen, H.E.; Bien-Möller, S.; Hubeny, A.; Vogelgesang, S.; Oswald, S.; Grube, M. OATP1A2 and OATP2B1 Are Interacting with Dopamine-Receptor Agonists and Antagonists. *Mol. Pharm.* **2020**, *17*, 1987–1995. [CrossRef] [PubMed]
62. Cheng, Z.; Liu, Q. Uptake Transport at the BBB—Examples and SAR. In *Blood-Brain Barrier in Drug Discovery: Optimizing Brain Exposure of CNS Drugs and Minimizing Brain Side Effects for Peripheral Drugs*; Wiley Online Library: Hoboken, NJ, USA, 2015; pp. 125–145. [CrossRef]
63. Di, L.; Kerns, E.H.; Bezar, I.F.; Petusky, S.L.; Huang, Y. Comparison of Blood-Brain Barrier Permeability Assays: In Situ Brain Perfusion, MDR1-MDCKII and PAMPA-BBB. *J. Pharm. Sci.* **2009**, *98*, 1980–1991. [CrossRef]
64. Kinzi, J.; Grube, M.; Zu Schwabedissen, H.E.M. OATP2B1–The Underrated Member of the Organic Anion Transporting Polypeptide Family of Drug Transporters? *Biochem. Pharmacol.* **2021**, *188*, 114534. [CrossRef] [PubMed]
65. Hussner, J.; Foletti, A.; Seibert, I.; Fuchs, A.; Schuler, E.; Malagnino, V.; Grube, M.; Meyer Zu Schwabedissen, H.E. Differences in Transport Function of the Human and Rat Orthologue of the Organic Anion Transporting Polypeptide 2B1 (OATP2B1). *Drug Metab. Pharmacokinet.* **2021**, *41*, 100418. [CrossRef] [PubMed]
66. Terao, J.; Murota, K.; Kawai, Y. Conjugated Quercetin Glucuronides as Bioactive Metabolites and Precursors of Aglycone in Vivo. *Food Funct.* **2011**, *2*, 11–17. [CrossRef] [PubMed]

pharmaceutics

Review

Uptake Transporters at the Blood–Brain Barrier and Their Role in Brain Drug Disposition

Md Masud Parvez [1], Armin Sadighi [1], Yeseul Ahn [2,3], Steve F. Keller [1] and Julius O. Enoru [1,*]

[1] Department of Quantitative, Translational & ADME Sciences (QTAS), AbbVie Biotherapeutics, San Francisco, CA 94080, USA; mdmasud.parvez@abbvie.com (M.M.P.)
[2] Department of Pharmaceutical Sciences, Jerry H. Hodge School of Pharmacy, Texas Tech University Health Sciences Center, 1300 S Coulter St., Amarillo, TX 79106, USA
[3] Center for Blood-Brain Barrier Research, Jerry H. Hodge School of Pharmacy, Texas Tech University Health Sciences Center, Amarillo, TX 79106, USA
* Correspondence: julius.enoru@abbvie.com

Abstract: Uptake drug transporters play a significant role in the pharmacokinetic of drugs within the brain, facilitating their entry into the central nervous system (CNS). Understanding brain drug disposition is always challenging, especially with respect to preclinical to clinical translation. These transporters are members of the solute carrier (SLC) superfamily, which includes organic anion transporter polypeptides (OATPs), organic anion transporters (OATs), organic cation transporters (OCTs), and amino acid transporters. In this systematic review, we provide an overview of the current knowledge of uptake drug transporters in the brain and their contribution to drug disposition. Here, we also assemble currently available proteomics-based expression levels of uptake transporters in the human brain and their application in translational drug development. Proteomics data suggest that in association with efflux transporters, uptake drug transporters present at the BBB play a significant role in brain drug disposition. It is noteworthy that a significant level of species differences in uptake drug transporters activity exists, and this may contribute toward a disconnect in inter-species scaling. Taken together, uptake drug transporters at the BBB could play a significant role in pharmacokinetics (PK) and pharmacodynamics (PD). Continuous research is crucial for advancing our understanding of active uptake across the BBB.

Keywords: uptake transporters; blood–brain barrier; pharmacokinetics; CNS drug delivery

Citation: Parvez, M.M.; Sadighi, A.; Ahn, Y.; Keller, S.F.; Enoru, J.O. Uptake Transporters at the Blood–Brain Barrier and Their Role in Brain Drug Disposition. *Pharmaceutics* **2023**, *15*, 2473. https://doi.org/10.3390/pharmaceutics15102473

Academic Editors: Gert Fricker and Elena Puris

Received: 13 September 2023
Revised: 3 October 2023
Accepted: 5 October 2023
Published: 16 October 2023

1. Introduction

Membrane transporters are expressed in several organs and play a significant role in pharmacokinetic (PK) drug disposition, pharmacodynamics (PD), and clinical drug–drug interactions [1]. The blood–brain barrier (BBB) always challenges the permeation of drugs into the brain, especially central nervous system (CNS)-targeted therapeutics [2,3]. Properties like unbound tissue partition coefficient (Kp,u) or, more specifically, the unbound brain-to-plasma drug concentration ratio ($K_{p,uu,brain}$) of a drug or chemical could be key determinant factors in drug penetration via BBB [4]. It is well established that drug transporters present at BBB could modulate drug delivery into the brain. To date, several in vitro, in vivo, and clinical studies have been conducted primarily to understand the role of efflux transporters in brain drug exposure. For example, brain exposure of raltegravir, an antiretroviral drug, was found to be strongly associated with p-glycoprotein (P-gp) and breast cancer-resistant protein (BCRP), with P-gp inhibitor PSC833 and BCRP inhibitor Ko143 significantly increasing raltegravir accumulation in human cerebral microvessel endothelial (hCMEC/D3) and mouse Sertoli TM4 cells [5]. On the contrary, similar investigations of the role of uptake transporters in brain drug disposition are poorly reported. However, a number of reports have demonstrated a solute carrier (SLC) family transporter expression in human microvessels, namely organic anion transporter polypeptide (OATP)-1B1, -1B3,

-2B1, -1A2, organic cation transporter (OCT), organic anion transporter (OAT), equilibrative nucleoside transporters (ENTs), concentrative nucleoside transporter (CNT), monocarboxylate transporters (MCTs), L-type amino acid transporter (LAT), and multidrug and toxin extrusion transporters (MATEs) [6–10]. There are also examples of clinical drugs that are well-known substrates of these uptake transporters, such as erythromycin, fexofenadine, imatinib, levofloxacin, methotrexate, pitavastatin, saquinavir for OATP1A2 [11–16], and atorvastatin, benzylpenicillin, bosentan, fexofenadine, glibenclamide, and rosuvastatin that are substrates of OATP2B1 [17–22]. Several studies using these compounds have suggested their potential contribution in PK and drug–drug interactions (DDIs), leading to speculations regarding their brain distribution as well. For example, OATP1A2-mediated DDIs were demonstrated using fexofenadine as a substrate for this transporter and the exposure (AUC) of fexofenadine was reduced by 25% and 40–70% due to flavonoids naringin and grapefruit juice (OATP1A2 inhibitor), respectively [23–25]. OATP1A2 has also been suggested to play a significant role in male hormone dehydroepiandrosterone sulfate (DHEAS) uptake into the brain and liver [23]. The coadministration of fexofenadine (a known substrate of P-gp), with terfenadine, increased brain penetration in mice by 25–27-fold, indicating that transporter-mediated disposition could be a key mechanism for fexofenadine brain exposure [26]. Despite the examples available in the literature, there is still a gap in quantifying the relative contribution of individual uptake transporters and their role in CNS exposure, efficacy, and safety. It should be noted that the clinical relevance of these uptake transporters is not limited to CNS-targeted therapeutic drugs. These uptake transporters also play a significant role in the PK and PD of non-CNS-targeted drugs such as statins [26–29].

Thus, understanding the mechanism and extent of drug penetration into the brain remains an unmet need in drug discovery and development. In this review, we summarize the current knowledge of the tools available, protein abundance, and clinical drug substrates of uptake transporters to facilitate our understanding and propose a roadmap on the significance of uptake transporters' contribution in brain drug disposition.

2. Study Highlights

We performed a comprehensive systematic review of publicly available literature data in the following areas: expression and tissue localization of uptake transporters in human, in vitro study models (cell line and rodent), in silico models, species differences, clinical PK/PD, and brain toxicity of therapeutic drugs. To capture available data, a through data mining was performed in the public domain (PubMed, Google Scholar, University of Washington DIDBW, Washington, DC, USA) by using key words (brain drug disposition, transporters in brain, transporter proteomics in brain, inter-species differences in brain drug disposition, drug-induced brain toxicity, pharmacokinetics in brain, blood–brain barrier, drug–drug interactions in brain). We included the reported studies in this review, considering study details, data quality, the concomitant use of the target drugs, pathophysiological conditions, protein quantification, and plausible data interpretation with minimal limitations.

3. Localization, Functions, and Expression of the Drug Uptake Transporters in the Brain

Results from our literature review showed that a large variety of uptake transporters are localized and significantly expressed at the BBB across species (Figure 1). There are very few studies that have demonstrated in vivo brain exposures in humans due to the difficulty in obtaining samples and a lack of in vitro-to-in vivo extrapolation (IVIVE) tools. Following recent advances in LC-MS/MS technology, the quantification of uptake transporter protein abundance is currently performed using LC-MS/MS-based proteomics. This enables translation via scaling or PBPK modeling and allows for estimations of their relative contribution in brain drug exposure. The human OATP family consists of 11 members, of which OATP1A2, 1B1, 1B3, and 2B1 play an important role in drug disposition and pharmacokinetics. Among these, OATP1A2 and OAT2B1 are highly expressed at the BBB and play a significant role in drug uptake into the brain. These OATP isoforms, namely OATP1A2, Oatp1a4, and OATP2B1 are expressed at the endothelial membrane, through which substrates enter the brain [30]. On the other hand, P-gp, BCRP, MRP1, MRP4, MRP5, and MRP2 are expressed at the luminal side that pumps substances out from the intracellular space [30]. Other uptake family transporters such as OCT1, OCT2, and OCTN2 are mainly expressed on the luminal side of brain microvessel endothelial cells and play a role in substrates' uptake into the brain [31]. OCTN2 and OCTN3 were reported to be expressed in rodent cell lines that uptake substrates into the brain that align with membrane potential and the proton gradient. However, these transporters need further investigation to confirm their localization in microvessels and choroid plexus epithelial cells in order to understand their contribution in CNS therapeutics [32–34]. MCT transporters are widely expressed in rat, mouse, or human brain endothelial cells, ependymocytes, and astrocytes, playing an important role in uptake into the brain [35]. MCTs facilitate lactate and monocarboxylates transport and cellular metabolism in a proton-dependent manner [36]. Apart from endogenous substances, MCTs also transport therapeutic drugs like atorvastatin and valproic acid [37]. Another uptake transporter at the BBB is LAT1, which is localized in both the apical and basolateral membrane of the brain capillary endothelial cells [38]. LAT1 is composed of a total 15 members and categorized into two subgroups: cationic amino acid transporters and LAT heterodimeric amino acid transporters [38]. LAT1 forms a heterodimeric amino acid transporter interacting with the glycoprotein CD98 to exert an uptake of a broad range of amino acids (tryptophan, phenylalanine, leucine, and histidine), prodrugs, and thyroid hormones T3 and T4 [39–41].

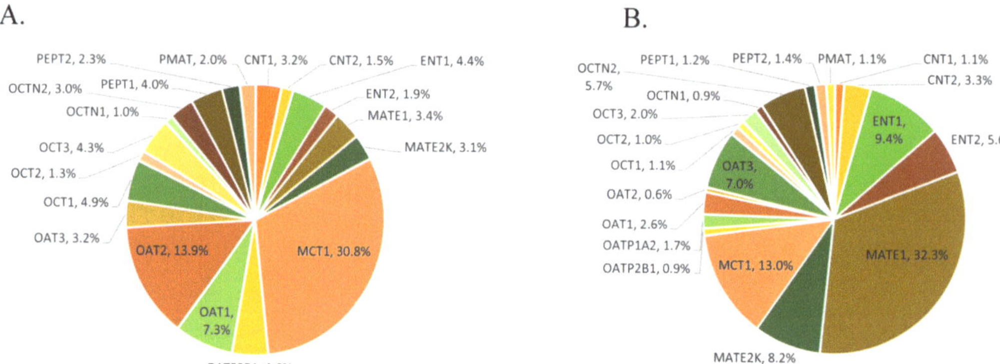

Figure 1. Uptake transporter proteomics in human brain cells. A summary of the literature reported proteomics-based quantitative abundance of clinically relevant major drug uptake transporters in human brain microvessels (**A**) and choroid plexus (**B**). Data represent an average percent of transporter protein absolute abundance (pmol/mg protein), summarized in Table 1. CNTs: concentrative nucleoside transporters; ENTs: equilibrative nucleoside transporters; MCTs: monocarboxylate transporters; OATPs: organic anion transporter polypeptides; PMAT: plasma membrane monoamine transporter; LATs: L-type amino acid transporters; OATs: organic anion transporters; OCTs: organic cation transporters; MATE: multidrug and toxic compound extrusion; PEPTs: peptide transporters. Graphical illustrations were made using Microsoft excel and Adobe IllustratorCC.

Table 1. Uptake transporters proteomics in human blood–brain barrier.

Transporter	Gene Name	Method	Protein Expression Level		Unit	Tissue/Cell	Reference
			Value	SD			
OCT1	SLC22A1	Targeted LC-MS/MS	ULQ < 0.289		fmol/µg protein	Choroid plexus (Plasma membrane fraction)	[9]
OCT1	SLC22A1	Targeted LC-MS/MS	ULQ < 0.288		fmol/µg protein	Brain microvessels	[8]
OCT1	SLC22A1	Targeted LC-MS/MS	0.58	0.11	pmol/mg protein	Brain microvessels	[6]
OCT1	SLC22A1	Targeted LC-MS/MS	0.54	0.06	pmol/mg protein	Brain microvessels	[6]
OCT2	SLC22A2	Targeted LC-MS/MS	ULQ < 0.254		fmol/µg protein	Choroid plexus (Plasma membrane fraction)	[9]
OCT2	SLC22A2	Targeted LC-MS/MS	ULQ < 0.123		fmol/µg protein	Brain microvessels	[8]
OCT3	SLC22A3	Targeted LC-MS/MS	ULQ < 0.534		fmol/µg protein	Choroid plexus (Plasma membrane fraction)	[9]
OCT3	SLC22A3	Targeted LC-MS/MS	ULQ < 0.207		fmol/µg protein	Brain microvessels	[8]
OCT3	SLC22A3	Targeted LC-MS/MS	0.62	0.08	pmol/mg protein	Brain microvessels	[6]
4F2hc	SLC3A2	Targeted LC-MS/MS	1.42	0.28	fmol/µg protein	Choroid plexus (Plasma membrane fraction)	[9]
4F2hc	SLC3A2	Targeted LC-MS/MS	3.47	0.83	fmol/µg protein	Brain microvessels	[8]
4F2hc	SLC3A2	Targeted LC-MS/MS	1.9	0.23	fmol/µg protein	hCMEC/D3 (plasma membrane fraction)	[7]
ASBT	SLC10A2	Targeted LC-MS/MS	ULQ < 0.288		fmol/µg protein	Choroid plexus (Plasma membrane fraction)	[9]
ASBT	SLC10A2	Targeted LC-MS/MS	ULQ < 0.12		fmol/µg protein	Brain microvessels	[8]
ASCT1	SLC1A4	Targeted LC-MS/MS	ULQ < 0.331		fmol/µg protein	Choroid plexus (Plasma membrane fraction)	[9]
ASCT2	SLC1A5	Targeted LC-MS/MS	ULQ < 1.38		fmol/µg protein	Choroid plexus (Plasma membrane fraction)	[9]
ASCT2	SLC1A5	Targeted LC-MS/MS	ULQ < 0.142		fmol/µg protein	Brain microvessels	[8]
ATA1	SLC38A1	Targeted LC-MS/MS	ULQ < 1.28		fmol/µg protein	Choroid plexus (Plasma membrane fraction)	[9]
ATA1	SLC38A1	Targeted LC-MS/MS	ULQ < 0.175		fmol/µg protein	Brain microvessels	[8]
ATA1	SLC38A1	Targeted LC-MS/MS	1.57	0.06	fmol/µg protein	hCMEC/D3 (plasma membrane fraction)	[7]
ATA2	SLC38A2	Targeted LC-MS/MS	ULQ < 0.497		fmol/µg protein	Choroid plexus (Plasma membrane fraction)	[9]
ATA2	SLC38A2	Targeted LC-MS/MS	ULQ < 0.143		fmol/µg protein	Brain microvessels	[8]

Table 1. *Cont.*

Transporter	Gene Name	Method	Protein Expression Level		Unit	Tissue/Cell	Reference
			Value	SD			
ATA3	*SLC38A4*	Targeted LC-MS/MS	ULQ < 0.823		fmol/μg protein	Choroid plexus (Plasma membrane fraction)	[9]
ATA3	*SLC38A4*	Targeted LC-MS/MS	ULQ < 0.0656		fmol/μg protein	Brain microvessels	[8]
BGT1	*SLC6A12*	Targeted LC-MS/MS	ULQ < 1.95		fmol/μg protein	Choroid plexus (Plasma membrane fraction)	[9]
BGT1	*SLC6A12*	Targeted LC-MS/MS	3.16	0.94	fmol/μg protein	Brain microvessels	[8]
BOCT	*SLC22A17*	Targeted LC-MS/MS	ULQ < 0.265		fmol/μg protein	Choroid plexus (Plasma membrane fraction)	[9]
BOIT	*SLC22A17*	Targeted LC-MS/MS	ULQ < 0.503		fmol/μg protein	Brain microvessels	[8]
CAT1	*SLC7A1*	Targeted LC-MS/MS	1.13	0.18	fmol/μg protein	Brain microvessels	[8]
CAT1	*SLC7A1*	Targeted LC-MS/MS	1.22	0.15	fmol/μg protein	Choroid plexus (Plasma membrane fraction)	[9]
CNT1	*SLC28A2*	Targeted LC-MS/MS	ULQ < 0.297		fmol/μg protein	Choroid plexus (Plasma membrane fraction)	[9]
CNT1	*SLC28A1*	Targeted LC-MS/MS	ULQ < 0.308		fmol/μg protein	Brain microvessels	[8]
CNT2	*SLC28A2*	Targeted LC-MS/MS	ULQ < 0.867		fmol/μg protein	Choroid plexus (Plasma membrane fraction)	[9]
CNT2	*SLC28A2*	Targeted LC-MS/MS	ULQ < 0.141		fmol/μg protein	Brain microvessels	[8]
CNT3	*SLC6A8*	Targeted LC-MS/MS	ULQ < 0.35		fmol/μg protein	Choroid plexus (Plasma membrane fraction)	[9]
CNT3	*SLC28A3*	Targeted LC-MS/MS	ULQ < 0.552		fmol/μg protein	Brain microvessels	[8]
CRT1	*SLC6A8*	Targeted LC-MS/MS	ULQ < 0.0915		fmol/μg protein	Brain microvessels	[8]
CT2	*SLC22A16*	Targeted LC-MS/MS	ULQ < 0.122		fmol/μg protein	Brain microvessels	[8]
CTL1	*SLC44A1*	Targeted LC-MS/MS	ULQ < 0.293		fmol/μg protein	Choroid plexus (Plasma membrane fraction)	[9]
CTL2	*SLC44A2*	Targeted LC-MS/MS	ULQ < 0.383		fmol/μg protein	Choroid plexus (Plasma membrane fraction)	[9]
EAAT1	*SLC1A3*	Targeted LC-MS/MS	5.04	0.18	fmol/μg protein	Choroid plexus (Plasma membrane fraction)	[9]
EAAT3	*SLC1A1*	Targeted LC-MS/MS	ULQ < 0.256		fmol/μg protein	Brain microvessels	[8]
EEAT1	*SLC1A3*	Targeted LC-MS/MS	25.4	12.5	fmol/μg protein	Brain microvessels	[8]
ENT1	*SLC29A1*	Targeted LC-MS/MS	0.568	0.134	fmol/μg protein	Brain microvessels	[8]
ENT1	*SLC29A1*	Targeted LC-MS/MS	5.94	0.35	fmol/μg protein	hCMEC/D3 (Plasma membrane fraction)	[7]

Table 1. *Cont.*

Transporter	Gene Name	Method	Protein Expression Level		Unit	Tissue/Cell	Reference
			Value	SD			
ENT1	SLC29A1	Targeted LC-MS/MS	0.27	0.1	pmol/mg protein	Brain microvessels	[6]
ENT1	SLC29A1	Targeted LC-MS/MS	2.49	0.12	fmol/μg protein	Choroid plexus (Plasma membrane fraction)	[9]
ENT2	SLC29A2	Targeted LC-MS/MS	ULQ < 1.49		fmol/μg protein	Choroid plexus (Plasma membrane fraction)	[9]
ENT2	SLC29A2	Targeted LC-MS/MS	ULQ < 0.18		fmol/μg protein	Brain microvessels	[8]
FATP1	SLC27A1	Targeted LC-MS/MS	ULQ < 1.04		fmol/μg protein	Choroid plexus (Plasma membrane fraction)	[9]
FATP2	SLC27A2	Targeted LC-MS/MS	ULQ < 0.199		fmol/μg protein	Choroid plexus (Plasma membrane fraction)	[9]
FATP3	SLC27A3	Targeted LC-MS/MS	ULQ < 0.44		fmol/μg protein	Choroid plexus (Plasma membrane fraction)	[9]
FLIPT1	SLC22A15	Targeted LC-MS/MS	ULQ < 0.245		fmol/μg protein	Choroid plexus (Plasma membrane fraction)	[9]
FLIPT1	SLC22A15	Targeted LC-MS/MS	ULQ < 0.101		fmol/μg protein	Brain microvessels	[8]
GAT2	SLC6A13	Targeted LC-MS/MS	ULQ < 1.3		fmol/μg protein	Choroid plexus (Plasma membrane fraction)	[9]
GAT2	SLC6A13	Targeted LC-MS/MS	ULQ < 0.374		fmol/μg protein	Brain microvessels	[8]
GLUT2	SLC2A2	Targeted LC-MS/MS	ULQ < 5.85		fmol/μg protein	Choroid plexus (Plasma membrane fraction)	[9]
GLUT4	SLC2A4	Targeted LC-MS/MS	ULQ < 2.52		fmol/μg protein	Choroid plexus (Plasma membrane fraction)	[9]
GLUT4	SLC2A4	Targeted LC-MS/MS	ULQ < 0.136		fmol/μg protein	Brain microvessels	[8]
LAT1	SLC7A5	Targeted LC-MS/MS	ULQ < 0.76		fmol/μg protein	Choroid plexus (Plasma membrane fraction)	[9]
LAT1	SLC7A5	Targeted LC-MS/MS	0.431	0.091	fmol/μg protein	Brain microvessels	[8]
LAT1	SLC7A5	Targeted LC-MS/MS	0.59	0.15	pmol/mg protein	Brain microvessels	[6]
LAT1	SLC7A5	Targeted LC-MS/MS	0.71	0.25	pmol/mg protein	Brain microvessels	[6]
LAT2	SLC7A6	Targeted LC-MS/MS	ULQ < 0.059		fmol/μg protein	Brain microvessels	[8]
LAT2	SLC7A6	Targeted LC-MS/MS	ULQ < 2.08		fmol/μg protein	Choroid plexus (Plasma membrane fraction)	[9]
MATE1	SLC47A1	Targeted LC-MS/MS	ULQ < 0.33		fmol/μg protein	Brain microvessels	[8]
MATE1	SLC47A2	Targeted LC-MS/MS	8.61	0.63	fmol/μg protein	Choroid plexus (Plasma membrane fraction)	[9]

Table 1. *Cont.*

Transporter	Gene Name	Method	Protein Expression Level		Unit	Tissue/Cell	Reference
			Value	SD			
MATE2-k	*SLC47A2*	Targeted LC-MS/MS	ULQ < 2.19		fmol/µg protein	Choroid plexus (Plasma membrane fraction)	[9]
MATE2-K	*SLC47A2*	Targeted LC-MS/MS	ULQ < 0.295		fmol/µg protein	Brain microvessels	[8]
MCT1	*SLC16A1*	Targeted LC-MS/MS	2.27	0.85	fmol/µg protein	Brain microvessels	[8]
MCT1	*SLC16A1*	Targeted LC-MS/MS	1.87	0.22	fmol/µg protein	hCMEC/D3 (Plasma membrane fraction)	[7]
MCT1	*SLC16A1*	Targeted LC-MS/MS	5.37	3.73	pmol/mg protein	Brain microvessels	[6]
MCT1	*SLC16A1*	Targeted LC-MS/MS	3.47	0.26	fmol/µg protein	Choroid plexus (Plasma membrane fraction)	[9]
MCT10	*SLC16A10*	Targeted LC-MS/MS	ULQ < 2.6		fmol/µg protein	Choroid plexus (Plasma membrane fraction)	[9]
MCT2	*SLC16A7*	Targeted LC-MS/MS	ULQ < 0.671		fmol/µg protein	Choroid plexus (Plasma membrane fraction)	[9]
MCT2	*SLC16A7*	Targeted LC-MS/MS	ULQ < 0.277		fmol/µg protein	Brain microvessels	[8]
MCT3	*SLC16A3*	Targeted LC-MS/MS	ULQ < 0.921		fmol/µg protein	Choroid plexus (Plasma membrane fraction)	[9]
MCT4	*SLC16A4*	Targeted LC-MS/MS	0.382	0.078	fmol/µg protein	Choroid plexus (Plasma membrane fraction)	[9]
MCT5	*SLC16A5*	Targeted LC-MS/MS	0.685	0.124	fmol/µg protein	Choroid plexus (Plasma membrane fraction)	[9]
MCT8	*SLC16A2*	Targeted LC-MS/MS	1.65	0.16	fmol/µg protein	Choroid plexus (Plasma membrane fraction)	[9]
MRP4	*ABCC4*	Targeted LC-MS/MS	0.818	0.14	fmol/µg protein	Choroid plexus (Plasma membrane fraction)	[9]
NET	*SLC6A2*	Targeted LC-MS/MS	ULQ < 0.361		fmol/µg protein	Choroid plexus (Plasma membrane fraction)	[9]
NET	*SLC6A2*	Targeted LC-MS/MS	ULQ < 0.441		fmol/µg protein	Brain microvessels	[8]
NTCP	*SLC10A1*	Targeted LC-MS/MS	ULQ < 0.771		fmol/µg protein	Choroid plexus (Plasma membrane fraction)	[9]
NTCP	*SLC10A1*	Targeted LC-MS/MS	ULQ < 0.454		fmol/µg protein	Brain microvessels	[8]
OAT1	*SLC22A6*	Targeted LC-MS/MS	ULQ < 0.687		fmol/µg protein	Choroid plexus (Plasma membrane fraction)	[9]
OAT1	*SLC22A6*	Targeted LC-MS/MS	ULQ < 0.909		fmol/µg protein	Brain microvessels	[8]
OAT1	*SLC22A6*	Targeted LC-MS/MS	0.48	0.11	pmol/mg protein	Brain microvessels	[6]

Table 1. *Cont.*

Transporter	Gene Name	Method	Protein Expression Level		Unit	Tissue/Cell	Reference
			Value	SD			
OAT2	SLC22A7	Targeted LC-MS/MS	ULQ < 0.152		fmol/μg protein	Choroid plexus (Plasma membrane fraction)	[9]
OAT2	SLC22A7	Targeted LC-MS/MS	ULQ < 0.153		fmol/μg protein	Brain microvessels	[8]
OAT2	SLC22A7	Targeted LC-MS/MS	7.9	3.8	pmol/mg protein	Brain microvessels	[6]
OAT3	SLC22A8	Targeted LC-MS/MS	1.87	0.12	fmol/μg protein	Choroid plexus (Plasma membrane fraction)	[9]
OAT3	SLC22A8	Targeted LC-MS/MS	ULQ < 0.348		fmol/μg protein	Brain microvessels	[8]
OAT3	SLC22A8	Targeted LC-MS/MS	0.27	0.03	pmol/mg protein	Brain microvessels	[6]
OAT4	SLC22A11	Targeted LC-MS/MS	ULQ < 0.534		fmol/μg protein	Choroid plexus (Plasma membrane fraction)	[9]
OAT4	SLC22A11	Targeted LC-MS/MS	ULQ < 0.243		fmol/μg protein	Brain microvessels	[8]
OAT5	SLC22A10	Targeted LC-MS/MS	ULQ < 3.27		fmol/μg protein	Choroid plexus (Plasma membrane fraction)	[9]
OAT5	SLC22A10	Targeted LC-MS/MS	ULQ < 0.0898		fmol/μg protein	Brain microvessels	[8]
OAT7	SLC22A9	Targeted LC-MS/MS	0.51	0.1	pmol/mg protein	Brain microvessels	[6]
OATP1	SLCO	Targeted LC-MS/MS	0.54	0.1	pmol/mg protein	Brain microvessels	[6]
OATP1A2	SLCO1A2	Targeted LC-MS/MS	ULQ < 0.452		fmol/μg protein	Choroid plexus (Plasma membrane fraction)	[9]
OATP1B1	SLCO1B1	Targeted LC-MS/MS	ULQ < 0.303		fmol/μg protein	Choroid plexus (Plasma membrane fraction)	[9]
OATP1B3	SLCO1B3	Targeted LC-MS/MS	ULQ < 0.619		fmol/μg protein	Choroid plexus (Plasma membrane fraction)	[9]
OATP1C1	SLCO1C1	Targeted LC-MS/MS	ULQ < 0.156		fmol/μg protein	Choroid plexus (Plasma membrane fraction)	[9]
OATP1C1	SLCO1C1	Targeted LC-MS/MS	0.27	0.03	pmol/mg protein	Brain microvessels	[6]
OATP2B1	SLCO2B1	Targeted LC-MS/MS	ULQ < 0.237		fmol/μg protein	Choroid plexus (Plasma membrane fraction)	[9]
OATP2B1	SLCO2B1	Targeted LC-MS/MS	0.4	0.04	pmol/mg protein	Brain microvessels	[6]
OATP2B1	SLCO2B1	Targeted LC-MS/MS	0.48	0.11	pmol/mg protein	Brain microvessels	[6]
OATP3A1	SLCO3A1	Targeted LC-MS/MS	0.641	12	fmol/μg protein	Choroid plexus (Plasma membrane fraction)	[9]

Table 1. *Cont.*

Transporter	Gene Name	Method	Protein Expression Level		Unit	Tissue/Cell	Reference
			Value	SD			
OATP4A1	SLCO4A1	Targeted LC-MS/MS	ULQ < 1.2		fmol/μg protein	Choroid plexus (Plasma membrane fraction)	[9]
OATP4C1	SLCO4C1	Targeted LC-MS/MS	ULQ < 0.283		fmol/μg protein	Choroid plexus (Plasma membrane fraction)	[9]
OATP5A1	SLCO5A1	Targeted LC-MS/MS	ULQ < 3.28		fmol/μg protein	Choroid plexus (Plasma membrane fraction)	[9]
OATP6A1	SLCO6A1	Targeted LC-MS/MS	ULQ < 0.545		fmol/μg protein	Choroid plexus (Plasma membrane fraction)	[9]
OATP8	SLCO1B3	Targeted LC-MS/MS	0.46	0.15	pmol/mg protein	Brain microvessels	[6]
OATP-8	SLCO1B3	Targeted LC-MS/MS	ULQ < 0.572		fmol/μg protein	Brain microvessels	[8]
OATP-A	SLCO1A2	Targeted LC-MS/MS	ULQ < 0.695		fmol/μg protein	Brain microvessels	[8]
OATP-B	SLCO2B1	Targeted LC-MS/MS	ULQ < 0.337		fmol/μg protein	Brain microvessels	[8]
OATP-C	SLCO1B1	Targeted LC-MS/MS	ULQ < 0.35		fmol/μg protein	Brain microvessels	[8]
OATP-D	SLCO3A1	Targeted LC-MS/MS	ULQ < 0.254		fmol/μg protein	Brain microvessels	[8]
OATP-E	SLCO4A1	Targeted LC-MS/MS	ULQ < 0.758		fmol/μg protein	Brain microvessels	[8]
OATP-F	SLCO1C1	Targeted LC-MS/MS	ULQ < 0.208		fmol/μg protein	Brain microvessels	[8]
OATP-H	SLCO4C1	Targeted LC-MS/MS	ULQ < 0.21		fmol/μg protein	Brain microvessels	[8]
OATP-I	SLCO	Targeted LC-MS/MS	ULQ < 0.082		fmol/μg protein	Brain microvessels	[8]
OATP-J	SLCO5A1	Targeted LC-MS/MS	ULQ < 0.061		fmol/μg protein	Brain microvessels	[8]
OCTL1	SLC22A13	Targeted LC-MS/MS	ULQ < 0.532		fmol/μg protein	Choroid plexus (Plasma membrane fraction)	[9]
OCTL1	SLC22A13	Targeted LC-MS/MS	ULQ < 0.699		fmol/μg protein	Brain microvessels	[8]
OCTL2	SLC22A14	Targeted LC-MS/MS	ULQ < 0.698		fmol/μg protein	Choroid plexus (Plasma membrane fraction)	[9]
OCTL2	SLC22A14	Targeted LC-MS/MS	ULQ < 0.527		fmol/μg protein	Brain microvessels	[8]
OCTN1	SLC22A4	Targeted LC-MS/MS	ULQ < 0.25		fmol/μg protein	Choroid plexus (Plasma membrane fraction)	[9]
OCTN1	SLC22A4	Targeted LC-MS/MS	ULQ < 0.123		fmol/μg protein	Brain microvessels	[8]
OCTN1	SLC22A4	Targeted LC-MS/MS	ULQ < 0.04	0.01	pmol/mg protein	Brain microvessels	[6]
OCTN2	SLC22A5	Targeted LC-MS/MS	ULQ < 0.907		fmol/μg protein	Choroid plexus (Plasma membrane fraction)	[9]
OCTN2	SLC22A5	Targeted LC-MS/MS	ULQ < 0.288		fmol/μg protein	Brain microvessels	[8]
OST-α	SLC51A	Targeted LC-MS/MS	0.45	0.13	pmol/mg protein	Brain microvessels	[6]

Table 1. *Cont.*

Transporter	Gene Name	Method	Protein Expression Level		Unit	Tissue/Cell	Reference
			Value	SD			
PCFT	SLC46A1	Targeted LC-MS/MS	ULQ < 0.419		fmol/µg protein	Brain microvessels	[8]
PCFT	SLC46A1	Targeted LC-MS/MS	1.78	0.17	fmol/µg protein	Choroid plexus (Plasma membrane fraction)	[9]
PEPT1	SLC15A1	Targeted LC-MS/MS	ULQ < 0.325		fmol/µg protein	Choroid plexus (Plasma membrane fraction)	[9]
PEPT1	SLC15A1	Targeted LC-MS/MS	ULQ < 0.379		fmol/µg protein	Brain microvessels	[8]
PEPT2	SLC15A2	Targeted LC-MS/MS	ULQ < 0.216		fmol/µg protein	Brain microvessels	[8]
PEPT2	SLC15A2	Targeted LC-MS/MS	ULQ < 0.37		fmol/µg protein	Choroid plexus (Plasma membrane fraction)	[9]
PGT	SLCO2A1	Targeted LC-MS/MS	ULQ < 0.233		fmol/µg protein	Choroid plexus (Plasma membrane fraction)	[9]
PGT	SLCO2A1	Targeted LC-MS/MS	ULQ < 0.186		fmol/µg protein	Brain microvessels	[8]
PHT2	SLC15A3	Targeted LC-MS/MS	ULQ < 0.456		fmol/µg protein	Choroid plexus (Plasma membrane fraction)	[9]
PMAT	SLC29A4	Targeted LC-MS/MS	ULQ < 0.191		fmol/µg protein	Brain microvessels	[8]
PMAT	SLC29A4	Targeted LC-MS/MS	0.288	0.041	fmol/µg protein	Choroid plexus (Plasma membrane fraction)	[9]
RFC	SLC19A	Targeted LC-MS/MS	0.76	0.04	fmol/µg protein	Brain microvessels	[8]
RFC	SLC19A	Targeted LC-MS/MS	0.76	0.04	fmol/µg protein	Brain microvessels	[8]
RFC1	SLC19A1	Targeted LC-MS/MS	3.68	0.09	fmol/µg protein	Choroid plexus (Plasma membrane fraction)	[9]
SERT	SLC6A4	Targeted LC-MS/MS	ULQ < 0.304		fmol/µg protein	Choroid plexus (Plasma membrane fraction)	[9]
SERT	SLC6A4	Targeted LC-MS/MS	ULQ < 0.116		fmol/µg protein	Brain microvessels	[8]
SLC22A18	SLC22A18	Targeted LC-MS/MS	ULQ < 0.375		fmol/µg protein	Choroid plexus (Plasma membrane fraction)	[9]
SLC22A18	SLC22A18	Targeted LC-MS/MS	ULQ < 0.345		fmol/µg protein	Brain microvessels	[9]
TAUT	SLC6A6	Targeted LC-MS/MS	ULQ < 0.169		fmol/µg protein	Choroid plexus (Plasma membrane fraction)	[9]
TAUT	SLC6A6	Targeted LC-MS/MS	ULQ < 0.0767		fmol/µg protein	Brain microvessels	[8]
TfR1	TFRC	Targeted LC-MS/MS	2.34	0.76	fmol/µg protein	Brain microvessels	[8]

Table 1. *Cont.*

Transporter	Gene Name	Method	Protein Expression Level		Unit	Tissue/Cell	Reference
			Value	SD			
URAT1	*SLC22A12*	Targeted LC-MS/MS	ULQ < 0.357		fmol/µg protein	Choroid plexus (Plasma membrane fraction)	[9]
URAT1	*SLC22A12*	Targeted LC-MS/MS	ULQ < 0.0566		fmol/µg protein	Brain microvessels	[8]
UST3	*SLC22A9*	Targeted LC-MS/MS	ULQ < 1.21		fmol/µg protein	Choroid plexus (Plasma membrane fraction)	[9]
UST3	*SLC22A9*	Targeted LC-MS/MS	ULQ < 0.326		fmol/µg protein	Brain microvessels	[8]
xCT	*SLC7A11*	Targeted LC-MS/MS	ULQ < 0.783		fmol/µg protein	Choroid plexus (Plasma membrane fraction)	[9]
xCT	*SLC7A11*	Targeted LC-MS/MS	ULQ < 0.429		fmol/µg protein	Brain microvessels	[8]

A summary of the uptake transporters proteomics-based absolute abundance (fmol/µg protein or pmol/mg protein) in human brain microvascular cells or choroid plexus. Data showing individual proteomics expression profiles of uptake transporters in the available literature.

BBB limits the brain penetration of therapeutics and reduces their efficacy in the treatment of brain malignancies and other CNS disorders [42]. Targeting uptake transporters to overcome the tightly integrated BBB and efficiently improve drug delivery to the site of action is desirable [43]. Thus, an accurate estimation of uptake transporters' absolute expression at the BBB will provide more insight into our mechanistic understanding of drug penetration [10]. This will help in predicting the bioavailability and disposition of current and future therapeutics [44]. Quantitative targeted absolute proteomics technique (QTAP) analyzed via targeted LC-MS/MS has been used to identify and quantify the expression of uptake transporters in the human brain. Human choroid plexus [9], human brain microvessels (BMVs) [8], and hCMEC/D3 cells [7] have been used for quantitative targeted absolute proteomics of human brain tissues and cells. Uchida and co-workers [9] stated that four selected/multiple reaction monitoring (SRM/MRM) transitions have been optimized for the identification and quantification of target peptide. To achieve accurate protein quantification with coefficients of variation (CV) of <20.0%, three or four positive peaks must be determined. When no positive peak is detected or only one or two SRM/MRM transitions occur for a specific protein, the protein expression is expressed as under the limit of quantification (ULQ). In other words, the sensitivity of the third most transition determines the LQ (fmol/µg protein). The absolute quantification of uptake transporters in the human brain (tissue and cells) has been reviewed here with more emphasis on FDA-recommended uptake transporters. Table 1 summarizes the list of uptake transporters quantified via targeted LC-MS/MS proteomics with their absolute quantitation in different regions of the human brain and plasma membrane fraction of hCMEC/D3 cells.

OATPs are responsible for the uptake of a wide range of structurally diverse substrates, from endogenous substances like steroid hormones and bile acids to statins and chemotherapeutics. The first attempt to find OATPs in human brain was accomplished by Kullak-Ublick and co-workers in search of a dehydroepiandrosterone sulfate (DHEAS) uptake protein [23]. In 1998, they provided the very first proof of the presence of OATPs in the brain based on the Northern blotting technique. Subsequently, the presence of OATP1A2 in the human BBB endothelial cells and brain capillaries was confirmed via immunohistochemical staining and Western blotting [45,46]. The expression and localization of six OATPs (i.e., OATP1A2, OATP1B1, OATP1B3, OATP1C1, OATP2B1, and OATP4A1) have been proven by Bronger et al. [47] in the endothelial cells of human gliomas. The localization of OATP1A2 and OATP2B1 at the BBB and the blood–tumor barrier has been identified at mRNA level and protein immunoblotting [47]. Uchida and co-workers [9] have identified and quantified OATP1A2 and OATP2B1 in a plasma membrane fraction of human choroid plexus (0.45 and 0.24 fmol/µg < ULQ) through a targeted LC-MS/MS approach. Moreover, Al-Majdoub and colleagues [6] quantified OATP2B1 as 0.40 to 0.48 fmol/µg total protein, using the same approach. Among the OATP family, OATP1B1 and OATP1B3 are recommended by the FDA for clinical DDI liability. Uchida and co-workers [9] have identified OATP1B1 and 1B3 as 0.30 and 0.62 fmol/µg, respectively, in a human choroid plexus; however, they were considered as ULQ (Table 1).

The SLC superfamily transporters, including OATs, OCTs, and novel organic cation transporters (OCTNs), are primarily responsible for the uptake of circulating solutes from blood to the brain [48]. Billington and co-workers have identified the inter-individual and inter-regional variability of drug transporters expression in the human brain using quantitative targeted proteomics [49]. According to their study, the abundance of OAT3, OCT1/2, and OCTN1/2 were estimated as below the limit of quantification. In another attempt, Al-Majdoub et al. [6] quantified the expression of OAT1, 2, 3, and 7 as 0.48, 7.9, 0.27, and 0.51 pmol/mg of the total protein in human BMVs, respectively. In addition, Uchida and co-workers [9] quantified OAT3 as 1.87 pmol/mg of plasma membrane fraction in human choroid plexus. Giacomini's group has identified the presence of OCT3 in human BMVs from two donors using the immunohistochemistry method [32]. This finding was supported by Al-Majdoub and co-workers, who identified the exact expression level of OCT3 as 0.62 ± 0.08 pmol/mg total protein in human BMVs [6]. In a more recent study, the absolute expression of OCT3 in human BMVs was estimated as 0.15 ± 0.056 fmol/μg of total protein using targeted LC-MS/MS-based proteomics [49]. This study has shown that OCT3 is the most highly expressed OCT in human BBB, while OCT1 and 2 are not detectable at the protein level. Multidrug and toxin extrusion protein 1 (MATE1) and MATE2-K have been found in the plasma membrane of human choroid plexus as 8.61 ± 0.63 and 2.19 fmol/μg protein (ULQ), respectively [9]. Moreover, MATE1 and MATE2-K have been quantified as ULQ as 0.33 and 0.29 fmol/μg protein in human BMVs (Table 1).

Other than FDA-recommended uptake transporters in the human brain, there are other uptake transporters essential for brain physiology [50]; LAT1(SLC 7A5/SLC2A3), which is responsible for the uptake of large neutral amino acids, thyroid hormones, and medicines, is one of them [51,52]. LAT1 is a light-chain amino acid uptake transporter connected by a disulfide bond to the heavy chain 4F2 cell-surface antigen heavy chain (4F2hc). The formation of the LAT1/4F2hc heteromeric complex is essential for the stabilization and localization of LAT1 in the BBB membrane [51,52]. Moreover, the overexpression of LAT1 in some tumor cells has been identified as a promising target in cancer therapy [51,53]. The absolute expression of LAT1 was quantified to be between 0.43 and 0.71 pmol/mg total plasma membrane protein fraction in healthy human BMVs [6,8]. Ohtsuki and co-workers have found 4F2hc, MCT1, ENT1 uptake transporters in both hCMEC/D3 cells and human BMVs [7]. MCTs play a crucial role in cellular metabolism by facilitating the transport of endogenous monocarboxylates like lactate into and out of brain cells [37]. ENTs are bidirectional, sodium-independent transporters involved in the inward and outward transport of nucleosides [54]. The expression levels of 4F2hc, MCT1, and ENT1 in the plasma membrane fraction of hCMEC/D3 cells were reported as 1.90 ± 0.23, 1.87 ± 0.22, and 5.94 ± 0.35 fmol/μg of protein, respectively (Table 1) [7]. The expression of ENT1 in human BMVs was estimated from 0.27 ± 0.1 [6] to 0.57 ± 0.13 fmol/μg protein [8]. The absolute expression of ENT1 was reported as 2.49 ± 0.12 fmol/μg protein in the plasma membrane fraction of human choroid plexus [9], while the expression of ENT2 in human BMVs and choroid plexus was ULQ (Table 1) [7,8].

3.1. Models Used to Study Uptake Transporters-Mediated Brain Drug Disposition

The BBB is a complex system that is made up of low-permeable brain capillary endothelial cells. This leads to a reduced transcytosis and tight junctions, resulting in very low paracellular transport [55]. Thus, many endogenous and exogenous compounds (e.g., drugs) need active transport processes to facilitate distribution in the brain [41]. In vitro brain cell models have been in use for over 50 years [56]. However, creating a BBB-like in vitro system is quite challenging. So far, several in vitro cell line models have been proposed in different species. Examples include: (i) immortalized and primary mouse brain endothelial cells, (ii) mono-culture rat brain capillary endothelial cells (BCEC), (iii) co-culture rat BCEC, (iv) triple-culture rat BCEC, (v) astrocytes co-culture with BCEC bovine cells, (vi) porcine monoculture cells with porcine brain endothelial cells (PBEC), (vii) human endothelial cells (hCMEC/D3), (viii) human BBB model with pluripotent stem cells (hPSCs), (ix) cord blood-derived endothelial progenitor cells, and (x) an hPSC-derived 3D spheroid system (Figure 2, an updated graphical presentation from HC Helms et al. [56]). Apart from these in vitro models, in vivo models have also been used, and include mouse, rat, monkey, and dog. Also very recently, proteomics-based IVIVE and physiologically based pharmacokinetic model (PBPK) data were published [41,42,57]. The isolation of these brain cell lines started since the 1970s and tremendous progress has been made toward optimizing isolation methods, as well as culture and phenotyping for transporter studies [27,58–61]. Since the establishment of monolayer culture models, several studies have shown progress on tight junction formation to study permeability and efflux transporters contribution. The majority of these studies used endothelial cells with astrocytes or pericytes [58,62–65]. Each of these models has its own specific advantage over the other. For example, triple-culture rat BCEC consists of BCEC cells with astrocyte/pericyte that facilitates the formation of monolayers having spindle shapes, that are known to express occludin and increase trans-epithelial electric resistance (TEER) [66,67]. The primary mouse BCEC–astrocytes co-culture system shows prominent tight junction (TEER up to 1000 $\Omega\cdot$cm^2); however, uptake transporter functions have not yet been characterized in this model [68–71]. Rat primary endothelial BCEC cells were characterized for P-gp, BCRP, MRP-1, and MRP1 functions. Rat BCEC cells express other uptake transporters such as Glut-1, LAT1, and PMAT [72–74]. Also, published results suggest that serum-free monolayer culture conditions are more sensitive to Glut-1-like uptake transporter downregulation than the efflux transporter expression [75,76]. hCMEC/D3 cell line first developed in 2005 represents a stable, easy growing, and maintenance line, with high translational capacity similar to brain microvascular endothelial cells in the BBB [77]. The hCMEC/D3 cells are derived from the hTERT/SV40-immortalized cloned cells from human temporal lobe microvessels isolated from an epileptic patient. On the other hand, hCMEC/D3 cells show a low TEER value, suggesting that an improvement in culture conditions is warranted. This led to the evaluation of a co-culture system with astrocytes and pericytes [78,79]. So far, over 140 uptake transporters (SLC family) have been identified in hCMEC/D3 cell lines, including Glut-1, LAT-1, MCT, and OATPs [5,77,80,81]. The hCMEC/D3 model has also been used for the discovery and development of antihistaminic drug candidates that need to reach the CNS [27]. A summary of the currently available opportunistic in vitro cell-based models is shown in Figure 2.

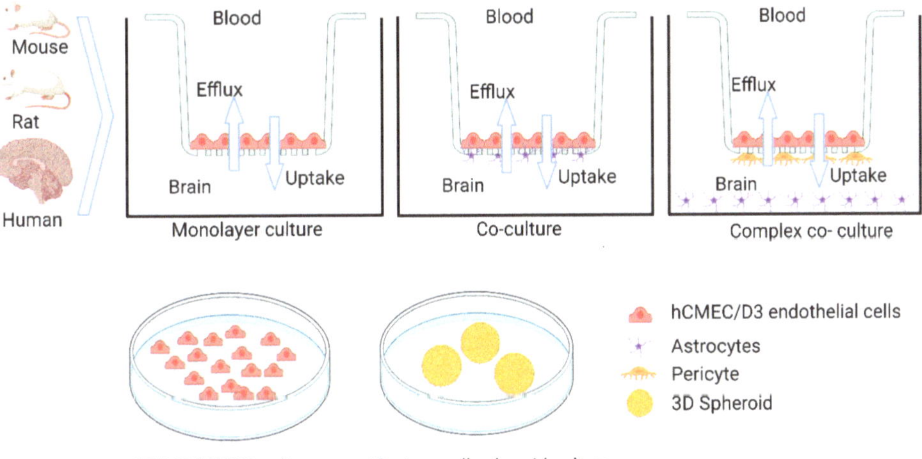

Figure 2. In vitro blood–brain barrier models for drug transporter studies. Graphical presentation of currently available in vitro models for drug transporter studies in different species (mouse, rat, human). In vitro models include brain endothelial microvascular cell-based culture for permeability into the brain (Kp,uu) using monolayer, co-culture with astrocytes; triple culture with pericytes and astrocytes (upper panel). Uptake transporter functional assay in vitro models showing 2D model with human microvascular endothelial cells (hCMEC/D3) and 3D culture with stem cell-derived spheroids. Graphical illustrations were made with BioRender and Adobe IllustratorCC.

3.2. Species Differences in Uptake Transporter Activity and Drug–Drug Interactions in Brain

Studies have identified a significant level of inter-species differences in brain drug dispositions. These species differences are known to be caused by physiological changes, especially a differential expression of the transporters and enzymes among species (i.e., mouse, rat, dog, monkey, and human). Table 2 summarizes a few examples of uptake transporters involved in rat, mouse, monkey, dog, and human brain uptake. As shown in this table, memantine brain uptake was 84.59 ± 9.73 pmol/mg brain tissue in control male rats, while it decreased by 36% in the presence of OCT1/2 inhibitor (cimetidine,

Table 2. Species differences in uptake drug transporters involved in brain drug exposure. A summary of the literature-reported brain drug exposure in different species.

Species	Transporter	Substrate	Perpetrator	Method	Exposure	DDI/Effect	Expression Level (fmol/µg Protein)	Reference
Mouse	Oatp1a4	Glyburide	Rifampicin	In situ brain perfusion	Kin (0.5 $\pm$ 0.11 µL/g/s)	No change (Kin: ~0.4)		[87]
Mouse	Oatp1a4	Rosuvastatin	-	In vivo brain uptake	57 $\pm$ 9 µL/min/g brain (brain/plasma = 22 µL/g)	-		[88]
		Pravastatin			24 $\pm$ 4 µL/min/g brain (brain/plasma = 8.3 µL/g)			
		Taurocholate			11 $\pm$ 2 µL/min/g brain			
		Ochratoxin A			11 $\pm$ 0 µL/min/g brain (brain/plasma = 7.6 µL/g)			
Mouse (Bend.3)	OCT1, OCT2, OCT3	Pentamidine	Amantadine (500 µM)	In vitro uptake (V_d)	V_d at different time points	59% Reduction		[96]
			Prazosin (100 µM)			No change (paracellular leakage increased)		
			N-methy-l nicotinamide (100 µM)			No change		
Rat	Oatp1a4	-	-		-	-	1.99	[49]
Rat	Oct1/2	Memantine (5 mh/kg i.v.)	Cimetidine (25 µM)	In vivo brain uptake	84.59 $\pm$ 9.73 pmol/mg brain tissue	37% Decreased (54.14 $\pm$ 8.35)	-	[82]
Rat	Oatp1a4	Atorvastatin	BMP-9 (1 µg/kg)	In vivo brain uptake	AUC: 987.9 $\pm$ 53.41 pmol $\times$ min/mg brain tissue	60% Increased (1581 $\pm$ 52.26)	-	[83]
			LDN (10 mg/kg) + BMP-9			Attenuated the BMP-9 effect		
		Pravastatin	BMP-9 (1 µg/kg)		AUC: 800.0 $\pm$ 47.41 pmol $\times$ min/mg brain tissue	69% Increased (1349.00 $\pm$ 48.00)		
			LDN (10 mg/kg) + BMP-9			Attenuated the BMP-9 effect		
		Rosuvastatin	BMP-9 (1 µg/kg)		AUC: 836.8 $\pm$ 50.53 pmol $\times$ min/mg brain tissue	74% Increased (1459.0 $\pm$ 53.51)		
			LDN (10 mg/kg) + BMP-9			Attenuated the BMP-9 effect		

Table 2. *Cont.*

Species	Transporter	Substrate	Perpetrator	Method	Exposure	DDI/Effect	Expression Level (fmol/µg Protein)	Reference
Rat	Oatp1a4	Atorvastatin	Fexofenadine (100 µM)	In situ brain perfusion	63.72 ± 9.78 pmol/mg brain tissue	39% reduced (24.89 ± 7.55)	-	[83]
		Pravastatin	Fexofenadine (100 µM)		54.98 ± 6.37 pmol/mg brain tissue	Reduced (12.39 ± 4.8)		
		Rosuvastatin	Fexofenadine (100 µM)		55.83 ± 7.84 pmol/mg brain tissue	Reduce (10.54 ± 3.65)		
Rat	Octs	SHY-01 (50 mg/kg)	-	In vivo	2.05 ± 0.18 (hr·µg/mL)	CL: 24.48 ± 2.25	-	[84]
		Metformin (50 mg/kg)	-	In vivo	1.89 ± 0.08 (hr·µg/mL)	CL: 26.46 ± 1.10		
Rat	Oatp	Digoxin (2 mg/kg, i.v.)	Rifampicin (30 mg/kg, oral)	In vivo	~0.07 ($K_{p,AUC,brain}$)	Increased (~1.8-fold)	-	[85]
					~0.02 ($K_{p,AUC,CSF}$)	Increased (~4-fold)		
Rat	Oatp1a4	Taurocholate	BMP-9 (1 µg/kg)	In vivo brain uptake	AUC: 1143.6 ± 57.92 pmol × min/mg brain tissue	79% Increased (2054.83 ± 66.13)	-	[89]
			E3S (100 µM)	In situ brain perfusion	65.31 ± 8.19 pmol/mg	59% Reduced (27.02 ± 7.56)		
			Fexofenadine (100 µM)			61% Reduced (25.61 ± 7.44)		
			BSP			No effect (66.81 ± 7.13)		
		Atorvastatin	E3S (100 µM)		34.07 ± 5.67 pmol/mg brain tissue	Reduced (17.67 ± 5.22 pmol/mg brain tissue)		
		Pravastatin			22.01 ± 6.27 pmol/mg brain tissue	Reduced (9.00 ± 4.98 pmol/mg brain tissue)		
Rat	Oatp1a4	Taurocholate	E3S (100 µM)	In situ brain perfusion	~55 pmol/g brain tissue	Reduced (2.2-fold)	-	[94]
			Digoxin (200 µM)			Reduced (2.4-fold)		
			Fexofenadine (100 µM)		$V_{brain} = 97.61 ±$ pmol/g	Reduced (2.2-fold)		
			BSP			No effect		

Table 2. *Cont.*

Species	Transporter	Substrate	Perpetrator	Method	Exposure	DDI/Effect	Expression Level (fmol/μg Protein)	Reference
Rat	Oats, Mrps, Oatps	Cefadroxil	Probenecid	Microdialysis	$AUC_{blood} = 1802 \pm 97$ ($\mu g \times min/mL$)	2873 ± 177 (Increased)	-	[95,97]
					$AUC_{ECF} = 40 \pm 7$	174 ± 35 (Increased)		
					$Kp_{uu,ECF} = 0.022 \pm 0.003$	0.058 ± 0.009 (Increased)		
					$AUC_{CSF} = 57 \pm 15$	117 ± 50 (Increased)		
					$Kp_{uu,CSF} = 0.031 \pm 0.007$	0.039 ± 0.015		
	Pept2	Cefadroxil	Ala-Ala	Brain slices	$V_{u,brain}$ (mL/g brain) = 3.67 ± 0.23	0.95 ± 0.45 (Reduced)		
			GlySar			1.10 ± 0.05 mL/g (Reduced)		
	Oats, Mrps, Oatps		Probenecid			6.06 ± 0.15 (Increased)		
Dog	OCT2		-	LC-MS/MS	-	-	<LOQ	[90,92,93]
		-	-		-	-	<LOQ	
	OAT3	-	-		-	-	<LOQ	
		-	-		-	-	<LOQ	
	OATP1A2	-	-		-	-	<LOQ	
		-	-		-	-	2.69 ± 0.78	
	OATP2B1	-	-		-	-	<LOQ	
		-	-		-	-	<LOQ	
	ENT1	-	-		-	-	0.581 ± 0.342	
		-	-		-	-	1.05 ± 0.47	
	LAT1	-	-		-	-	<LOQ	
		-	-		-	-	<LOQ	
	OCT3/P-gp	Quinidine (7.71 μmol/kg)	-	In vivo brain uptake	$K_{p,uu,brain} = 0.363 \pm 0.11$	-	-	
			-		$K_{p,uu,csf} = 0.131 \pm 0.036$	-		
	OAT2/BCRP	Dantrolene (1.59 μmol/kg)	-		$K_{p,uu,brain} = 0.0614 \pm 0.0021$	-	-	
			-		$K_{p,uu,csf} = 0.505 \pm 0.025$	-		

Table 2. *Cont.*

Species	Transporter	Substrate	Perpetrator	Method	Exposure	DDI/Effect	Expression Level (fmol/µg Protein)	Reference
Monkey (Baboon)	OATP2B1, OATP1A2	Glyburide	Rifampicin	PET	4.5 ± 1.0 ($AUC_{brain}/AUC_{blood} = 0.032$)	No change: 11.5 (0.018)	-	[87]
			Cyclosporine			No change: 17.2 (0.029)		
			Pantoprazole			No change: 8.1 (0.035)		
Monkey	OATP2B1	-	-		-	-	0.12	[49]
Human (hCMEC/D3)	OCT1, OCT2, OCT3	Pentamidine	Amantadine (500 µM)	In vitro uptake (V_d)	V_d at different time points	45% Reduction	-	[96]
			Prazosin (100 µM)			39% Reduction		
			N-methyl-1 nicotinamide (100 µM)			No change		
Human	OATP2B1, OATP1A2	Glyburide	Rifampicin (9 mg/kg i.v.)	PET	5.82 ± 0.74 ($AUC_{brain}/AUC_{blood} = 0.03$)	No change: 7.72 (0.03)	-	[86]

A summary of the literature-reported brain drug exposure in rodent models. The data show brain exposure, tissue partition coefficient ($k_{p,uu}$), and transporter-mediated drug–drug interaction on brain concentrations of these uptake and/or efflux transporter substrate drugs.

Table 3. Physicochemical properties and transporter interactions in brain disposition of therapeutic drugs.

Drug Name	Drug Class	Mw [a]	LogD [b]	Plasma Protein Binding (%) [a]	Efflux Transporter Substrate		Uptake Transporter Substrate															$K_{p,uu,brain}$			
					MDR1	BCRP	OATP1B1	OATP1B3	OATP2B1	OATP1A2	OCT1	OCT2	OCT3	OCTN1	OCTN2	LAT1	OAT1	OAT3	MATE1	MATE2k	Rat	Mouse	Monkey	Human
Dolutegravir	HIV-Integrase strand transfer inhibitor	419.38	1.10	98.90	Yes [98]	Yes [98]	NA	NA	NA	NA	NA	NA	NA	NA	NA	NA	NA	NA	NA	NA	0.02 [99]	NA	NA	NA

Pharmaceutics **2023**, 15, 2473

Table 3. *Cont.*

Gabapentin	Fexofenadine	Erlotinib	Efavirenz	Drug Name	
Anticonvulsant	H-1 Receptor antagonists	Kinase inhibitor	Non-nucleoside Reverse transcriptase inhibitors (NNRTI)	Drug Class	
171.20	501.66	393.40	315.68	Mw [a]	
−1.27	2.93	3.05	4.46	LogD [b]	
<3	65.00	93.00	99.60	Plasma Protein Binding (%) [a]	
No [119]	Yes [108]	Yes [104]	No [100]	MDR1	Efflux Transporter Substrate
NA	No [109]	Yes [104]	NA	BCRP	
NA	Yes [110]	No [105]	No [101]	OATP1B1	Uptake Transporter Substrate
NA	Yes [111]	No [105]	No [101]	OATP1B3	
NA	Yes [112]	Yes [54]	NA	OATP2B1	
No [120]	Yes [113]	NA	No [101]	OATP1A2	
No [121]	Yes [114]	No [106]	No [102]	OCT1	
Yes [122]	No [115]	Yes [106]	No [102]	OCT2	
No [123]	No [112]	NA	NA	OCT3	
Yes [124]	NA	NA	NA	OCTN1	
No [121]	No [112]	NA	NA	OCTN2	
Yes [122]	NA	NA	NA	LAT1	
No [122]	Yes [116]	No [106]	NA	OAT1	
No [122]	Yes [116]	Yes [106]	NA	OAT3	
NA	Yes [117]	NA	NA	MATE1	
NA	Yes [117]	NA	NA	MATE2k	
0.14 [125]	0.05 [118]	0.06 [104]	0.20 [103]	Rat	$K_{p,uu,brain}$
NA	0.22 [26]	NA	NA	Mouse	
NA	NA	0.05 [104]	NA	Monkey	
0.16 [125]	NA	0.08 [107]	NA	Human	

Table 3. *Cont.*

Category	Drug Name	Lamotrigine	Loperamide	Methotrexate	Pitavastatin
	Drug Class	Anticonvulsants	Antidiarrheal	Antimetabolite	HMG CoA Reductase Inhibitors (statin)
	Mw [a]	256.10	477.10	454.40	421.50
	LogD [b]	1.91	2.77	−6.56	0.89
	Plasma Protein Binding (%) [a]	55.00	95.00	50.25	>99
Efflux Transporter Substrate	MDR1	No [104]	Yes [104]	Yes [130]	Yes [104]
	BCRP	No [104]	No [104]	Yes [131]	Yes [104]
Uptake Transporter Substrate	OATP1B1	NA	NA	Yes [132]	Yes [138]
	OATP1B3	NA	NA	Yes [133]	Yes [139]
	OATP2B1	NA	NA	Yes [134]	Yes [140]
	OATP1A2	No [120]	NA	Yes [134]	Yes [141]
	OCT1	Yes [126]	NA	NA	No
	OCT2	Yes [126]	NA	Yes [135]	NA
	OCT3	Yes [126]	NA	NA	NA
	OCTN1	No [127]	NA	NA	NA
	OCTN2	No [127]	NA	NA	NA
	LAT1	NA	NA	NA	NA
	OAT1	NA	NA	Yes [136]	NA
	OAT3	NA	NA	Yes [137]	NA
	MATE1	NA	NA	Yes [135]	NA
	MATE2k	NA	NA	Yes [135]	NA
$K_{p,uu,brain}$	Rat	0.88 [125]	0.02 [128]	0.006 [125]	NA
	Mouse	NA	NA	NA	NA
	Monkey	0.86 [104]	0.04 [129]	0.04 [104]	0.24 [104]
	Human	2.80 [128]	NA	NA	NA

Table 3. *Cont.*

Drug Name	Drug Class	Mw [a]	$LogD$ [b]	Plasma Protein Binding (%) [a]	Efflux Transporter Substrate		Uptake Transporter Substrate															$K_{p,uu,brain}$			
					MDR1	BCRP	OATP1B1	OATP1B3	OATP2B1	OATP1A2	OCT1	OCT2	OCT3	OCTN1	OCTN2	LAT1	OAT1	OAT3	MATE1	MATE2k	Rat	Mouse	Monkey	Human
Quinidine	Antiarrhythmic	324.40	0.86	78.00	Yes [104]	No [104]	NA	NA	NA	NA	NA	NA	Yes [92]	Yes [142]	No [143]	NA	NA	NA	No [144]	No [144]	0.04 [104]	NA	0.10 [104]	NA
Raltegravir	HIV-Integrase Strand transfer inhibitor	444.42	−0.92	83.00	Yes [5]	Yes [5]	No [145]	No [145]	NA	No [145]	No [145]	NA	NA	NA	No [145]	NA	Yes [146]	NA	NA	NA	0.13 [147]	NA	0.12 [147]	NA
Rifampicin	Antibiotic	822.90	2.87	89.00	Yes [148]	No [149]	Yes [150]	Yes [150]	No [150]	No [150]	No [102]	No [102]	No [151]	NA	NA	NA	NA	NA	NA	NA	0.04 [125]	NA	NA	NA
Rosuvastatin	HMG CoA Reductase Inhibitors (statin)	481.54	−1.24	88.00	Yes [152]	Yes [153]	Yes [154]	Yes [154]	Yes [155]	Yes [156]	No [139]	NA	NA	NA	NA	NA	NA	NA	NA	NA	3.97 [157]	NA	NA	NA
Zidovudine	Nucleoside Reverse Transcriptase Inhibitors (NRTI)	267.20	−0.41	<38	Yes [104]	Yes [104]	NA	NA	NA	NA	No [158]	No [158]	No [159]	NA	NA	NA	Yes [160]	Yes [158]	NA	NA	0.09 [125]	NA	NA	NA

Data show potential interactions of drugs with efflux and uptake transporters and the effect on brain exposure. We summarized their reported Kp,uu,brain in rodents, monkey, and human to justify the role of the active transport in their brain disposition via BBB. [a] Retrieved information from DrugBank database. [b] Retrieved information from ChEMBL database. NA: Not available.

3.3. Significance of Uptake Transporters in Brain Drug Disposition

Table 3 summarizes the reported $K_{p,uu,brain}$ for different compounds, including dolutegravir (DTG), efavirenz (EFV), erlotinib (ERL), fexofenadine (FEX), gabapentin (GBP), lamotrigine (LMG), loperamide (LPM), methotrexate (MTX), pitavastatin (PTV), quinidine (QND), raltegravir (RLT), rifampicin (RFP), rosuvastatin (RSV), and zidovudine (ZDV). $K_{p,uu,brain}$ is defined as the unbound brain-to-plasma drug concentration ratio calculated using Equation (1):

$$K_{p,uu,brain} = \frac{f_{u,brain} \times C_{brain}}{f_{u,plasma} \times C_{plasma}} \tag{1}$$

where, $f_{u,brain}$ and $f_{u,plasma}$ stand for unbound fraction of drug in the brain and plasma, respectively. C_{brain} and C_{plasma} are the total drug concentration in brain and plasma, respectively.

In addition to the $K_{p,uu,brain}$ values, in vitro efflux and uptake brain transporters involvement in the disposition of drugs were retrieved from different literature studies to elucidate the probable role of transporters in brain exposure. MDR1 and BCRP are two major efflux transporters in the BBB that, along with uptake transporters OATPs, OCTs, OCTNs, OATs, LAT1, MATE 1, and MATE2K, are listed in Table 3. Among all the listed compounds in Table 3, RSV shows the highest $K_{p,uu,brain}$ of 3.97 in rat [157]. This might be due to the involvement of low-capacity OATP2B1 and OATP1A2 uptake transporters in BBB. This observation warrants additional validation in vitro using OATP2B1 and OATP1A2 orthologs in rat. Table 3 shows LMG with the highest $K_{p,uu,brain}$ in human at 2.8, suggesting the involvement of OCT1, OCT2, and OCT3 uptake transporters in human BBB [128]. Furthermore, the reported $K_{p,uu,brain}$ for MTX is very low (0.006 in rat and 0.04 in monkey) with a significant level of species difference [104,125]. The reason for this low brain penetration of MTX might be due to the efflux through MDR1 and BCRP, which may counter the MTX influx through BBB.

4. Conclusions and Future Directions

One major concern in the development of CNS target drugs is how the BBB affects brain exposure. CNS delivery of many compounds is greatly restricted by the BBB. To date, several strategies have been reported for its use in enhancing drug delivery to the brain. These include systemic and local routes of administration and comprise the following intranasal route, viral vectors, nanocarriers, and formulations (i.e., nanoparticles) [161]. Another attractive approach to brain drug delivery has been the linking of drugs to amino acids that actively cross the BBB, e.g., a methotrexate (MTX)–lysine conjugate enhances MTX brain uptake through the endogenous transporter system of lysine [162]. In this manuscript, we summarize the up to date information on uptake transporters' involvement in brain disposition, citing specific studies and examples of the role of uptake transporters. While a lot more information exists on the role of efflux transporters in brain drug disposition, the equivalent information for uptake transporters at the BBB is quite sparse, and this limits the success of research on CNS-targeted drugs. Assembling the brain localization (Figure 3) and functions, proteomic expression (Table 1), species differences (Table 2) in uptake transporter activity, and the currently available in vitro models (Figure 2) not only highlights the tools available, but also sheds light on the potential preclinical to clinical translation and evaluation of DDI potential. However, the clinical relevance of these transporter contributions remains to be elucidated in a broader sense. For example, the OCT1/2 inhibitor cimetidine decreased memantine brain concentration by 37% in rat [82]. On the other hand, rifampicin, an inhibitor of OATP2B1 and OATP1A2, had no effect on glyburide brain exposure in humans [86]. Therefore, with the information assembled in this review, it is clear that additional tools and investigations are needed to further our understanding of the clinical relevance of uptake transporters expressed at the BBB. Taken together, we believe that there are a few approaches that hold promise and could be helpful in investigations regarding the role of uptake transporters on CNS drug disposition. These include: (i) proteomics-based extrapolations of the fractional transport (ft) to in vivo levels,

(ii) PBPK modeling and simulations to predict brain compartmental concentrations of the drugs, (iii) transgenic/knock-out animal models for use in evaluating the role of these transporters on drug brain exposure, and (iv) a 3D stem cell-based assessment of low-clearance drugs. Furthermore, it will be highly beneficial if we can explore the rate-limiting steps involved in CNS drug exposure by investigating the interplay between uptake and efflux transport at the BBB.

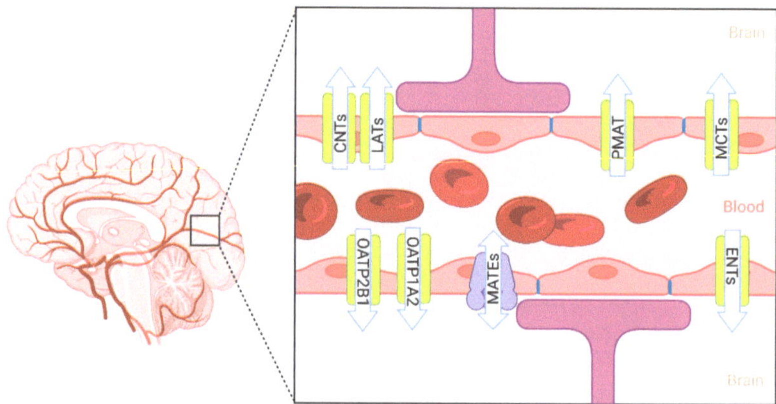

Figure 3. Localization of uptake transporters at the blood–brain barrier. A summary of the major drug and amino acids' uptake transporters localization at BBB endothelial cells. The graphical localization shows the uptake transporters and their functional direction for taking drugs and amino acids through the BBB into the brain. CNTs: concentrative nucleoside transporters; ENTs: equilibrative nucleoside transporters; MCTs: monocarboxylate transporters; OATPs: organic anion transporter polypeptides; PMAT: plasma membrane monoamine transporter; LATs: L-type amino acid transporters; MATE: multidrug and toxic compound extrusion. Graphical illustrations were made with BioRender and Adobe IllustratorCC.

Indeed, future successes in the development of CNS-targeted drugs will depend on improvements in our understanding of BBB uptake transport mechanisms. Research should focus on investigations of the relative contributions of different uptake processes and their role in human brain drug disposition. The identification of brain uptake transporter-specific inhibitors will enhance our understanding of how large (>450 Daltons) and water-soluble drug molecules enter the brain. Overall, this review improves our current knowledge of brain uptake transporters and recommends that additional investigations are warranted in this field. Information gained so far from BBB uptake transporter mechanistic studies can be applied throughout the drug development process, and will provide a better understanding of human brain drug disposition.

Author Contributions: Writing, review, and editing, M.M.P., A.S., Y.A., S.F.K. and J.O.E. All authors have read and agreed to the published version of the manuscript.

Funding: This research and APC was sponsored by AbbVie Biotherapeutics.

Data Availability Statement: All raw data was extracted from the literature that will be available upon request.

Acknowledgments: All authors namely M.M.P., A.S., S.F.K. and J.O.E. contributed in part or participated in the design, data analysis, interpretation, and writing of this manuscript. Y.A. was an experiential summer intern at AbbVie Biotherapeutics at the time this work was conducted.

Conflicts of Interest: The authors declare no conflict of interest.

Abbreviations

CNTs: concentrative nucleoside transporters; ENTs: equilibrative nucleoside transporters; MCTs: monocarboxylate transporters; OATPs: organic anion transporter polypeptides; PMAT: plasma membrane monoamine transporter; LATs: L-type amino acid transporters; OATs: organic anion transporters; OCTs: organic cation transporters; MATE: multidrug and toxic compound extrusion; PEPTs: peptide transporters; 4F2hc: several heterodimeric amino acids transporter; ASBT: apical sodium-dependent bile acid transporter; ASCT: alanine serine cysteine transporter; ATA: amino acid transporter; BGT: gamma-aminobutyric acid transporter; BOCT: brain organic cation transporter; CRT: creatine transporter; CT: human l-carnitine transporter; CTL: choline transporter; EAAT: excitatory amino acid transporter; FATP: long-chain fatty acid transport protein; GAT: gamma-aminobutyric acid (GABA) transporter; GLUT: glucose transporters; NET: norepinephrine transporter; OSTα: organic solute transporter alpha; PCFT: proton-coupled folate transporter; PGT: prostaglandin transporter; PHT2: peptide histidine transporter 2; RFC: reduced folate transporter; SERT: serotonin transporter; TAUT: sodium- and chloride-dependent taurine transporter; TfR1: transferrin receptor 1; URAT: urate transporter; UST: sodium-independent organic anion transporter; xCT: cystine/glutamate antiporter. ULQ represents the limit of quantification. In case of no positive peak or just one or two SRM/MRM transitions for a specific protein, the protein expression was determined as ULQ.

References

1. Giacomini, K.M.; Huang, S.-M.; Tweedie, D.J.; Benet, L.Z.; Brouwer, K.L.R.; Chu, X.; Dahlin, A.; Evers, R.; Fischer, V.; Hillgren, K.M.; et al. Membrane Transporters in Drug Development. *Nat. Rev. Drug Discov.* **2010**, *9*, 215–236. [CrossRef]
2. Alavijeh, M.S.; Chishty, M.; Qaiser, M.Z.; Palmer, A.M. Drug Metabolism and Pharmacokinetics, the Blood-Brain Barrier, and Central Nervous System Drug Discovery. *NeuroRx* **2005**, *2*, 554–571. [CrossRef] [PubMed]
3. Osborne, O.; Peyravian, N.; Nair, M.; Daunert, S.; Toborek, M. The Paradox of HIV Blood-Brain Barrier Penetrance and Antiretroviral Drug Delivery Deficiencies. *Trends Neurosci.* **2020**, *43*, 695–708. [CrossRef] [PubMed]
4. Loryan, I.; Reichel, A.; Feng, B.; Bundgaard, C.; Shaffer, C.; Kalvass, C.; Bednarczyk, D.; Morrison, D.; Lesuisse, D.; Hoppe, E.; et al. Unbound Brain-to-Plasma Partition Coefficient, $K_{p,Uu,Brain}$—A Game Changing Parameter for CNS Drug Discovery and Development. *Pharm. Res.* **2022**, *39*, 1321–1341. [CrossRef]
5. Hoque, M.T.; Kis, O.; Rosa, M.F.D.; Bendayan, R. Raltegravir Permeability across Blood-Tissue Barriers and the Potential Role of Drug Efflux Transporters. *Antimicrob. Agents Chemother.* **2015**, *59*, 2572–2582. [CrossRef]
6. Al-Majdoub, Z.M.; Feteisi, H.A.; Achour, B.; Warwood, S.; Neuhoff, S.; Rostami-Hodjegan, A.; Barber, J. Proteomic Quantification of Human Blood-Brain Barrier SLC and ABC Transporters in Healthy Individuals and Dementia Patients. *Mol. Pharm.* **2019**, *16*, 1220–1233. [CrossRef] [PubMed]
7. Ohtsuki, S.; Ikeda, C.; Uchida, Y.; Sakamoto, Y.; Miller, F.; Glacial, F.; Decleves, X.; Scherrmann, J.-M.; Couraud, P.-O.; Kubo, Y.; et al. Quantitative Targeted Absolute Proteomic Analysis of Transporters, Receptors and Junction Proteins for Validation of Human Cerebral Microvascular Endothelial Cell Line HCMEC/D3 as a Human Blood-Brain Barrier Model. *Mol. Pharm.* **2013**, *10*, 289–296. [CrossRef] [PubMed]
8. Uchida, Y.; Ohtsuki, S.; Katsukura, Y.; Ikeda, C.; Suzuki, T.; Kamiie, J.; Terasaki, T. Quantitative Targeted Absolute Proteomics of Human Blood-Brain Barrier Transporters and Receptors. *J. Neurochem.* **2011**, *117*, 333–345. [CrossRef]
9. Uchida, Y.; Zhang, Z.; Tachikawa, M.; Terasaki, T. Quantitative Targeted Absolute Proteomics of Rat Blood-Cerebrospinal Fluid Barrier Transporters: Comparison with a Human Specimen. *J. Neurochem.* **2015**, *134*, 1104–1115. [CrossRef] [PubMed]
10. Bao, X.; Wu, J.; Xie, Y.; Kim, S.; Michelhaugh, S.; Jiang, J.; Mittal, S.; Sanai, N.; Li, J. Protein Expression and Functional Relevance of Efflux and Uptake Drug Transporters at the Blood-Brain Barrier of Human Brain and Glioblastoma. *Clin. Pharmacol. Ther.* **2020**, *107*, 1116–1127. [CrossRef]
11. Su, Y.; Zhang, X.; Sinko, P.J. Human Organic Anion-Transporting Polypeptide OATP-A (SLC21A3) Acts in Concert with P-Glycoprotein and Multidrug Resistance Protein 2 in the Vectorial Transport of Saquinavir in Hep G2 Cells. *Mol. Pharm.* **2004**, *1*, 49–56. [CrossRef] [PubMed]
12. Maeda, T.; Takahashi, K.; Ohtsu, N.; Oguma, T.; Ohnishi, T.; Atsumi, R.; Tamai, I. Identification of Influx Transporter for the Quinolone Antibacterial Agent Levofloxacin. *Mol. Pharm.* **2007**, *4*, 85–94. [CrossRef] [PubMed]
13. Franke, R.M.; Baker, S.D.; Mathijssen, R.H.; Schuetz, E.G.; Sparreboom, A. Influence of Solute Carriers on the Pharmacokinetics of CYP3A4 Probes. *Clin. Pharmacol. Ther.* **2008**, *84*, 704–709. [CrossRef] [PubMed]
14. Hu, S.; Franke, R.M.; Filipski, K.K.; Hu, C.; Orwick, S.J.; de Bruijn, E.A.; Burger, H.; Baker, S.D.; Sparreboom, A. Interaction of Imatinib with Human Organic Ion Carriers. *Clin. Cancer Res.* **2008**, *14*, 3141–3148. [CrossRef]
15. Cvetkovic, M.; Leake, B.; Fromm, M.F.; Wilkinson, G.R.; Kim, R.B. OATP and P-Glycoprotein Transporters Mediate the Cellular Uptake and Excretion of Fexofenadine. *Drug Metab. Dispos.* **1999**, *27*, 866–871.

16. Fujino, H.; Saito, T.; Ogawa, S.-I.; Kojima, J. Transporter-Mediated Influx and Efflux Mechanisms of Pitavastatin, a New Inhibitor of HMG-CoA Reductase. *J. Pharm. Pharmacol.* **2005**, *57*, 1305–1311. [CrossRef]

17. Treiber, A.; Schneiter, R.; Häusler, S.; Stieger, B. Bosentan Is a Substrate of Human OATP1B1 and OATP1B3: Inhibition of Hepatic Uptake as the Common Mechanism of Its Interactions with Cyclosporin A, Rifampicin, and Sildenafil. *Drug Metab. Dispos.* **2007**, *35*, 1400–1407. [CrossRef]

18. Satoh, H.; Yamashita, F.; Tsujimoto, M.; Murakami, H.; Koyabu, N.; Ohtani, H.; Sawada, Y. Citrus Juices Inhibit the Function of Human Organic Anion-Transporting Polypeptide OATP-B. *Drug Metab. Dispos.* **2005**, *33*, 518–523. [CrossRef]

19. Ho, R.H.; Tirona, R.G.; Leake, B.F.; Glaeser, H.; Lee, W.; Lemke, C.J.; Wang, Y.; Kim, R.B. Drug and Bile Acid Transporters in Rosuvastatin Hepatic Uptake: Function, Expression, and Pharmacogenetics. *Gastroenterology* **2006**, *130*, 1793–1806. [CrossRef]

20. Nozawa, T.; Imai, K.; Nezu, J.-I.; Tsuji, A.; Tamai, I. Functional Characterization of PH-Sensitive Organic Anion Transporting Polypeptide OATP-B in Human. *J. Pharmacol. Exp. Ther.* **2004**, *308*, 438–445. [CrossRef]

21. Tamai, I.; Nezu, J.; Uchino, H.; Sai, Y.; Oku, A.; Shimane, M.; Tsuji, A. Molecular Identification and Characterization of Novel Members of the Human Organic Anion Transporter (OATP) Family. *Biochem. Biophys. Res. Commun.* **2000**, *273*, 251–260. [CrossRef] [PubMed]

22. Grube, M.; Köck, K.; Oswald, S.; Draber, K.; Meissner, K.; Eckel, L.; Böhm, M.; Felix, S.B.; Vogelgesang, S.; Jedlitschky, G.; et al. Organic Anion Transporting Polypeptide 2B1 Is a High-Affinity Transporter for Atorvastatin and Is Expressed in the Human Heart. *Clin. Pharmacol. Ther.* **2006**, *80*, 607–620. [CrossRef]

23. Kullak-Ublick, G.A.; Fisch, T.; Oswald, M.; Hagenbuch, B.; Meier, P.J.; Beuers, U.; Paumgartner, G. Dehydroepiandrosterone Sulfate (DHEAS): Identification of a Carrier Protein in Human Liver and Brain. *FEBS Lett.* **1998**, *424*, 173–176. [CrossRef]

24. Dresser, G.K.; Bailey, D.G.; Leake, B.F.; Schwarz, U.I.; Dawson, P.A.; Freeman, D.J.; Kim, R.B. Fruit Juices Inhibit Organic Anion Transporting Polypeptide-Mediated Drug Uptake to Decrease the Oral Availability of Fexofenadine. *Clin. Pharmacol. Ther.* **2002**, *71*, 11–20. [CrossRef] [PubMed]

25. Bailey, D.G.; Dresser, G.K.; Leake, B.F.; Kim, R.B. Naringin Is a Major and Selective Clinical Inhibitor of Organic Anion-Transporting Polypeptide 1A2 (OATP1A2) in Grapefruit Juice. *Clin. Pharmacol. Ther.* **2007**, *81*, 495–502. [CrossRef] [PubMed]

26. Zhao, R.; Kalvass, J.C.; Yanni, S.B.; Bridges, A.S.; Pollack, G.M. Fexofenadine Brain Exposure and the Influence of Blood-Brain Barrier P-Glycoprotein after Fexofenadine and Terfenadine Administration. *Drug Metab. Dispos.* **2009**, *37*, 529–535. [CrossRef] [PubMed]

27. Neuhaus, W.; Mandikova, J.; Pawlowitsch, R.; Linz, B.; Bennani-Baiti, B.; Lauer, R.; Lachmann, B.; Noe, C.R. Blood-Brain Barrier In Vitro Models as Tools in Drug Discovery: Assessment of the Transport Ranking of Antihistaminic Drugs. *Pharmazie* **2012**, *67*, 432–439. [PubMed]

28. Kalliokoski, A.; Niemi, M. Impact of OATP Transporters on Pharmacokinetics. *Br. J. Pharmacol.* **2009**, *158*, 693–705. [CrossRef]

29. Tahara, H.; Kusuhara, H.; Fuse, E.; Sugiyama, Y. P-Glycoprotein Plays a Major Role in the Efflux of Fexofenadine in the Small Intestine and Blood-Brain Barrier, but Only a Limited Role in Its Biliary Excretion. *Drug Metab. Dispos.* **2005**, *33*, 963–968. [CrossRef]

30. Chan, G.N.Y.; Hoque, M.T.; Bendayan, R. Role of Nuclear Receptors in the Regulation of Drug Transporters in the Brain. *Trends Pharmacol. Sci.* **2013**, *34*, 361–372. [CrossRef]

31. Lin, C.-J.; Tai, Y.; Huang, M.-T.; Tsai, Y.-F.; Hsu, H.-J.; Tzen, K.-Y.; Liou, H.-H. Cellular Localization of the Organic Cation Transporters, OCT1 and OCT2, in Brain Microvessel Endothelial Cells and Its Implication for MPTP Transport across the Blood-Brain Barrier and MPTP-Induced Dopaminergic Toxicity in Rodents. *J. Neurochem.* **2010**, *114*, 717–727. [CrossRef]

32. Geier, E.G.; Chen, E.C.; Webb, A.; Papp, A.C.; Yee, S.W.; Sadee, W.; Giacomini, K.M. Profiling Solute Carrier Transporters in the Human Blood-Brain Barrier. *Clin. Pharmacol Ther.* **2013**, *94*, 636–639. [CrossRef]

33. Friedrich, A.; Prasad, P.D.; Freyer, D.; Ganapathy, V.; Brust, P. Molecular Cloning and Functional Characterization of the OCTN2 Transporter at the RBE4 Cells, an In Vitro Model of the Blood-Brain Barrier. *Brain Res.* **2003**, *968*, 69–79. [CrossRef]

34. Morris, M.E.; Rodriguez-Cruz, V.; Felmlee, M.A. SLC and ABC Transporters: Expression, Localization, and Species Differences at the Blood-Brain and the Blood-Cerebrospinal Fluid Barriers. *AAPS J.* **2017**, *19*, 1317–1331. [CrossRef] [PubMed]

35. Halestrap, A.P.; Meredith, D. The SLC16 Gene Family-from Monocarboxylate Transporters (MCTs) to Aromatic Amino Acid Transporters and Beyond. *Pflugers Arch.* **2004**, *447*, 619–628. [CrossRef]

36. Zhang, M.; Wang, Y.; Bai, Y.; Dai, L.; Guo, H. Monocarboxylate Transporter 1 May Benefit Cerebral Ischemia via Facilitating Lactate Transport From Glial Cells to Neurons. *Front. Neurol.* **2022**, *13*, 781063. [CrossRef] [PubMed]

37. Vijay, N.; Morris, M.E. Role of Monocarboxylate Transporters in Drug Delivery to the Brain. *Curr. Pharm. Des.* **2014**, *20*, 1487–1498. [CrossRef] [PubMed]

38. Scalise, M.; Galluccio, M.; Console, L.; Pochini, L.; Indiveri, C. The Human SLC7A5 (LAT1): The Intriguing Histidine/Large Neutral Amino Acid Transporter and Its Relevance to Human Health. *Front. Chem.* **2018**, *6*, 243. [CrossRef]

39. Friesema, E.C.; Docter, R.; Moerings, E.P.; Verrey, F.; Krenning, E.P.; Hennemann, G.; Visser, T.J. Thyroid Hormone Transport by the Heterodimeric Human System L Amino Acid Transporter. *Endocrinology* **2001**, *142*, 4339–4348. [CrossRef]

40. del Amo, E.M.; Urtti, A.; Yliperttula, M. Pharmacokinetic Role of L-Type Amino Acid Transporters LAT1 and LAT2. *Eur. J. Pharm. Sci.* **2008**, *35*, 161–174. [CrossRef]

41. Murata, Y.; Neuhoff, S.; Rostami-Hodjegan, A.; Takita, H.; Al-Majdoub, Z.M.; Ogungbenro, K. In Vitro to In Vivo Extrapolation Linked to Physiologically Based Pharmacokinetic Models for Assessing the Brain Drug Disposition. *AAPS J.* **2022**, *24*, 28. [CrossRef]

42. Puris, E.; Fricker, G.; Gynther, M. Targeting Transporters for Drug Delivery to the Brain: Can We Do Better? *Pharm. Res.* **2022**, *39*, 1415–1455. [CrossRef]

43. Rihani, S.B.A.; Darakjian, L.I.; Deodhar, M.; Dow, P.; Turgeon, J.; Michaud, V. Disease-Induced Modulation of Drug Transporters at the Blood–Brain Barrier Level. *Int. J. Mol. Sci.* **2021**, *22*, 3742. [CrossRef]

44. Sanchez-Covarrubias, L.; Slosky, L.M.; Thompson, B.J.; Davis, T.P.; Ronaldson, P.T. Transporters at CNS Barrier Sites: Obstacles or Opportunities for Drug Delivery? *Curr. Pharm. Des.* **2014**, *20*, 1422–1449. [CrossRef]

45. Lee, W.; Glaeser, H.; Smith, L.H.; Roberts, R.L.; Moeckel, G.W.; Gervasini, G.; Leake, B.F.; Kim, R.B. Polymorphisms in Human Organic Anion-Transporting Polypeptide 1A2 (OATP1A2): Implications for Altered Drug Disposition and Central Nervous System Drug Entry. *J. Biol. Chem.* **2005**, *280*, 9610–9617. [CrossRef]

46. Gao, B.; Hagenbuch, B.; Kullak-Ublick, G.A.; Benke, D.; Aguzzi, A.; Meier, P.J. Organic Anion-Transporting Polypeptides Mediate Transport of Opioid Peptides across Blood-Brain Barrier. *J. Pharmacol. Exp. Ther.* **2000**, *294*, 73–79. [PubMed]

47. Bronger, H.; König, J.; Kopplow, K.; Steiner, H.-H.; Ahmadi, R.; Herold-Mende, C.; Keppler, D.; Nies, A.T. ABCC Drug Efflux Pumps and Organic Anion Uptake Transporters in Human Gliomas and the Blood-Tumor Barrier. *Cancer Res.* **2005**, *65*, 11419–11428. [CrossRef] [PubMed]

48. Betterton, R.D.; Davis, T.P.; Ronaldson, P.T. Organic Cation Transporters in the Central Nervous System. *Handb. Exp. Pharmacol.* **2021**, *266*, 301–328. [CrossRef]

49. Billington, S.; Salphati, L.; Hop, C.E.C.A.; Chu, X.; Evers, R.; Burdette, D.; Rowbottom, C.; Lai, Y.; Xiao, G.; Humphreys, W.G.; et al. Interindividual and Regional Variability in Drug Transporter Abundance at the Human Blood-Brain Barrier Measured by Quantitative Targeted Proteomics. *Clin. Pharmacol. Ther.* **2019**, *106*, 228–237. [CrossRef]

50. Bay, C.; Bajraktari-Sylejmani, G.; Haefeli, W.E.; Burhenne, J.; Weiss, J.; Sauter, M. Functional Characterization of the Solute Carrier LAT-1 (SLC7A5/SLC2A3) in Human Brain Capillary Endothelial Cells with Rapid UPLC-MS/MS Quantification of Intracellular Isotopically Labelled L-Leucine. *Int. J. Mol. Sci.* **2022**, *23*, 3637. [CrossRef] [PubMed]

51. Yan, R.; Zhao, X.; Lei, J.; Zhou, Q. Structure of the Human LAT1-4F2hc Heteromeric Amino Acid Transporter Complex. *Nature* **2019**, *568*, 127–130. [CrossRef]

52. Haining, Z.; Kawai, N.; Miyake, K.; Okada, M.; Okubo, S.; Zhang, X.; Fei, Z.; Tamiya, T. Relation of LAT1/4F2hc Expression with Pathological Grade, Proliferation and Angiogenesis in Human Gliomas. *BMC Clin. Pathol.* **2012**, *12*, 4. [CrossRef] [PubMed]

53. Watanabe, T.; Sanada, Y.; Hattori, Y.; Suzuki, M. Correlation between the Expression of LAT1 in Cancer Cells and the Potential Efficacy of Boron Neutron Capture Therapy. *J. Radiat. Res.* **2023**, *64*, 91–98. [CrossRef] [PubMed]

54. Bauer, M.; Matsuda, A.; Wulkersdorfer, B.; Philippe, C.; Traxl, A.; Özvegy-Laczka, C.; Stanek, J.; Nics, L.; Klebermass, E.-M.; Poschner, S.; et al. Influence of OATPs on Hepatic Disposition of Erlotinib Measured With Positron Emission Tomography. *Clin. Pharmacol. Ther.* **2018**, *104*, 139–147. [CrossRef] [PubMed]

55. Wolburg, H.; Lippoldt, A. Tight Junctions of the Blood-Brain Barrier: Development, Composition and Regulation. *Vascul. Pharmacol.* **2002**, *38*, 323–337. [CrossRef]

56. Helms, H.C.; Abbott, N.J.; Burek, M.; Cecchelli, R.; Couraud, P.-O.; Deli, M.A.; Förster, C.; Galla, H.J.; Romero, I.A.; Shusta, E.V.; et al. In Vitro Models of the Blood-Brain Barrier: An Overview of Commonly Used Brain Endothelial Cell Culture Models and Guidelines for Their Use. *J. Cereb. Blood Flow Metab.* **2016**, *36*, 862–890. [CrossRef]

57. Vallianatou, T.; Tsopelas, F.; Tsantili-Kakoulidou, A. Prediction Models for Brain Distribution of Drugs Based on Biomimetic Chromatographic Data. *Molecules* **2022**, *27*, 3668. [CrossRef]

58. Joó, F.; Karnushina, I. A Procedure for the Isolation of Capillaries from Rat Brain. *Cytobios* **1973**, *8*, 41–48. [PubMed]

59. Erdlenbruch, B.; Alipour, M.; Fricker, G.; Miller, D.S.; Kugler, W.; Eibl, H.; Lakomek, M. Alkylglycerol Opening of the Blood-Brain Barrier to Small and Large Fluorescence Markers in Normal and C6 Glioma-Bearing Rats and Isolated Rat Brain Capillaries. *Br. J. Pharmacol.* **2003**, *140*, 1201–1210. [CrossRef]

60. DeBault, L.E.; Kahn, L.E.; Frommes, S.P.; Cancilla, P.A. Cerebral Microvessels and Derived Cells in Tissue Culture: Isolation and Preliminary Characterization. *In Vitro* **1979**, *15*, 473–487. [CrossRef]

61. Cecchelli, R.; Dehouck, B.; Descamps, L.; Fenart, L.; Buée-Scherrer, V.V.; Duhem, C.; Lundquist, S.; Rentfel, M.; Torpier, G.; Dehouck, M. In Vitro Model for Evaluating Drug Transport across the Blood-Brain Barrier. *Adv. Drug Deliv. Rev.* **1999**, *36*, 165–178. [CrossRef]

62. Dehouck, M.P.; Méresse, S.; Delorme, P.; Fruchart, J.C.; Cecchelli, R. An Easier, Reproducible, and Mass-Production Method to Study the Blood-Brain Barrier In Vitro. *J. Neurochem.* **1990**, *54*, 1798–1801. [CrossRef] [PubMed]

63. Gaillard, P.J.; Voorwinden, L.H.; Nielsen, J.L.; Ivanov, A.; Atsumi, R.; Engman, H.; Ringbom, C.; de Boer, A.G.; Breimer, D.D. Establishment and Functional Characterization of an In Vitro Model of the Blood-Brain Barrier, Comprising a Co-Culture of Brain Capillary Endothelial Cells and Astrocytes. *Eur. J. Pharm. Sci.* **2001**, *12*, 215–222. [CrossRef]

64. DeBault, L.E.; Cancilla, P.A. Gamma-Glutamyl Transpeptidase in Isolated Brain Endothelial Cells: Induction by Glial Cells In Vitro. *Science* **1980**, *207*, 653–655. [CrossRef] [PubMed]

65. Tao-Cheng, J.H.; Nagy, Z.; Brightman, M.W. Tight Junctions of Brain Endothelium In Vitro Are Enhanced by Astroglia. *J. Neurosci.* **1987**, *7*, 3293–3299. [CrossRef] [PubMed]

66. Thomsen, L.B.; Burkhart, A.; Moos, T. A Triple Culture Model of the Blood-Brain Barrier Using Porcine Brain Endothelial Cells, Astrocytes and Pericytes. *PLoS ONE* **2015**, *10*, e0134765. [CrossRef]

67. Nakagawa, S.; Deli, M.A.; Kawaguchi, H.; Shimizudani, T.; Shimono, T.; Kittel, A.; Tanaka, K.; Niwa, M. A New Blood-Brain Barrier Model Using Primary Rat Brain Endothelial Cells, Pericytes and Astrocytes. *Neurochem. Int.* **2009**, *54*, 253–263. [CrossRef]

68. Burek, M.; Förster, C.Y. Cloning and Characterization of the Murine Claudin-5 Promoter. *Mol. Cell. Endocrinol.* **2009**, *298*, 19–24. [CrossRef] [PubMed]

69. Blecharz, K.G.; Drenckhahn, D.; Förster, C.Y. Glucocorticoids Increase VE-Cadherin Expression and Cause Cytoskeletal Rearrangements in Murine Brain Endothelial CEND Cells. *J. Cereb. Blood Flow Metab.* **2008**, *28*, 1139–1149. [CrossRef] [PubMed]

70. Harke, N.; Leers, J.; Kietz, S.; Drenckhahn, D.; Förster, C. Glucocorticoids Regulate the Human Occludin Gene through a Single Imperfect Palindromic Glucocorticoid Response Element. *Mol. Cell. Endocrinol.* **2008**, *295*, 39–47. [CrossRef]

71. Förster, C.; Silwedel, C.; Golenhofen, N.; Burek, M.; Kietz, S.; Mankertz, J.; Drenckhahn, D. Occludin as Direct Target for Glucocorticoid-Induced Improvement of Blood-Brain Barrier Properties in a Murine In Vitro System. *J. Physiol.* **2005**, *565*, 475–486. [CrossRef]

72. Hughes, C.C.; Lantos, P.L. Uptake of Leucine and Alanine by Cultured Cerebral Capillary Endothelial Cells. *Brain Res.* **1989**, *480*, 126–132. [CrossRef]

73. Pifferi, F.; Jouin, M.; Alessandri, J.M.; Haedke, U.; Roux, F.; Perrière, N.; Denis, I.; Lavialle, M.; Guesnet, P. n-3 Fatty Acids Modulate Brain Glucose Transport in Endothelial Cells of the Blood-Brain Barrier. *Prostaglandins Leukot. Essent. Fat. Acids* **2007**, *77*, 279–286. [CrossRef] [PubMed]

74. Pifferi, F.; Jouin, M.; Alessandri, J.-M.; Roux, F.; Perrière, N.; Langelier, B.; Lavialle, M.; Cunnane, S.; Guesnet, P. n-3 Long-Chain Fatty Acids and Regulation of Glucose Transport in Two Models of Rat Brain Endothelial Cells. *Neurochem. Int.* **2010**, *56*, 703–710. [CrossRef]

75. Calabria, A.R.; Shusta, E.V. A Genomic Comparison of In Vivo and In Vitro Brain Microvascular Endothelial Cells. *J. Cereb. Blood Flow Metab.* **2008**, *28*, 135–148. [CrossRef] [PubMed]

76. Garberg, P.; Ball, M.; Borg, N.; Cecchelli, R.; Fenart, L.; Hurst, R.D.; Lindmark, T.; Mabondzo, A.; Nilsson, J.E.; Raub, T.J.; et al. In Vitro Models for the Blood-Brain Barrier. *Toxicol. Vitr.* **2005**, *19*, 299–334. [CrossRef]

77. Weksler, B.; Romero, I.A.; Couraud, P.-O. The HCMEC/D3 Cell Line as a Model of the Human Blood Brain Barrier. *Fluids Barriers CNS* **2013**, *10*, 16. [CrossRef]

78. Gromnicova, R.; Davies, H.A.; Sreekanthreddy, P.; Romero, I.A.; Lund, T.; Roitt, I.M.; Phillips, J.B.; Male, D.K. Glucose-Coated Gold Nanoparticles Transfer across Human Brain Endothelium and Enter Astrocytes In Vitro. *PLoS ONE* **2013**, *8*, e81043. [CrossRef] [PubMed]

79. Steiner, O.; Coisne, C.; Engelhardt, B.; Lyck, R. Comparison of Immortalized BEnd5 and Primary Mouse Brain Microvascular Endothelial Cells as In Vitro Blood-Brain Barrier Models for the Study of T Cell Extravasation. *J. Cereb. Blood Flow Metab.* **2011**, *31*, 315–327. [CrossRef]

80. Carl, S.M.; Lindley, D.J.; Das, D.; Couraud, P.O.; Weksler, B.B.; Romero, I.; Mowery, S.A.; Knipp, G.T. ABC and SLC Transporter Expression and Proton Oligopeptide Transporter (POT) Mediated Permeation across the Human Blood--Brain Barrier Cell Line, HCMEC/D3 [Corrected]. *Mol. Pharm.* **2010**, *7*, 1057–1068. [CrossRef] [PubMed]

81. Lopez-Ramirez, M.A.; Male, D.K.; Wang, C.; Sharrack, B.; Wu, D.; Romero, I.A. Cytokine-Induced Changes in the Gene Expression Profile of a Human Cerebral Microvascular Endothelial Cell-Line, HCMEC/D3. *Fluids Barriers CNS* **2013**, *10*, 27. [CrossRef]

82. Stanton, J.A.; Williams, E.I.; Betterton, R.D.; Davis, T.P.; Ronaldson, P.T. Targeting Organic Cation Transporters at the Blood-Brain Barrier to Treat Ischemic Stroke in Rats. *Exp. Neurol.* **2022**, *357*, 114181. [CrossRef]

83. Betterton, R.D.; Abdullahi, W.; Williams, E.I.; Lochhead, J.J.; Brzica, H.; Stanton, J.; Reddell, E.; Ogbonnaya, C.; Davis, T.P.; Ronaldson, P.T. Regulation of Blood-Brain Barrier Transporters by Transforming Growth Factor-β/Activin Receptor-Like Kinase 1 Signaling: Relevance to the Brain Disposition of 3-Hydroxy-3-Methylglutaryl Coenzyme A Reductase Inhibitors (i.e., Statins). *Drug Metab. Dispos.* **2022**, *50*, 942–956. [CrossRef] [PubMed]

84. Zhang, G.; Chen, S.; Jia, J.; Liu, C.; Wang, W.; Zhang, H.; Zhen, X. Development and Evaluation of Novel Metformin Derivative Metformin Threonate for Brain Ischemia Treatment. *Front. Pharmacol.* **2022**, *13*, 879690. [CrossRef]

85. Taskar, K.S.; Mariappan, T.T.; Kurawattimath, V.; Gautam, S.S.; Mullapudi, T.V.R.; Sridhar, S.K.; Kallem, R.R.; Marathe, P.; Mandlekar, S. Unmasking the Role of Uptake Transporters for Digoxin Uptake Across the Barriers of the Central Nervous System in Rat. *J. Cent. Nerv. Syst. Dis.* **2017**, *9*, 1179573517693596. [CrossRef]

86. Marie, S.; Breuil, L.; Chalampalakis, Z.; Becquemont, L.; Verstuyft, C.; Lecoq, A.-L.; Caillé, F.; Gervais, P.; Lebon, V.; Comtat, C.; et al. [(11)C]Glyburide PET Imaging for Quantitative Determination of the Importance of Organic Anion-Transporting Polypeptide Transporter Function in the Human Liver and Whole-Body. *Biomed. Pharmacother.* **2022**, *156*, 113994. [CrossRef] [PubMed]

87. Tournier, N.; Saba, W.; Cisternino, S.; Peyronneau, M.-A.; Damont, A.; Goutal, S.; Dubois, A.; Dollé, F.; Scherrmann, J.-M.; Valette, H.; et al. Effects of Selected OATP and/or ABC Transporter Inhibitors on the Brain and Whole-Body Distribution of Glyburide. *AAPS J.* **2013**, *15*, 1082–1090. [CrossRef]

88. Ose, A.; Kusuhara, H.; Endo, C.; Tohyama, K.; Miyajima, M.; Kitamura, S.; Sugiyama, Y. Functional Characterization of Mouse Organic Anion Transporting Peptide 1a4 in the Uptake and Efflux of Drugs across the Blood-Brain Barrier. *Drug Metab. Dispos.* **2010**, *38*, 168–176. [CrossRef]

89. Abdullahi, W.; Brzica, H.; Hirsch, N.A.; Reilly, B.G.; Ronaldson, P.T. Functional Expression of Organic Anion Transporting Polypeptide 1a4 Is Regulated by Transforming Growth Factor-β/Activin Receptor-like Kinase 1 Signaling at the Blood-Brain Barrier. *Mol. Pharmacol.* **2018**, *94*, 1321–1333. [CrossRef]

90. Braun, C.; Sakamoto, A.; Fuchs, H.; Ishiguro, N.; Suzuki, S.; Cui, Y.; Klinder, K.; Watanabe, M.; Terasaki, T.; Sauer, A. Quantification of Transporter and Receptor Proteins in Dog Brain Capillaries and Choroid Plexus: Relevance for the Distribution in Brain and CSF of Selected BCRP and P-Gp Substrates. *Mol. Pharm.* **2017**, *14*, 3436–3447. [CrossRef] [PubMed]

91. Kodaira, H.; Kusuhara, H.; Fujita, T.; Ushiki, J.; Fuse, E.; Sugiyama, Y. Quantitative Evaluation of the Impact of Active Efflux by P-Glycoprotein and Breast Cancer Resistance Protein at the Blood-Brain Barrier on the Predictability of the Unbound Concentrations of Drugs in the Brain Using Cerebrospinal Fluid Concentration as a Surrogate. *J. Pharmacol. Exp. Ther.* **2011**, *339*, 935–944. [CrossRef]

92. Hasannejad, H.; Takeda, M.; Narikawa, S.; Huang, X.-L.; Enomoto, A.; Taki, K.; Niwa, T.; Jung, S.H.; Onozato, M.L.; Tojo, A.; et al. Human Organic Cation Transporter 3 Mediates the Transport of Antiarrhythmic Drugs. *Eur. J. Pharmacol.* **2004**, *499*, 45–51. [CrossRef] [PubMed]

93. Burckhardt, B.C.; Henjakovic, M.; Hagos, Y.; Burckhardt, G. Counter-Flow Suggests Transport of Dantrolene and 5-OH Dantrolene by the Organic Anion Transporters 2 (OAT2) and 3 (OAT3). *Pflugers Arch.* **2016**, *468*, 1909–1918. [CrossRef] [PubMed]

94. Ronaldson, P.T.; Finch, J.D.; Demarco, K.M.; Quigley, C.E.; Davis, T.P. Inflammatory Pain Signals an Increase in Functional Expression of Organic Anion Transporting Polypeptide 1a4 at the Blood-Brain Barrier. *J. Pharmacol. Exp. Ther.* **2011**, *336*, 827–839. [CrossRef]

95. Chen, X.; Loryan, I.; Payan, M.; Keep, R.F.; Smith, D.E.; Hammarlund-Udenaes, M. Effect of Transporter Inhibition on the Distribution of Cefadroxil in Rat Brain. *Fluids Barriers CNS* **2014**, *11*, 25. [CrossRef]

96. Sekhar, G.N.; Georgian, A.R.; Sanderson, L.; Vizcay-Barrena, G.; Brown, R.C.; Muresan, P.; Fleck, R.A.; Thomas, S.A. Organic Cation Transporter 1 (OCT1) Is Involved in Pentamidine Transport at the Human and Mouse Blood-Brain Barrier (BBB). *PLoS ONE* **2017**, *12*, e0173474. [CrossRef]

97. Mochizuki, T.; Mizuno, T.; Kurosawa, T.; Yamaguchi, T.; Higuchi, K.; Tega, Y.; Nozaki, Y.; Kawabata, K.; Deguchi, Y.; Kusuhara, H. Functional Investigation of Solute Carrier Family 35, Member F2, in Three Cellular Models of the Primate Blood-Brain Barrier. *Drug Metab. Dispos.* **2021**, *49*, 3–11. [CrossRef] [PubMed]

98. Reese, M.J.; Savina, P.M.; Generaux, G.T.; Tracey, H.; Humphreys, J.E.; Kanaoka, E.; Webster, L.O.; Harmon, K.A.; Clarke, J.D.; Polli, J.W. In Vitro Investigations into the Roles of Drug Transporters and Metabolizing Enzymes in the Disposition and Drug Interactions of Dolutegravir, a HIV Integrase Inhibitor. *Drug Metab. Dispos.* **2013**, *41*, 353–361. [CrossRef]

99. EMA. EMA Assessment Report. Available online: https://www.ema.europa.eu/en/documents/assessment-report/dovato-epar-public-assessment-report_en.pdf (accessed on 4 October 2023).

100. Janneh, O.; Chandler, B.; Hartkoorn, R.; Kwan, W.S.; Jenkinson, C.; Evans, S.; Back, D.J.; Owen, A.; Khoo, S.H. Intracellular Accumulation of Efavirenz and Nevirapine Is Independent of P-Glycoprotein Activity in Cultured CD4 T Cells and Primary Human Lymphocytes. *J. Antimicrob. Chemother.* **2009**, *64*, 1002–1007. [CrossRef]

101. Hartkoorn, R.C.; Kwan, W.S.; Shallcross, V.; Chaikan, A.; Liptrott, N.; Egan, D.; Sora, E.S.; James, C.E.; Gibbons, S.; Bray, P.G.; et al. HIV Protease Inhibitors Are Substrates for OATP1A2, OATP1B1 and OATP1B3 and Lopinavir Plasma Concentrations Are Influenced by SLCO1B1 Polymorphisms. *Pharmacogenet. Genom.* **2010**, *20*, 112–120. [CrossRef]

102. Jung, N.; Lehmann, C.; Rubbert, A.; Knispel, M.; Hartmann, P.; van Lunzen, J.; Stellbrink, H.-J.; Faetkenheuer, G.; Taubert, D. Relevance of the Organic Cation Transporters 1 and 2 for Antiretroviral Drug Therapy in Human Immunodeficiency Virus Infection. *Drug Metab. Dispos.* **2008**, *36*, 1616–1623. [CrossRef] [PubMed]

103. Curley, P.; Rajoli, R.K.R.; Moss, D.M.; Liptrott, N.J.; Letendre, S.; Owen, A.; Siccardi, M. Efavirenz Is Predicted To Accumulate in Brain Tissue: An In Silico, In Vitro, and In Vivo Investigation. *Antimicrob. Agents Chemother.* **2017**, *61*, e01841-16. [CrossRef] [PubMed]

104. Sato, S.; Matsumiya, K.; Tohyama, K.; Kosugi, Y. Translational CNS Steady-State Drug Disposition Model in Rats, Monkeys, and Humans for Quantitative Prediction of Brain-to-Plasma and Cerebrospinal Fluid-to-Plasma Unbound Concentration Ratios. *AAPS J.* **2021**, *23*, 81. [CrossRef]

105. Khurana, V.; Minocha, M.; Pal, D.; Mitra, A.K. Role of OATP-1B1 and/or OATP-1B3 in Hepatic Disposition of Tyrosine Kinase Inhibitors. *Drug Metab. Drug Interact.* **2014**, *29*, 179–190. [CrossRef]

106. Elmeliegy, M.A.; Carcaboso, A.M.; Tagen, M.; Bai, F.; Stewart, C.F. Role of ATP-Binding Cassette and Solute Carrier Transporters in Erlotinib CNS Penetration and Intracellular Accumulation. *Clin. Cancer Res.* **2011**, *17*, 89–99. [CrossRef]

107. Bauer, M.; Karch, R.; Wulkersdorfer, B.; Philippe, C.; Nics, L.; Klebermass, E.-M.; Weber, M.; Poschner, S.; Haslacher, H.; Jäger, W.; et al. A Proof-of-Concept Study to Inhibit ABCG2- and ABCB1-Mediated Efflux Transport at the Human Blood–Brain Barrier. *J. Nucl. Med.* **2019**, *60*, 486–491. [CrossRef] [PubMed]

108. Sampson, K.E.; Brinker, A.; Pratt, J.; Venkatraman, N.; Xiao, Y.; Blasberg, J.; Steiner, T.; Bourner, M.; Thompson, D.C. Zinc Finger Nuclease–Mediated Gene Knockout Results in Loss of Transport Activity for P-Glycoprotein, BCRP, and MRP2 in Caco-2 Cells. *Drug Metab. Dispos.* **2015**, *43*, 199–207. [CrossRef]

109. Matsushima, S.; Maeda, K.; Hayashi, H.; Debori, Y.; Schinkel, A.H.; Schuetz, J.D.; Kusuhara, H.; Sugiyama, Y. Involvement of Multiple Efflux Transporters in Hepatic Disposition of Fexofenadine. *Mol. Pharmacol.* **2008**, *73*, 1474–1483. [CrossRef]

110. Izumi, S.; Nozaki, Y.; Maeda, K.; Komori, T.; Takenaka, O.; Kusuhara, H.; Sugiyama, Y. Investigation of the Impact of Substrate Selection on In Vitro Organic Anion Transporting Polypeptide 1B1 Inhibition Profiles for the Prediction of Drug-Drug Interactions. *Drug Metab. Dispos.* **2015**, *43*, 235–247. [CrossRef]

111. Shimizu, M.; Fuse, K.; Okudaira, K.; Nishigaki, R.; Maeda, K.; Kusuhara, H.; Sugiyama, Y. Contribution of OATP (organic anion-transporting polypeptide) family transporters to the hepatic uptake of fexofenadine in humans. *Drug Metab. Dispos.* **2005**, *33*, 1477–1481. [CrossRef]

112. Ming, X.; Knight, B.M.; Thakker, D.R. Vectorial Transport of Fexofenadine across Caco-2 Cells: Involvement of Apical Uptake and Basolateral Efflux Transporters. *Mol. Pharm.* **2011**, *8*, 1677–1686. [CrossRef]

113. Morita, T.; Akiyoshi, T.; Sato, R.; Uekusa, Y.; Katayama, K.; Yajima, K.; Imaoka, A.; Sugimoto, Y.; Kiuchi, F.; Ohtani, H. Citrus Fruit-Derived Flavanone Glycoside Narirutin Is a Novel Potent Inhibitor of Organic Anion-Transporting Polypeptides. *J. Agric. Food Chem.* **2020**, *68*, 14182–14191. [CrossRef] [PubMed]

114. Glaeser, H.; Bailey, D.G.; Dresser, G.K.; Gregor, J.C.; Schwarz, U.I.; McGrath, J.S.; Jolicoeur, E.; Lee, W.; Leake, B.F.; Tirona, R.G.; et al. Intestinal Drug Transporter Expression and the Impact of Grapefruit Juice in Humans. *Clin. Pharmacol. Ther.* **2007**, *81*, 362–370. [CrossRef]

115. Tahara, H.; Kusuhara, H.; Maeda, K.; Koepsell, H.; Fuse, E.; Sugiyama, Y. Inhibition of OAT3-Mediated Renal Uptake as a Mechanism for Drug-Drug Interaction Between Fexofenadine and Probenecid. *Drug Metab. Dispos.* **2006**, *34*, 743–747. [CrossRef] [PubMed]

116. Mathialagan, S.; Piotrowski, M.A.; Tess, D.A.; Feng, B.; Litchfiled, J.; Varma, M.V. Quantitative Prediction of Human Renal Clearance and Drug-Drug Interactions of Organic Anion Transporter Substrates Using In Vitro Transport Data. *Drug Metab. Dispos.* **2017**, *45*, 409–417. [CrossRef]

117. Matsushima, S.; Maeda, K.; Inoue, K.; Ohta, K.; Yuasa, H.; Kondo, T.; Nakayama, H.; Horita, S.; Kusuhara, H.; Sugiyama, Y. The Inhibition of Human Multidrug and Toxin Extrusion 1 Is Involved in the Drug-Drug Interaction Caused by Cimetidine. *Drug Metab. Dispos.* **2009**, *37*, 555–559. [CrossRef] [PubMed]

118. Doan, K.M.; Wring, S.A.; Shampine, L.J.; Jordan, K.H.; Bishop, J.P.; Kratz, J.; Yang, E.; Serabjit-Singh, C.J.; Adkison, K.K.; Polli, J.W. Steady-State Brain Concentrations of Antihistamines in Rats. *Pharmacology* **2004**, *72*, 92–98. [CrossRef]

119. Crowe, A.; Teoh, Y.-K. Limited P-Glycoprotein Mediated Efflux for Anti-Epileptic Drugs. *J. Drug Target.* **2006**, *14*, 291–300. [CrossRef]

120. Cheng, Z.; Liu, H.; Yu, N.; Wang, F.; An, G.; Xu, Y.; Liu, Q.; Guan, C.; Ayrton, A. Hydrophilic Anti-Migraine Triptans Are Substrates for OATP1A2, a Transporter Expressed at Human Blood-Brain Barrier. *Xenobiotica* **2012**, *42*, 880–890. [CrossRef] [PubMed]

121. Dickens, D.; Webb, S.D.; Antonyuk, S.; Giannoudis, A.; Owen, A.; Rädisch, S.; Hasnain, S.S.; Pirmohamed, M. Transport of Gabapentin by LAT1 (SLC7A5). *Biochem. Pharmacol.* **2013**, *85*, 1672–1683. [CrossRef] [PubMed]

122. Feng, B.; Hurst, S.; Lu, Y.; Varma, M.V.; Rotter, C.J.; El-Kattan, A.; Lockwood, P.; Corrigan, B. Quantitative Prediction of Renal Transporter-Mediated Clinical Drug–Drug Interactions. *Mol. Pharm.* **2013**, *10*, 4207–4215. [CrossRef]

123. Gebauer, L.; Jensen, O.; Brockmöller, J.; Dücker, C. Substrates and Inhibitors of the Organic Cation Transporter 3 and Comparison with OCT1 and OCT2. *J. Med. Chem.* **2022**, *65*, 12403–12416. [CrossRef]

124. Pochini, L.; Galluccio, M.; Scalise, M.; Console, L.; Indiveri, C. OCTN: A Small Transporter Subfamily with Great Relevance to Human Pathophysiology, Drug Discovery, and Diagnostics. *SLAS Discov.* **2019**, *24*, 89–110. [CrossRef] [PubMed]

125. Fridén, M.; Winiwarter, S.; Jerndal, G.; Bengtsson, O.; Wan, H.; Bredberg, U.; Hammarlund-Udenaes, M.; Antonsson, M. Structure-Brain Exposure Relationships in Rat and Human Using a Novel Data Set of Unbound Drug Concentrations in Brain Interstitial and Cerebrospinal Fluids. *J. Med. Chem.* **2009**, *52*, 6233–6243. [CrossRef]

126. Redeker, K.-E.M.; Jensen, O.; Gebauer, L.; Meyer-Tönnies, M.J.; Brockmöller, J. Atypical Substrates of the Organic Cation Transporter 1. *Biomolecules* **2022**, *12*, 1664. [CrossRef] [PubMed]

127. Hasegawa, N.; Furugen, A.; Ono, K.; Koishikawa, M.; Miyazawa, Y.; Nishimura, A.; Umazume, T.; Narumi, K.; Kobayashi, M.; Iseki, K. Cellular Uptake Properties of Lamotrigine in Human Placental Cell Lines: Investigation of Involvement of Organic Cation Transporters (SLC22A1–5). *Drug Metab. Pharmacokinet.* **2020**, *35*, 266–273. [CrossRef] [PubMed]

128. Summerfield, S.G.; Lucas, A.J.; Porter, R.A.; Jeffrey, P.; Gunn, R.N.; Read, K.R.; Stevens, A.J.; Metcalf, A.C.; Osuna, M.C.; Kilford, P.J.; et al. Toward an Improved Prediction of Human In Vivo Brain Penetration. *Xenobiotica* **2008**, *38*, 1518–1535. [CrossRef]

129. Uchida, Y.; Ohtsuki, S.; Kamiie, J.; Terasaki, T. Blood-Brain Barrier (BBB) Pharmacoproteomics: Reconstruction of In Vivo Brain Distribution of 11 P-Glycoprotein Substrates Based on the BBB Transporter Protein Concentration, In Vitro Intrinsic Transport Activity, and Unbound Fraction in Plasma and Brain in Mice. *J. Pharmacol. Exp. Ther.* **2011**, *339*, 579–588. [CrossRef]

130. Jia, Y.; Liu, Z.; Wang, C.; Meng, Q.; Huo, X.; Liu, Q.; Sun, H.; Sun, P.; Yang, X.; Ma, X.; et al. P-Gp, MRP2 and OAT1/OAT3 Mediate the Drug-Drug Interaction between Resveratrol and Methotrexate. *Toxicol. Appl. Pharmacol.* **2016**, *306*, 27–35. [CrossRef] [PubMed]

131. Nozaki, Y.; Kusuhara, H.; Kondo, T.; Iwaki, M.; Shiroyanagi, Y.; Nakayama, H.; Horita, S.; Nakazawa, H.; Okano, T.; Sugiyama, Y. Species Difference in the Inhibitory Effect of Nonsteroidal Anti-Inflammatory Drugs on the Uptake of Methotrexate by Human Kidney Slices. *J. Pharmacol. Exp. Ther.* **2007**, *322*, 1162–1170. [CrossRef]

132. Ramsey, L.B.; Bruun, G.H.; Yang, W.; Treviño, L.R.; Vattathil, S.; Scheet, P.; Cheng, C.; Rosner, G.L.; Giacomini, K.M.; Fan, Y.; et al. Rare versus Common Variants in Pharmacogenetics: SLCO1B1 Variation and Methotrexate Disposition. *Genome Res.* **2012**, *22*, 1–8. [CrossRef] [PubMed]

133. Abe, T.; Unno, M.; Onogawa, T.; Tokui, T.; Kondo, T.N.; Nakagomi, R.; Adachi, H.; Fujiwara, K.; Okabe, M.; Suzuki, T.; et al. LST-2, A Human Liver-Specific Organic Anion Transporter, Determines Methotrexate Sensitivity in Gastrointestinal Cancers. *Gastroenterology* **2001**, *120*, 1689–1699. [CrossRef] [PubMed]

134. Visentin, M.; Chang, M.-H.; Romero, M.F.; Zhao, R.; Goldman, I.D. Substrate- and PH-Specific Antifolate Transport Mediated by Organic Anion-Transporting Polypeptide 2B1 (OATP2B1-SLCO2B1). *Mol. Pharmacol.* **2012**, *81*, 134–142. [CrossRef] [PubMed]

135. Badagnani, I.; Castro, R.A.; Taylor, T.R.; Brett, C.M.; Huang, C.C.; Stryke, D.; Kawamoto, M.; Johns, S.J.; Ferrin, T.E.; Carlson, E.J.; et al. Interaction of Methotrexate with Organic-Anion Transporting Polypeptide 1A2 and Its Genetic Variants. *J. Pharmacol. Exp. Ther.* **2006**, *318*, 521–529. [CrossRef]

136. Yuan, Y.; Yang, H.; Kong, L.; Li, Y.; Li, P.; Zhang, H.; Ruan, J. Interaction between Rhein Acyl Glucuronide and Methotrexate Based on Human Organic Anion Transporters. *Chem.-Biol. Interact.* **2017**, *277*, 79–84. [CrossRef]

137. Kurata, T.; Iwamoto, T.; Kawahara, Y.; Okuda, M. Characteristics of Pemetrexed Transport by Renal Basolateral Organic Anion Transporter HOAT3. *Drug Metab. Pharmacokinet.* **2014**, *29*, 148–153. [CrossRef]

138. Fujino, H.; Nakai, D.; Nakagomi, R.; Saito, M.; Tokui, T.; Kojima, J. Metabolic Stability and Uptake by Human Hepatocytes of Pitavastatin, a New Inhibitor of HMG-CoA Reductase. *Arzneimittelforschung* **2004**, *54*, 382–388. [CrossRef]

139. Bi, Y.; Costales, C.; Mathialagan, S.; West, M.; Eatemadpour, S.; Lazzaro, S.; Tylaska, L.; Scialis, R.; Zhang, H.; Umland, J.; et al. Quantitative Contribution of Six Major Transporters to the Hepatic Uptake of Drugs: "SLC-Phenotyping" Using Primary Human Hepatocytes. *J. Pharmacol. Exp. Ther.* **2019**, *370*, 72–83. [CrossRef]

140. Varma, M.V.; Rotter, C.J.; Chupka, J.; Whalen, K.M.; Duignan, D.B.; Feng, B.; Litchfield, J.; Goosen, T.C.; El-Kattan, A.F. PH-Sensitive Interaction of HMG-CoA Reductase Inhibitors (Statins) with Organic Anion Transporting Polypeptide 2B1. *Mol. Pharm.* **2011**, *8*, 1303–1313. [CrossRef]

141. SHIRASAKA, Y.; SUZUKI, K.; SHICHIRI, M.; NAKANISHI, T.; TAMAI, I. Intestinal Absorption of HMG-CoA Reductase Inhibitor Pitavastatin Mediated by Organic Anion Transporting Polypeptide and P-Glycoprotein/Multidrug Resistance 1. *Drug Metab. Pharmacokinet.* **2011**, *26*, 171–179. [CrossRef]

142. Yabuuchi, H.; Tamai, I.; Nezu, J.; Sakamoto, K.; Oku, A.; Shimane, M.; Sai, Y.; Tsuji, A. Novel Membrane Transporter OCTN1 Mediates Multispecific, Bidirectional, and PH-Dependent Transport of Organic Cations. *J. Pharmacol. Exp. Ther.* **1999**, *289*, 768–773. [PubMed]

143. Grigat, S.; Fork, C.; Bach, M.; Golz, S.; Geerts, A.; Schömig, E.; Gründemann, D. The Carnitine Transporter SLC22A5 Is Not a General Drug Transporter, but It Efficiently Translocates Mildronate. *Drug Metab. Dispos.* **2009**, *37*, 330–337. [CrossRef] [PubMed]

144. Tanihara, Y.; Masuda, S.; Sato, T.; Katsura, T.; Ogawa, O.; Inui, K. Substrate Specificity of MATE1 and MATE2-K, Human Multidrug and Toxin Extrusions/H+-Organic Cation Antiporters. *Biochem. Pharmacol.* **2007**, *74*, 359–371. [CrossRef] [PubMed]

145. Hashiguchi, Y.; Hamada, A.; Shinohara, T.; Tsuchiya, K.; Jono, H.; Saito, H. Role of P-Glycoprotein in the Efflux of Raltegravir from Human Intestinal Cells and CD4+ T-Cells as an Interaction Target for Anti-HIV Agents. *Biochem. Biophys. Res. Commun.* **2013**, *439*, 221–227. [CrossRef] [PubMed]

146. Moss, D.M.; Kwan, W.S.; Liptrott, N.J.; Smith, D.L.; Siccardi, M.; Khoo, S.H.; Back, D.J.; Owen, A. Raltegravir Is a Substrate for SLC22A6: A Putative Mechanism for the Interaction between Raltegravir and Tenofovir. *Antimicrob. Agents Chemother.* **2011**, *55*, 879–887. [CrossRef] [PubMed]

147. Srinivas, N.; Rosen, E.P.; Gilliland, W.M., Jr.; Kovarova, M.; Remling-Mulder, L.; Cruz, G.D.L.; White, N.; Adamson, L.; Schauer, A.P.; Sykes, C.; et al. Antiretroviral Concentrations and Surrogate Measures of Efficacy in the Brain Tissue and CSF of Preclinical Species. *Xenobiotica* **2019**, *49*, 1192–1201. [CrossRef]

148. Collett, A.; Tanianis-Hughes, J.; Hallifax, D.; Warhurst, G. Predicting P-Glycoprotein Effects on Oral Absorption: Correlation of Transport in Caco-2 with Drug Pharmacokinetics in Wild-Type and Mdr1a(-/-) Mice In Vivo. *Pharm. Res.* **2004**, *21*, 819–826. [CrossRef]

149. Wegler, C.; Gazit, M.; Issa, K.; Subramaniam, S.; Artursson, P.; Karlgren, M. Expanding the Efflux In Vitro Assay Toolbox: A CRISPR-Cas9 Edited MDCK Cell Line with Human BCRP and Completely Lacking Canine MDR1. *J. Pharm. Sci.* **2021**, *110*, 388–396. [CrossRef]

150. Vavricka, S.R.; Montfoort, J.V.; Ha, H.R.; Meier, P.J.; Fattinger, K. Interactions of Rifamycin SV and Rifampicin with Organic Anion Uptake Systems of Human Liver. *Hepatology* **2002**, *36*, 164–172. [CrossRef] [PubMed]

151. Te Brake, L.H.M.; van den Heuvel, J.J.M.W.; Buaben, A.O.; van Crevel, R.; Bilos, A.; Russel, F.G.; Aarnoutse, R.E.; Koenderink, J.B. Moxifloxacin Is a Potent In Vitro Inhibitor of OCT- and MATE-Mediated Transport of Metformin and Ethambutol. *Antimicrob. Agents Chemother.* **2016**, *60*, 7105–7114. [CrossRef] [PubMed]

152. Deng, F.; Tuomi, S.-K.; Neuvonen, M.; Hirvensalo, P.; Kulju, S.; Wenzel, C.; Oswald, S.; Filppula, A.M.; Niemi, M. Comparative Hepatic and Intestinal Efflux Transport of Statins. *Drug Metab. Dispos.* **2021**, *49*, 750–759. [CrossRef]

153. Wen, J.; Wei, X.; Sheng, X.; Zhou, D.; Peng, H.; Lu, Y.; Zhou, J. Effect of Ursolic Acid on Breast Cancer Resistance Protein-Mediated Transport of Rosuvastatin In Vivo and Vitro. *Chin. Med. Sci. J.* **2015**, *30*, 218–225. [CrossRef] [PubMed]

154. Kimoto, E.; Li, R.; Scialis, R.J.; Lai, Y.; Varma, M.V.S. Hepatic Disposition of Gemfibrozil and Its Major Metabolite Gemfibrozil 1-O-β-Glucuronide. *Mol. Pharm.* **2015**, *12*, 3943–3952. [CrossRef]

155. Bednarczyk, D.; Sanghvi, M.V. Organic Anion Transporting Polypeptide 2B1 (OATP2B1), an Expanded Substrate Profile, Does It Align with OATP2B1's Hypothesized Function? *Xenobiotica* **2020**, *50*, 1128–1137. [CrossRef]
156. Ronaldson, P.T.; Brzica, H.; Abdullahi, W.; Reilly, B.G.; Davis, T.P. Transport Properties of Statins by OATP1A2 and Regulation by Transforming Growth Factor-β (TGF-β) Signaling in Human Endothelial Cells. *J. Pharmacol. Exp. Ther.* **2020**, *376*. [CrossRef]
157. Harati, R.; Benech, H.; Villégier, A.S.; Mabondzo, A. P-Glycoprotein, Breast Cancer Resistance Protein, Organic Anion Transporter 3, and Transporting Peptide 1a4 during Blood-Brain Barrier Maturation: Involvement of Wnt/β-Catenin and Endothelin-1 Signaling. *Mol. Pharm.* **2013**, *10*, 1566–1580. [CrossRef]
158. Takeda, M.; Khamdang, S.; Narikawa, S.; Kimura, H.; Kobayashi, Y.; Yamamoto, T.; Cha, S.H.; Sekine, T.; Endou, H. Human Organic Anion Transporters and Human Organic Cation Transporters Mediate Renal Antiviral Transport. *J. Pharmacol. Exp. Ther.* **2002**, *300*, 918–924. [CrossRef]
159. Minuesa, G.; Volk, C.; Molina-Arcas, M.; Gorboulev, V.; Erkizia, I.; Arndt, P.; Clotet, B.; Pastor-Anglada, M.; Koepsell, H.; Martinez-Picado, J. Transport of Lamivudine [(-)-β-l-2′,3′-Dideoxy-3′-Thiacytidine] and High-Affinity Interaction of Nucleoside Reverse Transcriptase Inhibitors with Human Organic Cation Transporters 1, 2, and 3. *J. Pharmacol. Exp. Ther.* **2009**, *329*, 252–261. [CrossRef] [PubMed]
160. Parvez, M.M.; Kaisar, N.; Shin, H.J.; Jung, J.A.; Shin, J.-G. Inhibitory Interaction Potential of 22 Antituberculosis Drugs on Organic Anion and Cation Transporters of the SLC22A Family. *Antimicrob. Agents Chemother.* **2016**, *60*, 6558–6567. [CrossRef]
161. Bors, L.A.; Erdő, F. Overcoming the Blood–Brain Barrier. Challenges and Tricks for CNS Drug Delivery. *Sci. Pharm.* **2019**, *87*, 6. [CrossRef]
162. Singh, V.K.; Subudhi, B.B. Development and Characterization of Lysine-Methotrexate Conjugate for Enhanced Brain Delivery. *Drug Deliv.* **2016**, *23*, 2327–2337. [CrossRef] [PubMed]

 *pharmaceutics*

Review

Strategies for Drug Delivery into the Brain: A Review on Adenosine Receptors Modulation for Central Nervous System Diseases Therapy

Mercedes Fernandez [1], Manuela Nigro [1], Alessia Travagli [1], Silvia Pasquini [2], Fabrizio Vincenzi [1], Katia Varani [1], Pier Andrea Borea [3], Stefania Merighi [1,*] and Stefania Gessi [1,*]

[1] Department of Translational Medicine, University of Ferrara, 44121 Ferrara, Italy; mercedes.fernandez@unife.it (M.F.); manuela.nigro@unife.it (M.N.); trvlss@unife.it (A.T.); fabrizio.vincenzi@unife.it (F.V.); vrk@unife.it (K.V.)

[2] Department of Chemical, Pharmaceutical and Agricultural Science, University of Ferrara, 44121 Ferrara, Italy; silvia.pasquini@unife.it

[3] University of Ferrara, 44121 Ferrara, Italy; bpa@unife.it

* Correspondence: mhs@unife.it (S.M.); gss@unife.it (S.G.)

Abstract: The blood–brain barrier (BBB) is a biological barrier that protects the central nervous system (CNS) by ensuring an appropriate microenvironment. Brain microvascular endothelial cells (ECs) control the passage of molecules from blood to brain tissue and regulate their concentration-versus-time profiles to guarantee proper neuronal activity, angiogenesis and neurogenesis, as well as to prevent the entry of immune cells into the brain. However, the BBB also restricts the penetration of drugs, thus presenting a challenge in the development of therapeutics for CNS diseases. On the other hand, adenosine, an endogenous purine-based nucleoside that is expressed in most body tissues, regulates different body functions by acting through its G-protein-coupled receptors (A1, A2A, A2B and A3). Adenosine receptors (ARs) are thus considered potential drug targets for treating different metabolic, inflammatory and neurological diseases. In the CNS, A1 and A2A are expressed by astrocytes, oligodendrocytes, neurons, immune cells and ECs. Moreover, adenosine, by acting locally through its receptors A1 and/or A2A, may modulate BBB permeability, and this effect is potentiated when both receptors are simultaneously activated. This review showcases in vivo and in vitro evidence supporting AR signaling as a candidate for modifying endothelial barrier permeability in the treatment of CNS disorders.

Keywords: blood–brain barrier; brain microvascular endothelial cells; tight junction proteins; adenosine receptor signaling; adenosine receptor agonists; adenosine receptor antagonist

Citation: Fernandez, M.; Nigro, M.; Travagli, A.; Pasquini, S.; Vincenzi, F.; Varani, K.; Borea, P.A.; Merighi, S.; Gessi, S. Strategies for Drug Delivery into the Brain: A Review on Adenosine Receptors Modulation for Central Nervous System Diseases Therapy. *Pharmaceutics* **2023**, *15*, 2441. https://doi.org/10.3390/pharmaceutics15102441

Academic Editors: Gert Fricker and Elena Puris

Received: 7 September 2023
Revised: 29 September 2023
Accepted: 8 October 2023
Published: 10 October 2023

1. Introduction

Current trends in applied and translational neuropharmacology involve the development of new drugs for the treatment of specific central nervous system (CNS) diseases, including the development of strategies that can overcome the blood–brain barrier (BBB) and enable drugs to reach the brain's microenvironment by directing them to the specific/target cells involved in such diseases [1,2].

Compared to other therapeutic fields, drug discovery for brain diseases is very unsuccessful mainly due to the high complexity of the brain, the long time needed for drug development, the lack of effective approaches for drug delivery into the brain, and significant drug side effects. Furthermore, due to the BBB's limited permeability, only around 2% of small-molecule drugs can cross the BBB on the basis of their molecular size, charge and hydrophobicity, as well as multidrug resistant protein substrate activity [1,3–8].

Although BBB is the major obstacle for delivering drugs into the CNS, it plays a critical role in the brain's metabolic activity, as well as neuronal function. Under healthy conditions, the brain is constantly and efficiently supplied of oxygen, nutrients and essential amino acids, while neurotoxic substances are taken away to maintain its homeostatic balance [9].

The BBB also plays a pivotal role in the regulation of some neurotransmitters' trafficking between the peripheral and central nervous systems for which a specific transport system across the BBB exists. Such is the case for adenosine, a purine nucleoside that acts through purinergic P1 receptors that are selective for adenosine [10,11]. It functions as an important local signaling molecule and is involved in various physiological functions, including neurotransmission, cardiac pace and immune regulation [12,13]. Adenosine, produced both intracellularly and extracellularly, mediates its function by acting through its 7-transmembrane G-protein-coupled receptors, A1, A2A, A2B and A3 adenosine receptors (A1R, A2AR, A2BR and A3R), which are widely distributed on various cell types in the body [14–17]. Among them, BBB microvascular endothelial cells (ECs) express A1Rs and A2ARs.

On the other hand, it has been described that selective agonists of A1R and A2AR induce a transient increase in the BBB permeability, thus allowing the entry of macromolecules into the CNS. This effect is, in turn, counteracted by blocking adenosine signaling through selective ARs antagonists [18], which suggests that adenosine signaling might be considered a strategy to modify the BBB permeability.

This review focuses on purinergic signaling mediated by P1 ARs as a putative tool to modify endothelial barrier permeability to facilitate the flow of substances across the BBB. In this paper, we discuss recent in vivo and in vitro research performed to understand how the BBB can be modified in some brain pathological conditions and exploit the ARs signaling through the use of different types of AR agonists and antagonists. An overview of current state of the art of new strategies that can be applied to increase BBB permeability to enable the delivery of therapeutic drugs for the treatment of CNS diseases is provided.

2. Biological Brain Barriers and Brain Drug Delivery

As is well known, biological barriers that protect the CNS providing an appropriate microenvironment are: the blood–brain barrier (BBB) constituted by brain microvessel endothelial cells; the blood–cerebrospinal fluid barrier (BCSFB) or blood CSF barrier, formed by choroid plexus epithelial cells; and the meningeal barrier, formed by arachnoid epithelial cells [4] (Figure 1).

These barriers are responsible for controlling the trafficking of molecules to guarantee the function of neuronal circuits, synaptic transmission, angiogenesis and neurogenesis. Moreover, they prevent the entry of immune cells into the brain [4].

The BBB regulates the interaction between blood and the brain through an extended surface of capillaries involved in regulating the passage of molecules/drugs from blood to brain tissue and the concentration-versus-time profiles, i.e., pharmacokinetics (PK). At the same time, PK depends on drug physicochemical properties and, consequently, on its transport across the BBB, as well as on the physiological characteristics of the CNS, including the effect of aging [19,20].

The BBB prevents bloodborne and exogenous molecules from entering the central nervous system and is crucial to maintaining brain homeostasis. The differentiation of the endothelium into a barrier layer begins during embryonic angiogenesis and, in adults, is maintained by a very refined and complex structure made of (i) microvascular endothelial cells (ECs) that delineate the cerebral capillaries in the brain and the spinal cord, providing CNS with a highly complex vasculature; (ii) inter-endothelial junctions that are protein complexes of three types, tight junctions (TJs), adherens junctions and gap junctions; and (iii) pericytes (PCs), astrocytes and the basal membrane (BM), a thin extracellular matrix layer that surrounds ECs, thus giving the BBB its "impermeable" property [4].

ECs possess specific markers including gamma-glutamyl transpeptidase (GGTP), alkaline phosphatase, von Willebrand factor (vWf), glucose transporter-1 (GLUT-1), endothelial barrier antigen (EBA) and OX-47 antigen. Furthermore, ECs are also characterized by the presence of P-glycoprotein (Pgp) and multidrug resistance-associated protein (MRP), which are two N-glycosylated phosphoproteins [21,22].

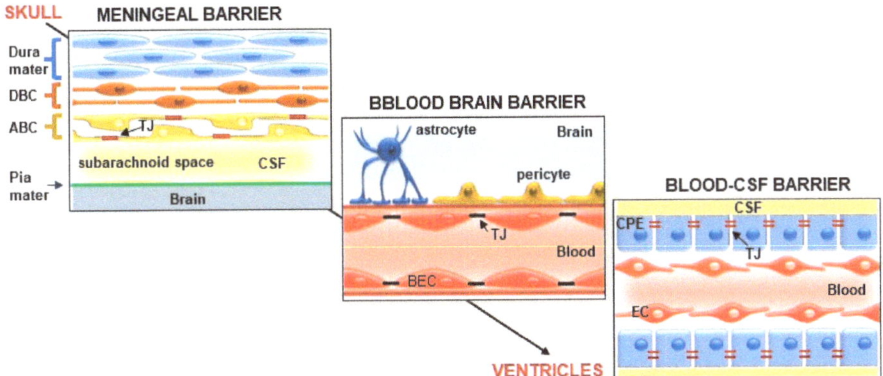

Figure 1. Biological barriers of the CNS. The meningeal barrier possesses the most complex structure of all the brain barriers. It is made of arachnoid barrier cells (ABC) and tight junctions (TJs) between adjacent cells forming a barrier between the cerebrospinal fluid (CSF) in the subarachnoid space (SAS) and the dural border cells (DBC) and the dura mater. The blood–brain barrier is made of brain endothelial cells (BECs), inter-endothelial junctions (TJs, adherens junctions and gap junctions), pericytes, astrocytes and the basal membrane. The blood–CSF barrier is situated in the choroid plexus within each brain ventricle. The barrier is made of cells of the choroid plexus epithelium (CPE) and tight junctions.

The permeability of the BBB is mainly controlled by inter-endothelial junctions, which link ECs in a coordinated way to form a continuous barrier capable of regulating its permeability and the communication between ECs in response to physiological and pathophysiological changes in the brain [9,23] (Figure 2).

Tight junction (TJ) complexes between brain capillary endothelial cells are important structural components of the BBB and consist of interconnected strands of transmembrane proteins and accessory proteins [3]. TJs possess associated transmembrane proteins, occludin and claudin (Cln), involved in the regulation of BBB integrity [24,25]. They act as junctional adhesion molecules that seal endothelial cells together, thus regulating the diffusion of fluid and small molecules across the BBB. Both proteins are essential for BBB formation. It has been reported that an increase in claudin-5 expression is associated with an increase in transendothelial electrical resistance (TEER), a functional parameter of paracellular permeability, and consequently a decrease in BBB permeability [26]. In the context of brain drug delivery, TEER values are indicators of cellular barrier integrity that should be considered when evaluating the transport of drugs into the brain. A decrease in TEER values correlates with increased paracellular space between adjacent endothelial cells and, hence, with increased permeability [27]. Accessory proteins, such as the zona occludens-1 (ZO-1) family, are cytoplasmatic proteins with regulatory roles and are involved in TJ structural organization through the linking of TJ complexes to the cytoskeleton, as well as in the support of protein signal transduction [28,29]. Ezrin/radixin/moesin (ERM) and myosin light chain (MLC) are actin-binding proteins present at TJs. Phosphorylated ERM proteins are involved in actin remodeling and vascular permeability control [30]. In addition, ERM proteins have been found to form adhesion complexes and actin–myosin stress fibers after the activation of Rho GTPases. Moreover, TJ barrier function can be regulated by signal transduction cascades that require the activation of Rho proteins and depend on the actin cytoskeleton. Furthermore, the phosphorylation of MLC by myosin light chain kinase (MLCK) provokes the activation of actomyosin contraction and the increased phosphorylation of MLC corresponds to increased BBB permeability [31–33].

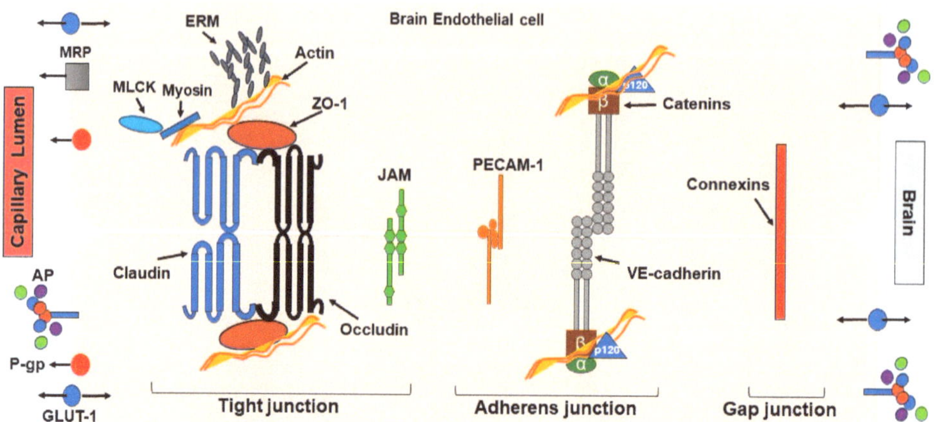

Figure 2. Brain microvascular endothelial cells (ECs)—inter-endothelial junctions complex of the BBB. The high vasculature that ECs provide to the CNS is made of inter-endothelial junctions that link the ECs in a coordinated way forming a continuous barrier. BBB permeability and the communication between ECs are regulated in response to physiological and pathophysiological conditions. Some of the ECs specific markers included in the cartoon are: Alkaline-phosphatase (AP) localized in the plasma membrane and highly expressed in ECs; the macromolecule transporter glucose transporter-1 (GLUT-1); and the two efflux transporter proteins P-glycoprotein (Pgp) and multidrug resistance-associated protein (MRP). There are three types of inter-endothelial junctions: tight junctions (TJs), adherens junctions and gap junctions. TJs consist of interconnected strands of the trans-membrane proteins occludin and claudins (Cln), which form the seal between endothelial cells. These proteins link with the actin cytoskeleton via the accessory proteins ZO-1, a family of cytoplasmatic proteins with regulatory and structural roles in the organization of TJs. Junction adhesion molecules (JAMs) are an immunoglobulin superfamily highly enriched at TJs that contribute to their properties. Ezrin/radixin/moesin (ERM) and myosin light chain (MLC) are actin-binding proteins present at TJs. The phosphorylation of ERM and the phosphorylation of MLC by the MLC kinase (MLCK) provokes the modulation of actin and actomyosin arrangement thus modifying the BBB permeability. Adherens junctions are characterized for the interaction between vascular endothelial (VE)-cadherin and catenin proteins (p120, β- and α-catenin). The platelet endothelial cell adhesion molecule-1 (PECAM-1) is a transmembrane protein that is also present at the adherens junctions and contributes to the regulation of vascular integrity. Together, these proteins control the formation, maintenance and function of adherens junctions. Gap junctions are composed of connexin protein subunits.

Adherens junctions possess multiple functions, including stabilization of cell–cell adhesion, regulation of the actin cytoskeleton, and intracellular signaling and transcription. The interaction between endothelial (E)-cadherin and the p120-, β- and α-catenin families of proteins occurs at adherens junctions [34]. Platelet endothelial cell adhesion molecule-1 (PECAM-1, or CD31) is a transmembrane protein that is also present at adherens junctions and contributes to the regulation of vascular integrity [35]. Connexins constitute a large family of transmembrane proteins that allow intercellular communication and the transfer of ions and small signaling molecules between cells.

Gap junctions are composed of connexin protein subunits. These junctions form channels between two adjacent cells, thus allowing direct intercellular communication [36].

Moreover, the structure that is made of ECs, the capillary BM, astrocytes, pericytes (PCs) embedded within the BM, and microglial and neuronal cells is known as the neurovascular unit (NVU) [4]. The maintenance of barrier function is guaranteed due to the high vascularity of the brain, which allows oxygen and nutrient supply from the peripheral blood to the brain [12,37].

Different types of transport across the BBB have been described, among them, passive diffusion, active efflux, carrier-mediated transport (CMT) and receptor-mediated transport (Figure 3).

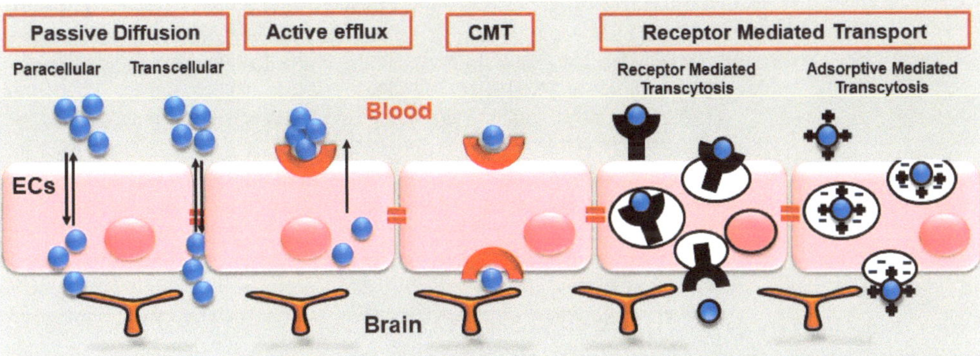

Figure 3. Types of transport across the BBB. Passive diffusion (paracellular and transcellular transport), active efflux, carrier-mediated transport (CMT) and receptor-mediated transport (transcytosis and adsorptive mediated transcytosis).

Passive diffusion consists of the passage of the molecules across the BBB by one of the following ways: (i) through the cells, i.e., transcellular transport; and (ii) between adjacent cells, i.e., paracellular transport [38]. In passive diffusion, lipid-soluble small molecules can diffuse passively across the BBB into the brain on the basis of their molecular parameters, such as lipid solubility and molecular weight (lower than 400 Da). As mentioned above, only 2% of small molecules can cross the BBB and only 5% of available drugs can be used to treat CNS diseases [3].

Active efflux is driven by different primary barrier interface proteins. Adenosine triphosphate (ATP)-binding cassette (ABC) transporters are BBB transporter proteins expressed on the luminal endothelial plasma membrane and limit the BBB permeability to toxins and therapeutic compounds. Their expression and functional activity have been found to decrease in neuropathological conditions such as Alzheimer's disease (AD) and Parkinson's disease (PD) [39]. Other efflux transporter proteins are P-glycoprotein (Pgp), multidrug resistance-associated proteins (MRPs) and breast cancer resistance protein (BCRP). Pgp and BCRP are expressed in the luminal membrane of the BBB and are responsible for the transport of substrates from the endothelium to blood, whereas MRPs are expressed in both luminal and abluminal membranes. When the primary barrier is dysfunctional, transport activity can be performed by pericytes and perivascular astrocytic endfeet.

Carrier-mediated transport (CMT) is an energy-dependent pathway generally used by small hydrophilic molecules. This type of transport is performed by carriers that possess specific receptors on their membrane that recognize the target molecules (amino acids, carbohydrates, monocarboxylic acids, fatty acids, hormones, nucleotides, organic anions, amines, choline and vitamins) and transport them across the cell membrane [40]. As an example, essential omega-3 fatty acids, such as docosahexaenoic acid (DHA), are transported into the brain via endothelial major facilitator superfamily (MSF) domain-containing protein 2a (MFSD2a) [41], a member of the MFS membrane carrier transporters known to play a critical role in maintaining the integrity of the BBB [42].

Receptor-mediated transport is a non-specific transcytotic mechanism to transport a variety of large bloodborne molecules and complexes across the BBB. In particular, it is used by specific neuroactive peptides, regulatory proteins, hormones and growth factors to cross the BBB. There are two types of receptor-mediated transport: receptor-mediated transcytosis (RMT) and adsorptive-mediated transcytosis (AMT) [43].

On the other hand, in the CNS there is a low infiltration of neutrophils into the brain and the interaction between BBB and immune cells is strictly regulated. It occurs at the endothelial cell junction through paracellular or transcellular transport and is known as diapedesis [44]. Moreover, during embryonic development and under physiological conditions, mononuclear cells enter the brain and become microglia, the resident immunological competent cells of the CNS [45].

On the other hand, different physiological functions have been ascribed to the BBB at the blood–brain interface:

(1) Regulate levels of neurotransmitters present in the central and peripheral nervous systems by keeping these different neurotransmitters' pools separate to minimize "crosstalk" and protect the brain from sudden changes in plasma levels. This is the case of the neuroexcitatory amino acid glutamate, whose levels vary significantly under specific physiological and/or pathological conditions (food intake and ischemic stroke). High levels of glutamate inside the brain significantly damage neuronal tissue [4].

(2) Maintain ionic homeostasis and brain nutrition via the action of specific ion channels and transporters, which guarantees the optimal conditions for neural and synaptic signaling functions. On the other hand, the BBB, together with the BCSFB, regulates pH levels and also calcium and magnesium homeostasis, which is essential to control neuronal excitability and the transmigration of macrophages across the BBB. Moreover, Ca^{2+} is involved in the modulation of BBB integrity and endothelial morphology [4].

(3) Protect the brain against increased access of neurotoxins of an endogenous and/or exogenous origin by regulating the entry of blood-circulating substances through the ABC energy-dependent efflux transporters (ATP-binding cassette transporters) present in the BBB luminal surface [46].

(4) Limit the outflow of plasma macromolecules into the brain under physiological conditions, thus preventing several pathological conditions that would be provoked when large-molecular-weight serum proteins cross the BBB [4].

Nevertheless, the BBB's integrity and physiological functions can be altered during pathological conditions. BBB dysfunction can be associated with either moderate and reversible changes due to TJ aperture or with severe and permanent changes in proteins, enzymes and other transport systems [4,47]. Therefore, microglial activation and infiltration of immune cells into the brain occur, provoking an alteration in CNS homeostasis and resulting in possible damage to the brain and thereby causing or exacerbating a pathological condition. The relative impermeable healthy BBB becomes more permeable in response to cytokines, vasoactive agents (for instance, adenosine) and chemical mediators (e.g., ATP, aspartate, nitric oxide and glutamate) that are released during the development of a pathological condition [4].

The blood–cerebrospinal fluid barrier (BCSFB) is formed by the epithelial cells of the choroid plexus (CPE) connected through TJs. These cells secrete CSF into the ventricular brain system in a controlled way [48] (Figure 1).

The meningeal barrier consists of an avascular arachnoid epithelium made of arachnoid barrier cells (ABC) and tight junctions (TJs). Situated under the dura mater, it wraps the CNS and keeps separate the extracellular fluids of the central nervous system and the fluids of the rest of the body. Due to its avascular nature and small surface area, the meningeal barrier contributes marginally to the exchange of substances between blood and the brain [4,49] (Figure 1).

3. Brain Drug Delivery Strategies

The BBB constitutes a physical semipermeable and dynamic interface between the blood and the brain and is mainly responsible for managing the exchanges between blood and brain tissue by allowing only certain molecules or ions to pass through via diffusion or via more specialized processes of facilitated diffusion, passive transport or active transport.

Keeping in mind the BBB's highly organized close-fitting structure, it is easy to understand why it also represents an obstacle for drug delivery to the brain. Thus, knowledge of the molecular and physiological mechanisms involved in the transport of compounds through the BBB represents an important key for CNS drug delivery, and further studies on brain drug pharmacokinetics and distribution into specific areas of the brain are needed [50]. The concept is to modify drugs that cannot cross the BBB in their original state by transporting them through the use of already-identified endogenous BBB carrier-mediated and receptor-mediated transport systems.

In the last forty years, a lot of new technologies aiming to modify the physicochemical properties of drugs to cross the BBB and deliver drugs into the brain have been developed, including BBB disruption, CSF delivery, trans-cranial delivery, lipid carriers, prodrugs, stem cells, exosomes, nanoparticles (NPs), gene therapy, recombinant proteins and nucleic-acid therapies [51,52].

One of the many classifications that can be found through the existing literature groups these brain drug delivery strategies into three categories, and the strategies included in these three categories can be invasive or non-invasive or miscellaneous for the brain depending on their method of action [51].

BBB disruption (BBBD) is a brain drug delivery strategy that can be invasive or miscellaneous. This technique consists of the use of substances that induce modifications of ECs and TJs, thus allowing the passage of certain molecules into the brain. The main disadvantage of this technique is that some substances (e.g., mannitol, glycerol and ethanol) can compromise BBB integrity and physiological functions; thus, noxious blood components that reach the brain tissue can cause injury [50,51,53].

CSF delivery strategies through intracerebroventricular or intrathecal infusion of drugs are invasive brain drug delivery strategies consisting of the injection of therapeutic proteins directly into the CSF. The advantage of these methods is the lower drug concentration of therapeutic drugs compared with regular intravenous administration, while also avoiding the risk of toxicity due to systemic exposure. Intrathecal drug administration can be performed via lumbar puncture or via an implanted intrathecal drug delivery device (IDDD) [54].

Trans-vascular brain drug delivery includes receptor-, carrier- and lipid-mediated transport and active efflux. In this context, transcytosis mediated by receptors (RMT) seems to be the easier way to bypass the BBB by allowing the drug being used to be first taken up via EC receptor-mediated endocytosis (RME), and then the targeting ligand binds to a receptor on the surface of ECs to transport the drug to the abluminal surface of the brain parenchyma via transcytosis [53]. The most commonly used targets in preclinical and clinical studies are transferrin, insulin, low-density lipoprotein, and diphtheria toxin receptors [55–59].

Nanotechnology applied to drug delivery into the brain is an approach mainly based on the use of nanosized technology for drug release in the brain. This delivery system uses a wide variety of nanoscale drug delivery platforms, mainly lipid- and polymer-based NPs, that assure a controlled and improved release of their cargo by protecting the loaded drugs from being metabolized [55]. Current efforts have mainly focused on increasing the ability of NPs to target their therapeutic site, thereby minimizing the doses of drugs released at undesired sites, to enhance the efficiency and, thus, the cost of these therapies [54,60].

Niosomes are a type of vesicular nanocarrier whose non-ionic, surfactant-based and non-toxic characteristics, together with their amphiphilic nature that is compatible for the encapsulation of lipophilic and hydrophilic molecules, make them a new delivery system for active drugs [61]. Many examples of NP applications have been reviewed in detail [4,51,53,60].

4. Adenosine, Adenosine Receptors and BBB Modulation: Implications in CNS Homeostasis in Both Physiological and Pathological Conditions

4.1. Physiological Conditions

Adenosine is an endogenous purine nucleoside that, under physiological conditions, is mainly generated inside cells from the substrate S-Adenosyl-L-homocysteine through the catalytic action of the enzyme S-Adenosyl-L-homocysteine hydrolase, leading to adenosine and homocysteine as the products. Adenosine is a basic constituent of nucleic acids and can also be present in the forms of AMP, ADP and ATP; these purine molecules play a crucial role in cellular energy supply and in the transfer of information within and across cells [20,62,63].

In the brain, adenosine can also be generated outside cells from ATP via sequential dephosphorylation reactions that are catalyzed mainly by nucleoside triphosphate diphosphohydrolase-1 (CD39), which transforms ATP into ADP and successively into AMP, and by the action of $5'$-ectonucleotidase (CD73), which transforms AMP into adenosine. Then, extracellular produced adenosine can be transported inside cells by nucleoside transporters (ENTs or CNTs) or transformed into AMP through a phosphorylation reaction catalyzed by the adenosine kinase (AK) enzyme [64]. Alternatively, extracellular adenosine can be deaminated to inosine by adenosine deaminase (ADA). In the extracellular space, adenosine acts as a signaling molecule with different actions, including regulating the supply of oxygen on a demand basis, involving in immunomodulation, inducing repair and stimulating angiogenesis, and acting as a neuromodulator/neurotransmitter in the CNS [63,65] (Figure 4).

The cycle of adenosine being released and then uptaken, together with its extracellular metabolism, is conditioned on the state of the cells being in a healthy or pathological condition [66–68].

Adenosine acts through its "7-transmembrane G-protein coupled" purinergic P1 receptor subtypes, namely A1, A2A, A2B and A3, which are functionally linked to adenylyl cyclase (AC), to inhibit (A1R and A3R) or stimulate (A2AR and A2BR) the production of the second messenger cAMP. A2A and A2B receptors interact with Gs proteins to activate adenylate cyclase, while A1 and A3 interact with Gi/0 proteins, which reduce the activity of adenylate cyclase. Moreover, ARs can be involved in other cAMP-mediated pathways, including protein kinase C (PKC), phosphoinositide 3 kinase (PI3K), PI3K/Akt activation and mitogen-activated protein kinase (MAPK) pathways [12,14,69,70] (Figure 4).

It has been reported that under physiological conditions, the intra- and extracellular concentrations of adenosine in the brain are in the same range, between 20 and 30 nM, and that these concentrations are equilibrated by ENTs and CNTs, which, respectively, mediate the efflux and influx of adenosine across the cell membrane [18].

The action of adenosine through its receptors is different depending on the AR expression levels in specific tissues and the adenosine concentration. In fact, AR subtypes differ in their binding affinity to their natural ligand adenosine. A1R and A3R are high-affinity receptors that bind adenosine at concentrations of 0.1–3 nM, whereas A2AR is a high-affinity receptor that binds adenosine at 1–20 nM and A2BR is a low-affinity AR that is activated at an adenosine concentration higher than 1 µM. The receptors A1R and A2AR are widely expressed in the brain. Thus, under physiological conditions, adenosine signaling in the brain is mainly mediated via A1R and A2AR, whereas A2BR and A3R are more active in other body compartments where adenosine concentration is much higher compared to the brain [15,18,71,72]. It should be noted that the moderate expression of these adenosine receptors, as shown by glial cells under physiological conditions, is upregulated during CNS insults, such as neuroinflammation [73].

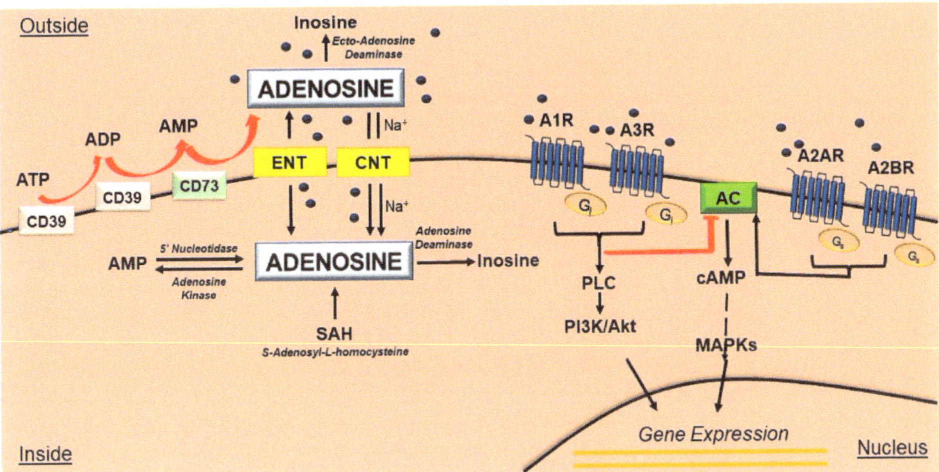

Figure 4. Adenosine metabolism and signal transduction. The nucleoside adenosine is generated from ATP and converted to ADP and AMP by ectonucleoside triphosphate diphosphohydrolase 1 (CD39) followed by conversion to adenosine by ecto-5-nucleotidase (CD73). Inside the cell, the generation of adenosine can occur either from ATP, ADP and AMP via the activity of cytoplasmic 5′-nucleotidases, or from the hydrolysis of S-Adenosyl-L-homocysteine (SAH) via the action of the enzyme S-Adenosyl-L-homocysteine hydrolase. Adenosine is then transported and metabolized to either inosine or AMP by the enzymes adenosine deaminase (AD) or adenosine kinase, respectively. Extracellular adenosine can be metabolized to inosine via the ecto-adenosine deaminase and/or uptaken by either the nucleoside transporters system (ENT) or the concentrative nucleoside transporters system CNT, a sodium-dependent Na+ transporter. Extracellular adenosine activates the adenosine receptors A1, A2A, A2B and A3. A1R and A3 are coupled to an inhibitory G protein (Gi). A2A and A2B are coupled to the stimulatory G protein (Gs). Signaling pathways mediated by A1 and A3 include adenyl cyclase inhibition, phospholipase C (PLC) activation and PI3K/Akt activation. A2A and A2B induce adenyl cyclase activity. ARs stimulate PI3K/Akt and the mitogen-activated protein kinase (MAPK) family which trigger specific cellular responses including the regulation of gene transcription. Abbreviations: ADP, adenosine diphosphate; AMP, adenosine monophosphate; Akt, serine-threonine protein kinase; ATP, adenosine triphosphate; CD39, cluster of differentiation 39; CD73, cluster of differentiation 73; CNT, Na+-dependent concentrative nucleoside transporter; ENT, Na+-independent equilibrative nucleoside transporter; PI3K, phosphoinositide 3-kinase, SAH, S-Adenosyl-L-homocysteine.

A1R and A2ARs are highly expressed in the brain. A1R is more abundant compared to A2AR, which exhibits a lower distribution under physiological conditions. A1R is expressed by neurons and is located presynaptically and postsynaptically. Adenosine acts as a neuromodulator that is involved in the control of transcellular messenger homeostasis and neurotransmitter release [14]. Moreover, adenosine is known to be a physiological neuromodulator that acts temporally at the place of production due to its short lifetime of around 10 s [74,75].

A1Rs and A2ARs are expressed by ECs, neurons, glial cells and immune cells [62,76]. Since ECs also express CD73, it is hypothesized that these cells are sensitive to the potential damage induced by extracellular ATP, which is known to be an injury signal in the brain [77,78], and might convert ATP into adenosine via the action of the ectoenzyme CD73. In this way, the resulting adenosine, acting locally through its receptor A1 and/or A2A, may modulate BBB permeability, with this effect being potentiated when both receptors are simultaneously activated. The literature shows evidence supporting AR signaling as a candidate for modifying endothelial barrier permeability [12,18,79,80].

Carman et al. extended on these studies by using an in vitro model of mouse brain endothelial cells (BEC), the Bend.3 cell line, an ex vivo model of mouse primary brain ECs, and in vivo mouse and rat models as well. The mouse Bend.3 cell line expresses A1R and A2AR mRNAs but does not express A2B and A3 receptors, whereas they express CD73. These authors demonstrated that the activation of A1R and A2AR by the selective agonists NECA (N-Ethylcarboxamidoadenosine) and Lexiscan (CVT-3146, also known as Regadenoson), respectively, induced a decrease in TEER, thus increasing the paracellular space between endothelial cells where TJs are located and, consequently, enhancing BBB permeability. These results suggest the involvement of the actin cytoskeleton in the maintenance of endothelial cell shape and barrier integrity, thus providing evidence for the involvement of adenosine signaling. The activation of A1Rs/A2aRs induced the formation of "actomyosin stress fibers", which induced cytoskeleton changes in ECs that increased BBB permeability. As mentioned above, claudin-5, occludin and ZO-1 are TJ-associated proteins involved in the control of BBB structure and function. The treatment of Bend.3 cells with NECA or Lexiscan induced a decrease in the expression of claudin-5 and ZO-1 proteins, suggesting that the change in BBB permeability could be modulated by AR signaling through a change in TJ protein expression. Moreover, Western blot analysis of primary mouse BEC lysate showed the expression of A1Rs and A2ARs, thus suggesting that ECs of this in vitro model were capable of directly responding to adenosine [79].

On the other hand, Lexiscan or NECA administration (i.v.) to mice and rats increased the CNS entry of intravenously administered fluorescently labeled dextrans (10 kDa and 70 kDa); this effect was observed to be dose dependent, and the duration of this BBB permeability increase depended on the adenosine agonist's lifetime, with the effect of NECA lasting longer (around hours) than that of Lexiscan (around minutes). The same effect was observed for macromolecules, such as antibodies, with this effect being significant for therapeutic agents that crossed the BBB. Moreover, immunofluorescent staining and fluorescent in situ hybridization methodologies applied to brain cortical slices demonstrated that ARs (A1 and A2A) co-stained with the CD31 endothelial marker [12]. From these in vivo results, it can be stated that adenosine, binding to its receptors, induces a transient and reversible BBB disruption that enhances BBB permeability.

In another study using an in vitro co-culture BBB model of a human cerebral microvascular endothelial cell line (hCMEC/D3) and a human normal glial cell line (HEB), it was demonstrated that the extract of Ginkgo biloba (EGb) induced a reversible increase in BBB permeability via the A1R signaling pathway. This effect was associated with an alteration in TJ ultrastructure, causing a reduction in the TEER parameter. In particular, the expression of phosphorylated TJ proteins, ezrin/radixin/moesin (ERM) and myosin light chain (MLC) (both ERM and MLC are actin-binding proteins) increased, whereas the expression of ZO-1, occludin and claudin-3 did not change. These observed effects were counteracted by the A1R antagonist DPCPX (1,3-dipropyl-8-cyclopentyl xanthine) [33].

All of the abovementioned results further suggest that adenosine, acting through A1Rs or A2ARs, could modulate BBB permeability to facilitate the entry of molecules into the CNS.

Other studies performed confirmed these results and expanded the investigation. Kim et al. demonstrated that BBB permeability could be controlled through adenosine signaling in a rapid, reversible and time-dependent way via endogenous mechanisms in which morphological changes in actin cytoskeletal reorganization, as well as claudin-5 and vascular endothelial (VE)-cadherin proteins, are involved. To perform their studies, these authors used the primary human brain microvascular ECs, HBMVECs, and the human EC cell line, hCMEC/D3. Both models highly express the ecto-enzymes CD73, CD39 and A2ARs. The authors demonstrated that the blockade of CD73 or A2AR inhibition hinders the migration of white blood cells into the CNS, whereas the activation of ARs by Lexiscan or NECA induces a decrease in the TEER parameter and, hence, an increase in BBB permeability. More specifically, the authors studied the kinetics of these effects and concluded that the duration of the increase in BBB permeability is proportional to the AR agonists' lifetime, with this effect being reverted afterward. On the other hand, the authors described a rapid increase in cAMP after the treatment of HBMVECs with Lexiscan, which, at the same time, increases RhoA activity, a GTPase known to regulate cell structure and morphology through the re-organization of actin cytoskeletal proteins and the disruption of cell-to-cell junctions [12,80].

Moreover, it has been reported that the activation of ARs in mouse Bend.3 cells with the agonists Lexiscan and NECA induces a downregulation in the expression levels of claudin-5 and VE-cadherin, with Lexiscan's effects being higher and faster than NECA's effects. The A2AR specific antagonist, SCH58261, reverses NECA's effects [12], demonstrating that the observed effects are mediated via A2AR. As already mentioned, claudin-5 is a TJ protein, whereas VE-cadherin is an adherens junction protein. Both proteins play a crucial role in maintaining endothelial barrier integrity.

Since A2BR and A3R are expressed at a lower level in the CNS compared to A1R and A2AR, there are not many reports on their role in the BBB. The A2BR agonist, BAY 60-656583, has been found to be involved in the regulation of cerebrovascular integrity in a rat transient middle cerebral artery occlusion model (tMCAO), inducing a decrease in tissue lesion volume and a decrease in the expression of TJ-associated proteins, such as ZO-1, thus protecting the BBB [81]. On the other hand, the A3R agonist, AST-004, has been found to play a neuroprotective role in a mouse model of traumatic brain injury (TBI). Mice treated with AST-004 showed less severe secondary brain injury (measured as cell death and loss of BBB breakdown) and a lower expression of neuroinflammatory markers compared to untreated mice. Moreover, the AST-004 treatment prevented the impairment of spatial memory in male mice [82]. More studies involving these two A2BRs and A3Rs should be performed.

Therefore, previous studies have identified a reversible paracellular increase in BBB permeability through AR signaling, which can be exploited in translational medicine to modulate the entry of molecules into the brain [12,50,62].

A summary of the principal proposed mechanisms underlying adenosine modulation of BBB permeability acting through its receptors is shown in Figure 5.

4.2. Pathological Conditions

As already mentioned, the proper physiological function of the brain is guaranteed by regulating the transport of molecules into the brain (good molecules such as nutrients and bad molecules such as toxic metabolites), which is mediated by the BBB that plays a protective role for the CNS and maintains its homeostasis. In addition to its low permeability, the BBB endothelium is characterized by the presence of specific transport systems and drug efflux systems, including Pgp, MRPs and BCRP [80,83]. Therefore, the BBB has a friend role as well as a foe role, since it presents an obstacle for the entry of therapeutic substances into the brain, in particular in the treatment of neurological diseases and primary brain tumors where a limited penetration of drugs has been reported [84,85].

The BBB is a dynamic structure that is continually regulated by physiological and pathological factors. BBB dysfunction has been reported in many CNS pathological conditions, such as multiple sclerosis (MS) [86–88], hypoxic and ischemic insults, PD, AD [89,90], brain cancer [83] and inflammation [91]. Increased BBB permeability has also been described in normal aging [18–20]. BBB impairment in various pathological conditions has also been reported to be caused by structural changes due to the altered expression of TJ proteins. Moreover, the loss of TJ proteins has been associated with neuroinflammatory and neurodegenerative disorders [92,93]. Thus, the identification of the altered expression of proteins in the cerebral vascular endothelium during CNS pathological conditions may be used as a potential therapeutic target in the treatment of CNS diseases [18].

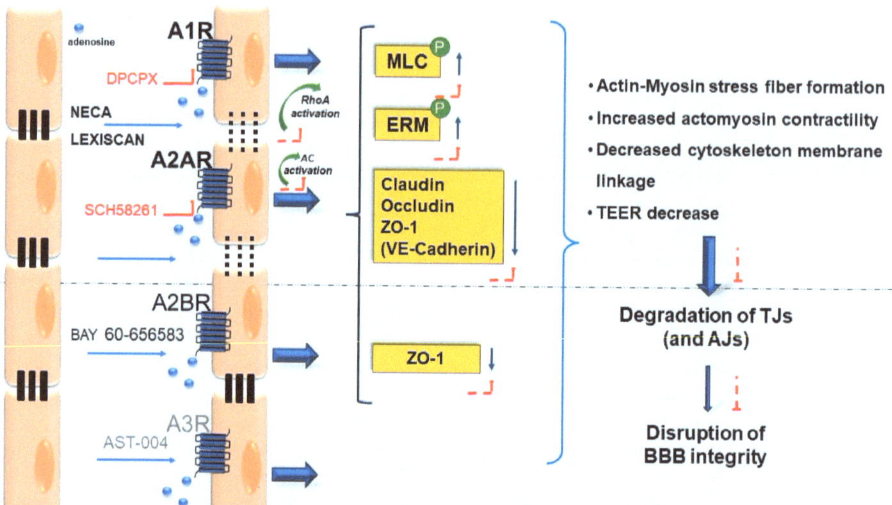

Figure 5. Adenosine modulation of BBB permeability proposed mechanisms. The activation of adenosine receptors after agonist binding induces specific signal transduction pathways depending on the AR subtype involved. The most expressed AR in brain microvascular endothelial cells (BECs) are A1 and A2A receptors, whereas A2B and A3 receptors are much less expressed. CD39 and CD73 are also expressed in BECs. Possible mechanisms of action mediated by ARs inducing the modification of BBB permeability are included in the figure and have been discussed in the text. In a proposed mechanism, agonists' activation of ARs results in the activation of AC and RhoA-GTPase enzymes which induces the increase in phosphorylated MCL and ERM proteins, causing the activation of actomyosin contraction and subsequent BBB disruption. In another proposed mechanism, agonist activation of ARs induces alteration in the expression of TJ and AJs proteins (claudin, occludin, ZO-1, and VE-cadherin) and morphological changes in the actin cytoskeleton which induces a sequence of effects leading to BBB permeability alteration. In the same studies, ARs antagonists contrasted the effects caused by ARs activation. Abbreviations: AJs, adherens junctions; DPCPX (1,3-dipropyl-8-cyclopentyl xanthine; ERM, ezrin/radixin/moesin; MLC, myosin light chain; Neca, N-Ethylcarboxamidoadenosine; Lexiscan; TEER, transendothelial resistance parameter paracellular permeability; TJs, tight junctions; ZO-1, zona occludens-1.

Much attention has been paid to how adenosine modulation of BBB permeability can be exploited to deliver drugs into the CNS to treat neurological diseases ranging from AD to brain tumors. To enable this, we need to first determine whether AR signaling regulates human BBB permeability in physiological and/or pathological conditions, and second, we need to understand the mechanisms that regulate brain endothelial barrier permeability and determine the role of AR signaling in the modulation of BBB permeability.

Several AR agonists and antagonists have been tested for their effects on BBB permeability, which differ in their affinity, time of action, and selectivity to ARs. For instance, the A2AR agonist, Lexiscan, is a low-affinity, short-acting and selective molecule, whereas CGS21680 is a high-affinity, selective molecule with a long duration of action and NECA is a nonselective agonist. In the case of low-affinity agonists, their potency of action depends on the AR expression levels in the targeted tissues and species, thus contributing to a higher tissue selectivity. Overall, previous studies have delineated the importance of receptor tissues and species differences, as these factors are related to drug responsivity, in an effort to optimize preclinical models that can be used for future clinical applications [83].

In some preclinical studies, the A2AR antagonist, caffeine, has been found to increase cognitive performance in AD animal models, demonstrating the effects of targeting A2AR in the therapy of AD [62,94,95]. Moreover, other clinical trials of neurodegenerative diseases, for instance, AD and PD, exploit the use of BBB-permeable adenosine-related compounds, mainly A2AR agonists and antagonists, alone or in combination with other drugs [96,97].

As has already been demonstrated, A2AR agonists increase BBB permeability via actin cytoskeletal reorganization, which is an integral part of the intercellular junction system. The proposed mechanisms to explain the modification in the functions of tight and adhesive junctions are (i) the increase in RhoA signaling activity and the consequent formation of actin stress fibers; (ii) the downregulation of VE-cadherin, ZO-1 and claudin-5 proteins; (iii) the reduced phosphorylation of adhesion-related factors; and (iv) the downregulation of P-glycoprotein expression in ECs. All of these alterations lead to an increase in the paracellular gap and an extended flow of substances across the BBB [18,80].

5. Adenosine Signaling under Pathological Conditions

An increase in the expression of A2AR in glial cells during hypoxia and inflammation has been described. In fact, A2AR agonists have been used to reduce inflammation during ischemic hemorrhagic shock to protect tissues from damage, mitigate inflammatory processes and protect neuronal cells [98,99].

Sleep restriction is another pathological condition in which increased BBB permeability and inflammatory responses have been described, with this effect being reverted after A2AR antagonism through SCH58261, which demonstrates the implication of A2AR in BBB dysfunction [100].

Gliomas are known to modify vascular permeability via BBB disruption. Consequently, to limit this pathological process, one tentative approach to repair the damage is by regulating certain proteins' expression and function [101]. Interestingly, endothelial cells in cerebral vessels are characterized by the presence of A1R and A2AR. It has been shown that substances affecting these receptors regulate the permeability of the BBB. More specifically, it has been demonstrated that A2AR agonists increase BBB permeability, and the proposed mechanisms involve actin cytoskeletal reorganization at TJs [83]. Therefore, A2AR agonists may be used as potential therapeutic strategies in the co-treatment of brain cancer to increase the effectiveness of oncological therapy.

The A2AR agonist, Lexiscan, has been demonstrated to increase BBB permeability in in vivo models of mice and rats through the downregulation of TJ proteins and P-glycoprotein. However, Lexiscan is known to have a short circulating lifetime and show important systemic side effects, such as stroke, shortness of breath and nausea. Moreover, it has shown no efficacy in some clinical studies [80,83,94].

Interestingly, strategies to vehicle drugs into the brain, such as the application of NPs, have also been used in the treatment of brain tumors in preclinical studies; more specifically, NPs are conjugated with A2AR agonists, such as Lexiscan [102]. Compared to the use of A2AR agonists alone, Lexiscan-conjugated NPs had an enhanced selectivity and reduced systemic side effect, which improved targeted drug delivery to the CNS and prolonged its time of action upon opening the BBB [102,103]. Nevertheless, they showed some disadvantages such as potential neuro- and systemic toxicity, and only small-sized NPs (<12 nm) can cross the BBB. Thus, NP technology must be further studied [83].

It has been demonstrated that high concentrations of ATP are released during ischemia, inflammation or cell damage as a "danger signal". Adenosine functions mainly as an anti-inflammatory and pro-resolving signal to suppress neuroinflammatory reaction and to reinforce the structural integrity of the CNS barriers [104]. ECs are capable of directly responding to extracellular adenosine. Since extracellular adenosine mediates many of its functions in inflammation or injury [105], we hypothesize that adenosine acting via ARs on ECs induces the recruitment of cells and/or molecules into the CNS to sites of damage or inflammation [62,91].

As already mentioned, conventional nanomedicine approaches take advance of ligands and NPs, whose physiochemical characteristics are employed to enhance drug delivery across the BBB. As a step up of this technology, some authors have created a targeted nanomedicine capable of bypassing the BBB and BCSFB to deliver specific A1R antagonists (theophylline and 1,3-dipropyl-8-cyclopentyl xanthine-DPCPX-) selectively to specific areas of the spinal cord by utilizing a retrograde neural tracer (wheat germ agglutinin-horseradish peroxidase, WGA-HRP). This targeted nanomedicine is able to revert the respiratory dysfunction caused by respiratory muscle paralysis during spinal cord injury (SCI) [106].

6. Conclusions

Despite its invaluable protective and functional role, the BBB is a major obstacle in the delivery of drugs for the treatment of CNS diseases.

The development of drugs for treating CNS diseases should be committed to both CNS drug discovery and CNS drug delivery.

Research has provided interesting outcomes in the development of new therapeutic strategies for effectively targeting a drug to the brain compartment, to minimize systemic side effects and to achieve the highest drug selectivity.

In this context, adenosine acting on distinct subtypes of P1 receptors expressed on BBB cells modulates the permeability of the BBB, as A1R and A2AR activation increases its permeability while A2BR and A3R stimulation maintains its integrity. It should be remarked that A1R and A2AR agonists, by briefly opening the BBB, allow drugs' passage, thereby presenting a significant potential for the delivery of drugs to the CNS. Moreover, the use of appropriate selective AR antagonists could, in turn, limit the entry of harmful substances and inflammatory immune cells into the brain and contribute to the maintenance of CNS homeostasis. In addition, since ECs express ARs and the ectonucleotidases CD73 and CD79, signaling at ECs may represent a druggable endogenous mechanism for modulating BBB permeability to treat CNS diseases.

Author Contributions: M.F.: conceptualization, writing—original draft preparation; S.G. and S.M.: writing—review and editing, supervision; M.N. and A.T.: writing—original draft preparation, figure preparation; S.P., F.V., K.V. and P.A.B.: supervision. All authors have read and agreed to the published version of the manuscript.

Funding: Work supported by #NEXTGENERATIONEU (NGEU) and funded by the Ministry of University and Research (MUR) National Recovery and Resilience Plan (NRRP), Project MNESYS (PE0000006)—A multiscale integrated approach to the study of the nervous system in health and disease (DN. 1553 11.10.2022).

Institutional Review Board Statement: Not applicable.

Informed Consent Statement: Not applicable.

Data Availability Statement: Data sharing not applicable.

Conflicts of Interest: The authors declare no conflict of interest.

Abbreviations

A1R, adenosine A1 receptor; A2AR, adenosine A2A receptor; A2BR, adenosine A2B receptor; A3R, adenosine A3 receptor; ABC, adenosine triphosphate (ATP)-binding cassette; AC, adenylyl cyclase; ABCs, arachnoid barrier cells; AD, Alzheimer's disease; ADA, adenosine deaminase; ADP, adenosine diphosphate; AK, adenosine kinase; Akt, serine-threonine protein kinase; AMP, adenosine monophosphate; AMT, adsorptive-mediated transcytosis; AR, adenosine receptors; ATP, adenosine triphosphate; BBB, blood–brain barrier; BBBD, blood–brain barrier disruption; BCRP, breast cancer resistant protein; BCSFB, blood–cerebrospinal fluid barrier; BECs, brain endothelial cells; BM, basal membrane; CD31, cluster of differentiation 31; CD39, cluster differentiation 39 (ectonucleoside triphosphate diphosphohydrolase-1); CD73, cluster differentiation 73 (ecto-5′-nucleotidase); Cln, claudins; CMT, carrier-mediated transport; CNS, central nervous system; CNTs, concentrative nucleoside transporters; CPE, choroid plexus epithelial cells; CSF, cerebrospinal fluid; DBC, dural border cells; DHA, docosahexaenoic acid; CSF, cerebrospinal fluid; DPCPX, 1,3-dipropyl-8-cyclopentylxanthine; EBA, endothelial barrier antigen; E-cadherin, endothelial-cadherin; ECs, brain microvessel endothelial cells; EGb, extract Ginkgo biloba; ENTs, equilibrative nucleoside transporters; ERM, ezrin/radixin/moesin; GGTP, gamma-glutamyl-transpeptidase; Gi, inhibitory G protein; GLUT-1, glucose transporter-1; Go, other G protein; Gs, stimulatory G protein; GTP, guanosine triphosphate; HBMVECs, human brain microvascular endothelial cells; hCMEC/D3, human cerebral microvascular endothelial cell line; HEB, human normal glial cell line; IDDD, intrathecal drug delivery device; JAMs, Junction adhesion molecules; MAPK, mitogen-activated protein kinase; MFSD2a, major facilitator superfamily domain-containing protein 2a; MLC, myosin light chain; MRP, multidrug resistance-associated protein; MS, multiple sclerosis; NECA, N-Ethylcarboxamidoadenosine; NPs, nanoparticles; NVU, neurovascular unit; P1, Purinergic receptor type 1; PCs, pericytes; PD, Parkinson's disease; PECAM-1, platelet endothelial cell adhesion molecule-1; Pgp, P-glycoprotein; PI3K, phosphoinositide 3 kinase; PK, pharmacokinetics; PKC, protein kinase C; PLC, phospholipase C; RME, receptor-mediated endocytosis; RMT, receptor-mediated transcytosis; SAH, S-Adenosyl-L-homocysteine; SAS, subarachnoid space; TBI, traumatic brain injury; TJs, tight junctions; tMCAO, transient middle cerebral artery occlusion model; VE, vascular endothelial; vWf, von Willebrand factor; WGA-HRP, wheat germ agglutinin-horseradish peroxidase; ZO-1, zona occludens-1.

References

1. Pardridge, W.M. CSF, Blood-Brain Barrier, and Brain Drug Delivery. *Expert Opin. Drug Deliv.* **2016**, *13*, 963–975. [CrossRef] [PubMed]
2. Tajes, M.; Ramos-Fernández, E.; Weng-Jiang, X.; Bosch-Morató, M.; Guivernau, B.; Eraso-Pichot, A.; Salvador, B.; Fernàndez-Busquets, X.; Roquer, J.; Muñoz, F.J. The Blood-Brain Barrier: Structure, Function and Therapeutic Approaches to Cross It. *Mol. Membr. Biol.* **2014**, *31*, 152–167. [CrossRef] [PubMed]
3. Ghose, A.K.; Viswanadhan, V.N.; Wendoloski, J.J. A Knowledge-Based Approach in Designing Combinatorial or Medicinal Chemistry Libraries for Drug Discovery. 1. A Qualitative and Quantitative Characterization of Known Drug Databases. *J. Comb. Chem.* **1999**, *1*, 55–68. [CrossRef] [PubMed]
4. Kadry, H.; Noorani, B.; Cucullo, L. A Blood–Brain Barrier Overview on Structure, Function, Impairment, and Biomarkers of Integrity. *Fluids Barriers CNS* **2020**, *17*, 69. [CrossRef]
5. Pardridge, W.M. Drug and Gene Delivery to the Brain. *Neuron* **2002**, *36*, 555–558. [CrossRef]
6. Pardridge, W.M. The Blood-Brain Barrier: Bottleneck in Brain Drug Development. *NeuroRX* **2005**, *2*, 3–14. [CrossRef]
7. Mehta, D.C.; Short, J.L.; Hilmer, S.N.; Nicolazzo, J.A. Drug Access to the Central Nervous System in Alzheimer's Disease: Preclinical and Clinical Insights. *Pharm. Res.* **2015**, *32*, 819–839. [CrossRef]
8. Dong, X. Current Strategies for Brain Drug Delivery. *Theranostics* **2018**, *8*, 1481–1493. [CrossRef]
9. Schaeffer, S.; Iadecola, C. Revisiting the Neurovascular Unit. *Nat. Neurosci.* **2021**, *24*, 1198–1209. [CrossRef]
10. Burnstock, G. Purine and Purinergic Receptors. *Brain Neurosci. Adv.* **2018**, *2*, 239821281881749. [CrossRef]
11. Linden Joel, M. Purinergic Receptors. In *Basic Neurochemistry*; Molecular, Cellular and Medical Aspects; Lippincott-Raven: Philadelphia, PA, USA, 1999; ISBN 10: 0-397-51820-X.
12. Kim, D.-G.; Bynoe, M.S. A2A Adenosine Receptor Regulates the Human Blood-Brain Barrier Permeability. *Mol. Neurobiol.* **2015**, *52*, 664–678. [CrossRef]
13. Borea, P.A.; Gessi, S.; Merighi, S.; Vincenzi, F.; Varani, K. Pharmacology of Adenosine Receptors: The State of the Art. *Physiol. Rev.* **2018**, *98*, 1591–1625. [CrossRef] [PubMed]
14. Fredholm, B.B.; Chen, J.-F.; Cunha, R.A.; Svenningsson, P.; Vaugeois, J.-M. Adenosine and Brain Function. In *International Review of Neurobiology*; Elsevier: Amsterdam, The Netherlands, 2005; Volume 63, pp. 191–270. ISBN 978-0-12-366864-6.
15. Borea, P.A.; Gessi, S.; Merighi, S.; Varani, K. Adenosine as a Multi-Signalling Guardian Angel in Human Diseases: When, Where and How Does It Exert Its Protective Effects? *Trends Pharmacol. Sci.* **2016**, *37*, 419–434. [CrossRef] [PubMed]
16. Fredholm, B.B.; IJzerman, A.P.; Jacobson, K.A.; Linden, J.; Müller, C.E. International Union of Basic and Clinical Pharmacology. LXXXI. Nomenclature and Classification of Adenosine Receptors—An Update. *Pharmacol. Rev.* **2011**, *63*, 1–34. [CrossRef]
17. IJzerman, A.P.; Jacobson, K.A.; Müller, C.E.; Cronstein, B.N.; Cunha, R.A. International Union of Basic and Clinical Pharmacology. CXII: Adenosine Receptors: A Further Update. *Pharmacol. Rev.* **2022**, *74*, 340–372. [CrossRef] [PubMed]
18. Bynoe, M.S.; Viret, C.; Yan, A.; Kim, D.-G. Adenosine Receptor Signaling: A Key to Opening the Blood–Brain Door. *Fluids Barriers CNS* **2015**, *12*, 20. [CrossRef]
19. Saleh, M.A.A.; De Lange, E.C.M. Impact of CNS Diseases on Drug Delivery to Brain Extracellular and Intracellular Target Sites in Human: A "WHAT-IF" Simulation Study. *Pharmaceutics* **2021**, *13*, 95. [CrossRef]
20. Zimmerman, B.; Rypma, B.; Gratton, G.; Fabiani, M. Age-related Changes in Cerebrovascular Health and Their Effects on Neural Function and Cognition: A Comprehensive Review. *Psychophysiology* **2021**, *58*, e13796. [CrossRef]
21. Frampton, J.; Shuler, M.; Shain, W.; Hynd, M. Biomedical Technologies for In Vitro Screening and Controlled Delivery of Neuroactive Compounds. *Cent. Nerv. Syst. Agents Med. Chem.* **2008**, *8*, 203–219. [CrossRef]

22. Brichacek, A.L.; Brown, C.M. Alkaline Phosphatase: A Potential Biomarker for Stroke and Implications for Treatment. *Metab. Brain Dis.* **2019**, *34*, 3–19. [CrossRef]

23. Lochhead, J.J.; Yang, J.; Ronaldson, P.T.; Davis, T.P. Structure, Function, and Regulation of the Blood-Brain Barrier Tight Junction in Central Nervous System Disorders. *Front. Physiol.* **2020**, *11*, 914. [CrossRef] [PubMed]

24. Feldman, G.; Mullin, J.; Ryan, M. Occludin: Structure, Function and Regulation. *Adv. Drug Deliv. Rev.* **2005**, *57*, 883–917. [CrossRef] [PubMed]

25. Piorntek, J.; Winkler, L.; Wolburg, H.; Müller, S.L.; Zuleger, N.; Piehl, C.; Wiesner, B.; Krause, G.; Blasig, I.E. Formation of Tight Junction: Determinants of Homophilic Interaction between Classic Claudins. *FASEB J.* **2008**, *22*, 146–158. [CrossRef] [PubMed]

26. Honda, M.; Nakagawa, S.; Hayashi, K.; Kitagawa, N.; Tsutsumi, K.; Nagata, I.; Niwa, M. Adrenomedullin Improves the Blood–Brain Barrier Function through the Expression of Claudin-5. *Cell. Mol. Neurobiol.* **2006**, *26*, 109–118. [CrossRef] [PubMed]

27. Li, G.; Simon, M.J.; Cancel, L.M.; Shi, Z.-D.; Ji, X.; Tarbell, J.M.; Morrison, B.; Fu, B.M. Permeability of Endothelial and Astrocyte Cocultures: In Vitro Blood–Brain Barrier Models for Drug Delivery Studies. *Ann. Biomed. Eng.* **2010**, *38*, 2499–2511. [CrossRef]

28. Jiao, H.; Wang, Z.; Liu, Y.; Wang, P.; Xue, Y. Specific Role of Tight Junction Proteins Claudin-5, Occludin, and ZO-1 of the Blood–Brain Barrier in a Focal Cerebral Ischemic Insult. *J. Mol. Neurosci.* **2011**, *44*, 130–139. [CrossRef]

29. Arpin, M.; Chirivino, D.; Naba, A.; Zwaenepoel, I. Emerging Role for ERM Proteins in Cell Adhesion and Migration. *Cell Adhes. Migr.* **2011**, *5*, 199–206. [CrossRef]

30. García-Ponce, A.; Citalán-Madrid, A.F.; Velázquez-Avila, M.; Vargas-Robles, H.; Schnoor, M. The Role of Actin-Binding Proteins in the Control of Endothelial Barrier Integrity. *Thromb. Haemost.* **2015**, *113*, 20–36. [CrossRef]

31. Taverner, A.; Dondi, R.; Almansour, K.; Laurent, F.; Owens, S.-E.; Eggleston, I.M.; Fotaki, N.; Mrsny, R.J. Enhanced Paracellular Transport of Insulin Can Be Achieved via Transient Induction of Myosin Light Chain Phosphorylation. *J. Control. Release* **2015**, *210*, 189–197. [CrossRef]

32. Brunner, J.; Ragupathy, S.; Borchard, G. Target Specific Tight Junction Modulators. *Adv. Drug Deliv. Rev.* **2021**, *171*, 266–288. [CrossRef]

33. Liang, W.; Xu, W.; Zhu, J.; Zhu, Y.; Gu, Q.; Li, Y.; Guo, C.; Huang, Y.; Yu, J.; Wang, W.; et al. Ginkgo Biloba Extract Improves Brain Uptake of Ginsenosides by Increasing Blood-Brain Barrier Permeability via Activating A1 Adenosine Receptor Signaling Pathway. *J. Ethnopharmacol.* **2020**, *246*, 112243. [CrossRef] [PubMed]

34. Greene, C.; Campbell, M. Tight Junction Modulation of the Blood Brain Barrier: CNS Delivery of Small Molecules. *Tissue Barriers* **2016**, *4*, e1138017. [CrossRef] [PubMed]

35. Wimmer, I.; Tietz, S.; Nishihara, H.; Deutsch, U.; Sallusto, F.; Gosselet, F.; Lyck, R.; Muller, W.A.; Lassmann, H.; Engelhardt, B. PECAM-1 Stabilizes Blood-Brain Barrier Integrity and Favors Paracellular T-Cell Diapedesis Across the Blood-Brain Barrier During Neuroinflammation. *Front. Immunol.* **2019**, *10*, 711. [CrossRef] [PubMed]

36. Beyer, E.C.; Berthoud, V.M. Gap Junction Gene and Protein Families: Connexins, Innexins, and Pannexins. *Biochim. Biophys. Acta BBA Biomembr.* **2018**, *1860*, 5–8. [CrossRef]

37. Gao, L.; Pan, X.; Zhang, J.H.; Xia, Y. Glial Cells: An Important Switch for the Vascular Function of the Central Nervous System. *Front. Cell. Neurosci.* **2023**, *17*, 1166770. [CrossRef]

38. Puris, E.; Fricker, G.; Gynther, M. Targeting Transporters for Drug Delivery to the Brain: Can We Do Better? *Pharm. Res.* **2022**, *39*, 1415–1455. [CrossRef]

39. Zlokovic, B.V. Neurovascular Pathways to Neurodegeneration in Alzheimer's Disease and Other Disorders. *Nat. Rev. Neurosci.* **2011**, *12*, 723–738. [CrossRef]

40. Lin, L.; Yee, S.W.; Kim, R.B.; Giacomini, K.M. SLC Transporters as Therapeutic Targets: Emerging Opportunities. *Nat. Rev. Drug Discov.* **2015**, *14*, 543–560. [CrossRef]

41. Nguyen, L.N.; Ma, D.; Shui, G.; Wong, P.; Cazenave-Gassiot, A.; Zhang, X.; Wenk, M.R.; Goh, E.L.K.; Silver, D.L. Mfsd2a Is a Transporter for the Essential Omega-3 Fatty Acid Docosahexaenoic Acid. *Nature* **2014**, *509*, 503–506. [CrossRef]

42. Huang, B.; Li, X. The Role of Mfsd2a in Nervous System Diseases. *Front. Neurosci.* **2021**, *15*, 730534. [CrossRef]

43. Barar, J.; Rafi, M.A.; Pourseif, M.M.; Omidi, Y. Blood-Brain Barrier Transport Machineries and Targeted Therapy of Brain Diseases. *BioImpacts* **2016**, *6*, 225–248. [CrossRef] [PubMed]

44. Filippi, M.-D. Mechanism of Diapedesis. In *Advances in Immunology*; Elsevier: Amsterdam, The Netherlands, 2016; Volume 129, pp. 25–53. ISBN 978-0-12-804799-6.

45. Glezer, I.; Simard, A.R.; Rivest, S. Neuroprotective Role of the Innate Immune System by Microglia. *Neuroscience* **2007**, *147*, 867–883. [CrossRef] [PubMed]

46. Abbott, N.J.; Patabendige, A.A.K.; Dolman, D.E.M.; Yusof, S.R.; Begley, D.J. Structure and Function of the Blood–Brain Barrier. *Neurobiol. Dis.* **2010**, *37*, 13–25. [CrossRef] [PubMed]

47. Moya, M.L.; Triplett, M.; Simon, M.; Alvarado, J.; Booth, R.; Osburn, J.; Soscia, D.; Qian, F.; Fischer, N.O.; Kulp, K.; et al. A Reconfigurable In Vitro Model for Studying the Blood–Brain Barrier. *Ann. Biomed. Eng.* **2020**, *48*, 780–793. [CrossRef] [PubMed]

48. Czarniak, N.; Kamińska, J.; Matowicka-Karna, J.; Koper-Lenkiewicz, O.M. Cerebrospinal Fluid–Basic Concepts Review. *Biomedicines* **2023**, *11*, 1461. [CrossRef]

49. Kandel, E.R.; Schwartz, J.H.; Jessell, T.M. *Principles of Neural Science*, 4th ed.; McGraw-Hill Health Professions Division: New York, NY, USA, 2000.

50. Bellettato, C.M.; Scarpa, M. Possible Strategies to Cross the Blood–Brain Barrier. *Ital. J. Pediatr.* **2018**, *44*, 131. [CrossRef]

51. Pardridge, W.M. A Historical Review of Brain Drug Delivery. *Pharmaceutics* **2022**, *14*, 1283. [CrossRef]
52. Sethi, B.; Kumar, V.; Mahato, K.; Coulter, D.W.; Mahato, R.I. Recent Advances in Drug Delivery and Targeting to the Brain. *J. Control. Release* **2022**, *350*, 668–687. [CrossRef]
53. Ndemazie, N.B.; Inkoom, A.; Morfaw, E.F.; Smith, T.; Aghimien, M.; Ebesoh, D.; Agyare, E. Multi-Disciplinary Approach for Drug and Gene Delivery Systems to the Brain. *AAPS PharmSciTech* **2022**, *23*, 11. [CrossRef]
54. Barua, S.; Mitragotri, S. Challenges Associated with Penetration of Nanoparticles across Cell and Tissue Barriers: A Review of Current Status and Future Prospects. *Nano Today* **2014**, *9*, 223–243. [CrossRef]
55. Johnsen, K.B.; Burkhart, A.; Melander, F.; Kempen, P.J.; Vejlebo, J.B.; Siupka, P.; Nielsen, M.S.; Andresen, T.L.; Moos, T. Targeting Transferrin Receptors at the Blood-Brain Barrier Improves the Uptake of Immunoliposomes and Subsequent Cargo Transport into the Brain Parenchyma. *Sci. Rep.* **2017**, *7*, 10396. [CrossRef] [PubMed]
56. Storck, S.E.; Pietrzik, C.U. Endothelial LRP1—A Potential Target for the Treatment of Alzheimer's Disease: Theme: Drug Discovery, Development and Delivery in Alzheimer's Disease Guest Editor: Davide Brambilla. *Pharm. Res.* **2017**, *34*, 2637–2651. [CrossRef] [PubMed]
57. Lapointe, S.; Perry, A.; Butowski, N.A. Primary Brain Tumours in Adults. *Lancet* **2018**, *392*, 432–446. [CrossRef] [PubMed]
58. Pulgar, V.M. Transcytosis to Cross the Blood Brain Barrier, New Advancements and Challenges. *Front. Neurosci.* **2019**, *12*, 1019. [CrossRef] [PubMed]
59. Johnsen, K.B.; Burkhart, A.; Thomsen, L.B.; Andresen, T.L.; Moos, T. Targeting the Transferrin Receptor for Brain Drug Delivery. *Prog. Neurobiol.* **2019**, *181*, 101665. [CrossRef]
60. Grabrucker, A.M.; Chhabra, R.; Belletti, D.; Forni, F.; Vandelli, M.A.; Ruozi, B.; Tosi, G. Nanoparticles as Blood–Brain Barrier Permeable CNS Targeted Drug Delivery Systems. In *The Blood Brain Barrier (BBB)*; Fricker, G., Ott, M., Mahringer, A., Eds.; Topics in Medicinal Chemistry; Springer: Berlin/Heidelberg, Germany, 2013; Volume 10, pp. 71–89. ISBN 978-3-662-43786-5.
61. Izhar, M.P.; Hafeez, A.; Kushwaha, P.; Simrah. Drug Delivery Through Niosomes: A Comprehensive Review with Therapeutic Applications. *J. Clust. Sci.* **2023**, *34*, 2257–2273. [CrossRef]
62. Wang, Y.; Zhu, Y.; Wang, J.; Dong, L.; Liu, S.; Li, S.; Wu, Q. Purinergic Signaling: A Gatekeeper of Blood-Brain Barrier Permeation. *Front. Pharmacol.* **2023**, *14*, 1112758. [CrossRef]
63. Nedeljkovic, N. Complex Regulation of Ecto-5′-Nucleotidase/CD73 and A2AR-Mediated Adenosine Signaling at Neurovascular Unit: A Link between Acute and Chronic Neuroinflammation. *Pharmacol. Res.* **2019**, *144*, 99–115. [CrossRef]
64. Zimmermann, H. History of Ectonucleotidases and Their Role in Purinergic Signaling. *Biochem. Pharmacol.* **2021**, *187*, 114322. [CrossRef]
65. Burnstock, G. Introduction to Purinergic Signalling in the Brain. In *Glioma Signaling*; Barańska, J., Ed.; Advances in Experimental Medicine and Biology; Springer International Publishing: Cham, Switzerland, 2020; Volume 1202, pp. 1–12, ISBN 978-3-030-30650-2.
66. Boison, D. Adenosine Kinase, Epilepsy and Stroke: Mechanisms and Therapies. *Trends Pharmacol. Sci.* **2006**, *27*, 652–658. [CrossRef]
67. Tescarollo, F.C.; Rombo, D.M.; DeLiberto, L.K.; Fedele, D.E.; Alharfoush, E.; Tomé, Â.R.; Cunha, R.A.; Sebastião, A.M.; Boison, D. Role of Adenosine in Epilepsy and Seizures. *J. Caffeine Adenosine Res.* **2020**, *10*, 45–60. [CrossRef]
68. Garcia-Gil, M.; Camici, M.; Allegrini, S.; Pesi, R.; Tozzi, M.G. Metabolic Aspects of Adenosine Functions in the Brain. *Front. Pharmacol.* **2021**, *12*, 672182. [CrossRef]
69. Daré, E.; Schulte, G.; Karovic, O.; Hammarberg, C.; Fredholm, B.B. Modulation of Glial Cell Functions by Adenosine Receptors. *Physiol. Behav.* **2007**, *92*, 15–20. [CrossRef]
70. Zimmermann, H.R.; Yang, W.; Kasica, N.P.; Zhou, X.; Wang, X.; Beckelman, B.C.; Lee, J.; Furdui, C.M.; Keene, C.D.; Ma, T. Brain-Specific Repression of AMPKα1 Alleviates Pathophysiology in Alzheimer's Model Mice. *J. Clin. Investig.* **2020**, *130*, 3511–3527. [CrossRef] [PubMed]
71. Vincenzi, F.; Pasquini, S.; Borea, P.A.; Varani, K. Targeting Adenosine Receptors: A Potential Pharmacological Avenue for Acute and Chronic Pain. *Int. J. Mol. Sci.* **2020**, *21*, 8710. [CrossRef] [PubMed]
72. Fredholm, B.; Cunha, R.; Svenningsson, P. Pharmacology of Adenosine A2A Receptors and Therapeutic Applications. *Curr. Top. Med. Chem.* **2003**, *3*, 413–426. [CrossRef]
73. Haskó, G.; Pacher, P.; Sylvester Vizi, E.; Illes, P. Adenosine Receptor Signaling in the Brain Immune System. *Trends Pharmacol. Sci.* **2005**, *26*, 511–516. [CrossRef] [PubMed]
74. Klabunde, R.E. Dipyridamole Inhibition of Adenosine Metabolism in Human Blood. *Eur. J. Pharmacol.* **1983**, *93*, 21–26. [CrossRef] [PubMed]
75. Liu, Y.; Chen, J.; Li, X.; Zhou, X.; Hu, Y.; Chu, S.; Peng, Y.; Chen, N. Research Progress on Adenosine in Central Nervous System Diseases. *CNS Neurosci. Ther.* **2019**, *25*, 899–910. [CrossRef]
76. Selmi, C.; Barin, J.G.; Rose, N.R. Current Trends in Autoimmunity and the Nervous System. *J. Autoimmun.* **2016**, *75*, 20–29. [CrossRef]
77. Davalos, D.; Grutzendler, J.; Yang, G.; Kim, J.V.; Zuo, Y.; Jung, S.; Littman, D.R.; Dustin, M.L.; Gan, W.-B. ATP Mediates Rapid Microglial Response to Local Brain Injury in Vivo. *Nat. Neurosci.* **2005**, *8*, 752–758. [CrossRef] [PubMed]
78. Di Virgilio, F.; Vultaggio-Poma, V.; Falzoni, S.; Giuliani, A.L. Extracellular ATP: A Powerful Inflammatory Mediator in the Central Nervous System. *Neuropharmacology* **2023**, *224*, 109333. [CrossRef] [PubMed]

79. Carman, A.J.; Mills, J.H.; Krenz, A.; Kim, D.-G.; Bynoe, M.S. Adenosine Receptor Signaling Modulates Permeability of the Blood–Brain Barrier. *J. Neurosci.* **2011**, *31*, 13272–13280. [CrossRef] [PubMed]
80. Kim, D.-G.; Bynoe, M.S. A2A Adenosine Receptor Modulates Drug Efflux Transporter P-Glycoprotein at the Blood-Brain Barrier. *J. Clin. Investig.* **2016**, *126*, 1717–1733. [CrossRef]
81. Li, Q.; Han, X.; Lan, X.; Hong, X.; Li, Q.; Gao, Y.; Luo, T.; Yang, Q.; Koehler, R.C.; Zhai, Y.; et al. Inhibition of tPA-Induced Hemorrhagic Transformation Involves Adenosine A2b Receptor Activation after Cerebral Ischemia. *Neurobiol. Dis.* **2017**, *108*, 173–182. [CrossRef]
82. Bozdemir, E.; Vigil, F.A.; Chun, S.H.; Espinoza, L.; Bugay, V.; Khoury, S.M.; Holstein, D.M.; Stoja, A.; Lozano, D.; Tunca, C.; et al. Neuroprotective Roles of the Adenosine A3 Receptor Agonist AST-004 in Mouse Model of Traumatic Brain Injury. *Neurotherapeutics* **2021**, *18*, 2707–2721. [CrossRef]
83. Wala, K.; Szlasa, W.; Saczko, J.; Rudno-Rudzińska, J.; Kulbacka, J. Modulation of Blood–Brain Barrier Permeability by Activating Adenosine A2 Receptors in Oncological Treatment. *Biomolecules* **2021**, *11*, 633. [CrossRef]
84. Deeken, J.F.; Löscher, W. The Blood-Brain Barrier and Cancer: Transporters, Treatment, and Trojan Horses. *Clin. Cancer Res.* **2007**, *13*, 1663–1674. [CrossRef]
85. Vézina, A.; Manglani, M.; Morris, D.; Foster, B.; McCord, M.; Song, H.; Zhang, M.; Davis, D.; Zhang, W.; Bills, J.; et al. Adenosine A2A Receptor Activation Enhances Blood–Tumor Barrier Permeability in a Rodent Glioma Model. *Mol. Cancer Res.* **2021**, *19*, 2081–2095. [CrossRef]
86. Correale, J.; Villa, A. The Blood–Brain-Barrier in Multiple Sclerosis: Functional Roles and Therapeutic Targeting. *Autoimmunity* **2007**, *40*, 148–160. [CrossRef]
87. Bynoe, M.S. Adenosine:Adenosine Receptor (AR) Axis in Blood Brain Barrier (BBB) Permeability Regulation. *J. Immunol.* **2019**, *202*, 116.11. [CrossRef]
88. Nishihara, H.; Perriot, S.; Gastfriend, B.D.; Steinfort, M.; Cibien, C.; Soldati, S.; Matsuo, K.; Guimbal, S.; Mathias, A.; Palecek, S.P.; et al. Intrinsic Blood–Brain Barrier Dysfunction Contributes to Multiple Sclerosis Pathogenesis. *Brain* **2022**, *145*, 4334–4348. [CrossRef] [PubMed]
89. Pan, Y.; Nicolazzo, J.A. Impact of Aging, Alzheimer's Disease and Parkinson's Disease on the Blood-Brain Barrier Transport of Therapeutics. *Adv. Drug Deliv. Rev.* **2018**, *135*, 62–74. [CrossRef] [PubMed]
90. Candelario-Jalil, E.; Dijkhuizen, R.M.; Magnus, T. Neuroinflammation, Stroke, Blood-Brain Barrier Dysfunction, and Imaging Modalities. *Stroke* **2022**, *53*, 1473–1486. [CrossRef]
91. Takata, F.; Nakagawa, S.; Matsumoto, J.; Dohgu, S. Blood-Brain Barrier Dysfunction Amplifies the Development of Neuroinflammation: Understanding of Cellular Events in Brain Microvascular Endothelial Cells for Prevention and Treatment of BBB Dysfunction. *Front. Cell. Neurosci.* **2021**, *15*, 661838. [CrossRef]
92. Zenaro, E.; Piacentino, G.; Constantin, G. The Blood-Brain Barrier in Alzheimer's Disease. *Neurobiol. Dis.* **2017**, *107*, 41–56. [CrossRef]
93. Liebner, S.; Dijkhuizen, R.M.; Reiss, Y.; Plate, K.H.; Agalliu, D.; Constantin, G. Functional Morphology of the Blood–Brain Barrier in Health and Disease. *Acta Neuropathol.* **2018**, *135*, 311–336. [CrossRef]
94. Jackson, S.; Weingart, J.; Nduom, E.K.; Harfi, T.T.; George, R.T.; McAreavey, D.; Ye, X.; Anders, N.M.; Peer, C.; Figg, W.D.; et al. The Effect of an Adenosine A2A Agonist on Intra-Tumoral Concentrations of Temozolomide in Patients with Recurrent Glioblastoma. *Fluids Barriers CNS* **2018**, *15*, 2. [CrossRef]
95. Laurent, C.; Eddarkaoui, S.; Derisbourg, M.; Leboucher, A.; Demeyer, D.; Carrier, S.; Schneider, M.; Hamdane, M.; Müller, C.E.; Buée, L.; et al. Beneficial Effects of Caffeine in a Transgenic Model of Alzheimer's Disease-like Tau Pathology. *Neurobiol. Aging* **2014**, *35*, 2079–2090. [CrossRef]
96. Torti, M.; Vacca, L.; Stocchi, F. Istradefylline for the Treatment of Parkinson's Disease: Is It a Promising Strategy? *Expert Opin. Pharmacother.* **2018**, *19*, 1821–1828. [CrossRef]
97. Shang, P.; Baker, M.; Banks, S.; Hong, S.-I.; Choi, D.-S. Emerging Nondopaminergic Medications for Parkinson's Disease: Focusing on A2A Receptor Antagonists and GLP1 Receptor Agonists. *J. Mov. Disord.* **2021**, *14*, 193–203. [CrossRef]
98. Bobermin, L.D.; Roppa, R.H.A.; Quincozes-Santos, A. Adenosine Receptors as a New Target for Resveratrol-Mediated Glioprotection. *Biochim. Biophys. Acta BBA Mol. Basis Dis.* **2019**, *1865*, 634–647. [CrossRef] [PubMed]
99. Sachdeva, S.; Gupta, M. Adenosine and Its Receptors as Therapeutic Targets: An Overview. *Saudi Pharm. J.* **2013**, *21*, 245–253. [CrossRef] [PubMed]
100. Hurtado-Alvarado, G.; Domínguez-Salazar, E.; Velázquez-Moctezuma, J.; Gómez-González, B. A2A Adenosine Receptor Antagonism Reverts the Blood-Brain Barrier Dysfunction Induced by Sleep Restriction. *PLoS ONE* **2016**, *11*, e0167236. [CrossRef] [PubMed]
101. Gao, F.; Cui, Y.; Jiang, H.; Sui, D.; Wang, Y.; Jiang, Z.; Zhao, J.; Lin, S. Circulating Tumor Cell Is a Common Property of Brain Glioma and Promotes the Monitoring System. *Oncotarget* **2016**, *7*, 71330–71340. [CrossRef]
102. Teleanu, D.; Chircov, C.; Grumezescu, A.; Volceanov, A.; Teleanu, R. Blood-Brain Delivery Methods Using Nanotechnology. *Pharmaceutics* **2018**, *10*, 269. [CrossRef]
103. Zou, Y.; Liu, Y.; Yang, Z.; Zhang, D.; Lu, Y.; Zheng, M.; Xue, X.; Geng, J.; Chung, R.; Shi, B. Effective and Targeted Human Orthotopic Glioblastoma Xenograft Therapy via a Multifunctional Biomimetic Nanomedicine. *Adv. Mater.* **2018**, *30*, 1803717. [CrossRef]

104. Cunha, R.A. How Does Adenosine Control Neuronal Dysfunction and Neurodegeneration? *J. Neurochem.* **2016**, *139*, 1019–1055. [CrossRef]
105. Blackburn, M.R.; Vance, C.O.; Morschl, E.; Wilson, C.N. Adenosine Receptors and Inflammation. In *Adenosine Receptors in Health and Disease*; Wilson, C.N., Mustafa, S.J., Eds.; Handbook of Experimental Pharmacology; Springer: Berlin/Heidelberg, Germany, 2009; Volume 193, pp. 215–269, ISBN 978-3-540-89614-2.
106. Hassan, M.M.; Hettiarachchi, M.; Kilani, M.; Gao, X.; Sankari, A.; Boyer, C.; Mao, G. Sustained A1 Adenosine Receptor Antagonist Drug Release from Nanoparticles Functionalized by a Neural Tracing Protein. *ACS Chem. Neurosci.* **2021**, *12*, 4438–4448. [CrossRef]

*pharmaceutics*

Review

Endocrine Regulation of Microvascular Receptor—Mediated Transcytosis and Its Therapeutic Opportunities: Insights by PCSK9—Mediated Regulation

Alexander D. Mazura and Claus U. Pietrzik *

Institute of Pathobiochemistry, University Medical Center of the Johannes Gutenberg, University Mainz, Duesbergweg 6, 55128 Mainz, Germany
* Correspondence: pietrzik@uni-mainz.de; Tel.: +49-6131-39-25390

Abstract: Currently, many neurological disorders lack effective treatment options due to biological barriers that effectively separate the central nervous system (CNS) from the periphery. CNS homeostasis is maintained by a highly selective exchange of molecules, with tightly controlled ligand-specific transport systems at the blood–brain barrier (BBB) playing a key role. Exploiting or modifying these endogenous transport systems could provide a valuable tool for targeting insufficient drug delivery into the CNS or pathological changes in the microvasculature. However, little is known about how BBB transcytosis is continuously regulated to respond to temporal or chronic changes in the environment. The aim of this mini-review is to draw attention to the sensitivity of the BBB to circulating molecules derived from peripheral tissues, which may indicate a fundamental endocrine-operating regulatory system of receptor-mediated transcytosis at the BBB. We present our thoughts in the context of the recent observation that low-density lipoprotein receptor-related protein 1 (LRP1)-mediated clearance of brain amyloid-β (Aβ) across the BBB is negatively regulated by peripheral proprotein convertase subtilisin/kexin type 9 (PCSK9). We hope that our conclusions will inspire future investigations of the BBB as dynamic communication interface between the CNS and periphery, whose peripheral regulatory mechanisms could be easily exploited for therapeutic purposes.

Keywords: blood brain barrier; receptor-mediated transcytosis; low-density lipoprotein receptor family; low-density lipoprotein receptor-related protein 1; proprotein convertase subtilisin/kexin type 9; central nervous system drug delivery; therapeutic blood–brain barrier modification

check for updates

Citation: Mazura, A.D.; Pietrzik, C.U. Endocrine Regulation of Microvascular Receptor—Mediated Transcytosis and Its Therapeutic Opportunities: Insights by PCSK9—Mediated Regulation. *Pharmaceutics* **2023**, *15*, 1268. https://doi.org/10.3390/pharmaceutics15041268

Academic Editors: Gert Fricker and Elena Puris

Received: 1 March 2023
Revised: 12 April 2023
Accepted: 14 April 2023
Published: 18 April 2023

1. Introduction

Many disorders of the central nervous system (CNS) are lacking effective therapeutic treatment options due to the insufficiency of conventional drugs to enter the brain following peripheral administration [1]. Despite this fact, peripheral drug administration, in contrast to transcranial approaches that deliver drugs directly into certain brain segments [1], is still the application route of choice when it comes to handling, cost, and safety.

The majority of the extravascular CNS is separated from the providing vasculature by the blood brain barrier (BBB) and the blood–cerebrospinal fluid barrier (BCSFB) [2]. The BBB is formed of specific endothelial cells lining the inside of brain microvasculature vessels [3], while the BCSFB is established by the epithelial layer of the choroid plexus of each ventricle [4]. Although both are physical obstacles, the morphological and structural differences determine the central barrier functions, which are considered to be fundamentally different; whereas the BCSFB controls the composition of the cerebrospinal fluid (CSF), which is constantly produced by plasma ultrafiltration at a high turnover rate [4], the BBB ensures a highly selective exchange of ions, molecules, and immune cells between the compartments while maintaining a potent protection against detrimental blood-borne substances, toxins, or pathogens [3]. The heavily restrictive character of the BBB is due to a variety of proteinaceous junctions—in particular tight junctions—connecting adjacent brain microvascular endothelial cells to a close-mesh network of cells, which eliminates paracellular flow [5], has a very low pinocytosis rate compared to other endothelial cells of the vascular system [6], and has a presence of multiple efflux transporters within the endothelial plasma membrane, capable of rapidly excluding endogenous and drug metabolites that diffuse into the brain microvascular endothelium [7]. This clear distinction of the separate compartments at the brain microvasculature level is the foundation for the blood- (luminal) and brain-faced (abluminal) endothelial membrane terminology and their significantly different composition [8]. The vascular basement membrane surrounds the abluminal surface of the brain microvascular tubes and embeds pericytes in a comparatively high pericyte-to-endothelial cell ratio [9,10]. This is followed by the tight envelopment of the basal lamina by end-feet of perivascular astrocytes, providing a cellular link between the microvasculature and neuronal circuitry [11]. Following increasing evidence that the endothelial monolayer is in close exchange with these perivascular cells as well as proximal neurons and microglia, the concept of the neurovascular unit has been proposed, emphasizing the importance of extra-endothelial cells in the CNS in the maintenance and regulation of BBB integrity and function in a paracrine manner [12].

Since the extremely dense brain microcapillary system represents the largest blood–brain interface, regulating the immediate microenvironment of cerebral cells throughout the brain, and possesses the structural prerequisites for controlled but efficient transport, crossing the BBB is the primary, and in certain circumstances the exclusive route of entry for peripheral substances [13]. However, a fully functional BBB excludes nearly 100% of available large and more than 98% of small molecule drugs due to the size limitations (<0.4 kDa) and lack of high lipid solubility (<8 hydrogen bonds) required for non-specific diffusion across the BBB [1]. An alternative non-invasive approach to bypass the highly selective BBB features intranasal drug administration, where the therapeutic agent can be delivered across the nasal mucosa along the olfactory or trigeminal nerve pathways directly into the interstitial fluid of the CNS, but may also end up in the olfactory bulbs, vascular system, lymph nodes, or CSF, depending on factors such as substance properties, formulation characteristics, and handling [14]. Although promising for some therapeutic agents, several clinical trials have produced disappointing data [15–18], indicating that further optimization is needed to increase the brain parenchyma penetration and distribution for a variety of therapeutics intended to use the nasal-brain delivery route [19,20]. In contrast to BBB circumvention, several approaches [21], including microbubble-assisted focused-ultrasound or the use of hyperosmotic adjuvants [22,23], induce the temporary opening or disruption of BBB integrity to various extents, providing non-specific BBB permeability for a wide range of drugs. However, compromising BBB integrity, even slightly, is accompanied by an unregulated influx of plasma constituents that affect the sensitive CNS milieu, usually associated, among others, with inflammation, immune responses, and neuronal damage [24]. Even though some approaches show promising preclinical and early clinical evidence [22,23,25,26], long-term randomized clinical trials will be required to demonstrate whether these approaches are applicable for human therapy in different pathological contexts beyond the treatment of severe brain tumours.

Due to the absence of paracellular or transcellular channels at the brain microvasculature, the required transfer of larger and/or polar solutes through the semi-permeable BBB is physiologically performed by ligand-specific endogenous transport systems such as carrier- or receptor-mediated transcytosis [27,28]. The ligand-binding carriers or receptors can be heterogeneously embedded in the luminal or abluminal endothelial plasma membrane, thereby controlling the direction of transcellular transport [29–31]. Carrier transport systems facilitate the transport of small molecules, especially nutrients, down their concentration gradient (generally directed from blood-to-brain) [32], whereas transmembrane receptors specifically recognize diverse extracellular macromolecules and initiate the transcytosis of the specific ligand to the opposite plasma membrane [27,28]. A prominent ligand-independent endogenous transport system for polycationic molecules is the process of adsorptive-mediated transcytosis, in which positively charged substances attach to negatively charged microdomains of the endothelial membrane, followed by vesicle formation, transendothelial trafficking, and exocytosis at the opposite cell surface (generally directed from blood to brain) [33,34].

Exploiting endogenous transcellular transport systems for drug delivery into the brain would ideally allow for extensive drug distribution within the CNS by utilizing the widely branched brain microcapillary system without interfering with the structure and function of the BBB [35,36]. Individual structural reengineering for efficient brain penetration is possible for certain substances, including small molecules mimicking endogenous nutrient structures or macromolecules fused to receptor-binding motifs [37,38], but structural modifications of the therapeutic agent must still conform in terms of metabolism, pharmacokinetics, and therapeutic functionality, which have to be assessed on a cost- and time-intense substance-by-substance basis. Several nanocarrier systems, including liposomes, polymersomes, and metal-based nanoparticles, permit a greater structural flexibility of therapeutics [39–41]. Depending on the nanoparticle type and cargo properties, therapeutics such as monoclonal antibodies, antisense drugs, or short interfering RNA, are entrapped by, integrated in, or bound to a nanocarrier shell, which provides multiple advantages including the introduction of unique features (sensitive to magnetic fields or specific pH values), improved stability, solubility, bioavailability, biocompatibility, and permeability towards physiological barriers [39–41]. For example, liposome formulations with cationic gemini amphiphiles, depending on their stereochemistry, promote efficient transport across an in vitro human BBB model compared to free tracer or tracer-loaded neutral liposomes, thought to be due to enhanced adsorptive-mediated transcytosis [42]. In another example, nanocarriers labelled with dual ligands targeting the glucose transporter GLUT1, which is highly expressed at the BBB, showed significantly increased carrier-mediated transport into the mouse brain compared to single-ligand labelled or untargeted nanocarriers [43].

However, a well-known procedure to achieve efficient BBB transcytosis is the conjugation of short peptides or monoclonal antibodies mimicking ligand epitopes (peptidomimetics) to the nanocarrier surface, which are recognized by the respective endogenous receptors ("Trojan horse" strategy) [44]. Since enriched protein levels available at the luminal surface as well as rapid and efficient blood-to-brain transcytosis rates across the BBB are important characteristics of a potential vehicle receptor, the common targets are non-microvascular-specific metabolic receptors for insulin [45], transferrin [46], insulin-like growth factors [47], leptin [48], or low-density lipoprotein (LDL) [49]. For example, fluorescent nanoparticles conjugated to transferrin or LDL exhibit enhanced receptor-mediated transcytosis rates across an in vitro murine BBB model compared to unconjugated nanoparticles [50]. In addition, the transcytosis efficiency of transferrin-modified nanoparticles can be further enhanced by the introduction of a short cell-penetrating peptides onto the nanocarrier surface, which reduces endosomal entrapment and subsequent degradation of the delivery vehicles during receptor-mediated transcytosis [51]. Although drug delivery systems paired with moieties for receptor-mediated transcytosis are constantly being improved, brain penetration efficiencies are insufficient so far due to the potential competition with endogenous ligands and receptor availability on off-target cells, but also due to the tight regulation of receptor activity, especially at the brain microvasculature [41,52].

2. Receptor-Mediated Transcytosis at the BBB: Regulation of LDL Receptors

Transcytosis across an intact brain endothelial microvasculature is strictly downregulated [27]. Despite its relevance, little is known about the cellular regulatory mechanisms [53–55], but it has become clear that brain microvascular endothelial cells are sensitive to signals derived from extra-endothelial cells, particularly well-demonstrated by pericytes, part of the neurovascular unit, that significantly regulate BBB transcytosis [56–58]. However, how transcytosis is constantly modified due to temporal or chronic changes in the brain homeostasis and the peripheral environment, especially during aging or pathophysiological conditions, still raises many questions [28,59]. The scaling of transporter levels at the endothelial surface in response to individual cellular and environmental demands might be a central element in the dynamic transcytosis regulation at the BBB.

Members of the LDL receptor family, including the eponym LDL receptor and the prominent low-density lipoprotein receptor-related protein 1 (LRP1), are essential cell surface receptors participating in diverse physiological processes [60]. Receptors of the LDL receptor family share a modular domain organization through the use of repetitive characteristic substructures, and are divided into a cytoplasmic and extracellular section of variable length, connected by a single-pass transmembrane domain [61]. The cytoplasmic tail can contain several recognition motifs for cellular adaptor or scaffold molecules, responsible for clathrin-mediated internalization, cellular trafficking, or cell signalling [60]. The extracellular part harbours at least one cluster of ligand-binding repeats (cysteine-rich complement-type repeats) and an epidermal growth factor (EGF) precursor homology domain, which is essential for the pH-sensitive conformational change following endocytosis [61]. Differences in the number, composition, and position of each substructure provide the basis for the functional diversity of LDL receptor family members [62]. Interactions with other cell surface proteins (co-receptors) further modulate the ligand profile and downstream activities of these receptors [60]. The ubiquitously expressed LDL receptor is well known to mediate the cellular internalization of lipoprotein particles by binding to the associated apolipoproteins (Apo)B100 and E [63,64], primarily ensuring cellular cholesterol supply and thereby controlling the content of LDL, the most abundant cholesterol-carrying lipoprotein, in extracellular fluids [65]. In contrast to the modest quantity of potential ligands and functions of the LDL receptor, LRP1 is a multifunctional receptor participating in numerous physiological processes besides lipoprotein metabolism [66,67], including blood coagulation [68–70], clearance of various proteases and protease inhibitors [71,72], and immune responses [73–78], but also plays an essential role in various pathophysiological conditions, such as Alzheimer's Disease (AD) [79–81]), atherosclerosis [82–84]), or cancer [85–87] through its interaction with at least 12 cytosolic adaptor proteins and the binding and internalization of more than 75 ligands, including ApoE, that vary significantly in structure and function [88].

LRP1 is expressed in certain cell types, such as hepatocytes, fibroblasts, macrophages, smooth muscle cells, neurons, and astrocytes, making it present in most tissues of the body [64,89]. In contrast to the high expression in most cell types [89], LRP1 is found at comparatively low levels in the brain microvasculature and supposed to be located predominantly in the abluminal (and to smaller quantities in the luminal) endothelial membrane [90–92].

Due to the structural prerequisites, members of the LDL receptor family are inherently endocytic receptors, which bind to specific ligands from the extracellular space and translocate in complex with the ligand from the plasma membrane into an intracellular vesicle. However, when it comes to extremely thin (~200 nm) and polarized endothelial cells of the brain microvasculature [6,8,37], the process of receptor-mediated endocytosis can be extended to receptor-mediated transcytosis, where complex vesicular sorting mechanisms and exocytosis at the opposite plasma membrane translocate the ligand from one extracellular fluid to another. Whereas the possibility of LDL receptor-mediated transcytosis is discussed controversially, brain-to-blood and blood-to-brain BBB transcytosis mediated by LRP1 has been documented for several ligands [91,93–96]. The high transcytosis rate of LRP1 and the tightly regulated low brain microvascular protein levels [92], which are significantly reduced with age or in AD [97–99], might reflect the significance and performance of LRP1 transcytosis at the BBB.

The receptor surface levels of members of the LDL receptor family can be regulated at different cellular levels in response to physiological or pathophysiological conditions. A central physiological feedback loop is transcriptional regulation due to cellular levels of a ligand. The cellular depletion of cholesterol initiates the two-step proteolysis of the transcription factor sterol-regulatory element binding protein (SREBP)-2 in the Golgi apparatus, which allows its translocation from the endoplasmic reticulum (ER) membrane into the nucleus [100]. While nuclear SREBP-2 enhances the transcription of the LDL receptor, it inhibits LRP1 transcription [101]. As cholesterol levels rise in the ER membrane, SREBP-2 translocation is prohibited, thereby returning the effect on the receptor transcription [100].

In addition to transcriptional regulation, cell surface activity of transmembrane receptors is limited by proteolytic shedding. Membrane-anchored metalloproteases, including ADAM10, ADAM17, and BACE1 are capable of cleaving adjacent LRP1 within the extracellular chain, releasing a large fragment into the extracellular space [102,103]. In contrast to secreted soluble LRP1, the remaining truncated receptor is unable to bind to extracellular ligands, which limits cell surface LRP1 activity.

Over the past two decades [104], an entirely different regulatory mechanism has moved into the centre of attention: the proprotein convertase subtilisin/kexin type 9 (PCSK9)-mediated degradation of several LDL receptor family members [105–107]. The serine protease PCSK9, a member of the proteinase K subfamily of subtilases, is expressed as a soluble zymogen primarily in the liver and to a lesser extent in the kidney, small intestine, and the brain [104], but not in brain microvascular endothelial cells [108]. Following the signal peptide (amino acids (aa) 1–30), pro-PCSK9 exhibits an N-terminal prodomain (aa 31–152) that precedes the catalytic subunit (aa 153–404), which contains the classical serine protease catalytic triad (Asp186, His226, Ser386) and the oxyanion hole (Asn317) [109–112]. The catalytic subunit is connected by a hinge region (aa 422–452) to a C-terminal Cys/His-rich domain (CHRD) (aa 453–692), which is composed of three tandem repeats (M1: aa 453–529, M2: aa 530–603, M3: aa 604–692) [109–112]. Maturation of pro-PCSK9 is achieved by Ca^{2+}-independent autocatalysis of its prodomain early in the ER, which is required for its secretion into the extracellular space [104,113,114]. The separated prodomain remains non-covalently attached to the subtilisin-like catalytic site, preventing further enzymatic activity [104,109–111]. The mode of extracellular PCSK9 regulation, which is independent of its proteolytic activity, is well studied for the LDL receptor (Figure 1). Usually, cell surface LDL receptors bind to extracellular cholesterol-rich lipoprotein particles and internalize their ligands through clathrin-coated pits [65]. The LDL receptor/ligand complex enters the endosomal/lysosomal degradation pathway and dissociates pH-dependently (pH < 6) in sorting endosomes due to an EGF homology domain-induced conformational change in the LDL receptor, disrupting the lipoprotein binding sites [115–117]. While the released LDL receptor is recycled back to the cell surface, restoring the initial receptor level, the lipoprotein particle remains in the degradation pathway until lysosomal disintegration, releasing the transported cholesterol [65]. Analogous to receptor-mediated ligand endocytosis, extracellular PCSK9 is able to bind with its catalytic subunit to the EGF-A repeat within the EGF homology domain of the LDL receptor in a Ca^{2+}-dependent manner, and enters the endosomal/lysosomal system in association with the receptor [118,119]. However, instead of complex dissociation, PCSK9-LDL receptor binding is enhanced with decreasing pH value, which, through an as-yet-unidentified mechanism(s), redirects the LDL receptor from the endosome to the lysosome for degradation rather than allowing recycling back to the cell surface [118,120]. Consequently, the LDL receptor level at the cell surface is reduced at the expense of extracellular PCSK9 molecules. Although not required for direct binding, the availability of the PCSK9 CHRD domain (aa 425–692), as well as at least three ligand-binding repeats and the β-propeller domain of the LDL receptor, are essential for subsequent LDL receptor degradation, indicating a more complex mechanism [120,121].

The PCSK9-mediated regulatory pathway extends the cholesterol-responsive negative feedback loop within cellular cholesterol homeostasis to the post-transcriptional level. In response to low cholesterol levels, PCSK9 and the LDL receptor are transcriptionally activated by SREBP-2 binding to the sterol-regulatory element site of the promotor sequences [122,123], are produced side by side as precursor proteins, and promote the efficient processing of each other [124]. Maturation of pro-PCSK9 via autocatalytic intramolecular cleavage in the ER is promoted by the binding to the EGF-A repeat of the pro-LDL receptor, whereas processed PCSK9 subsequently acts as an LDL receptor chaperone, supporting the transport of pro-LDL receptor molecules from the ER to the Golgi apparatus and towards the cell membrane after completion of glycosylation [124,125].

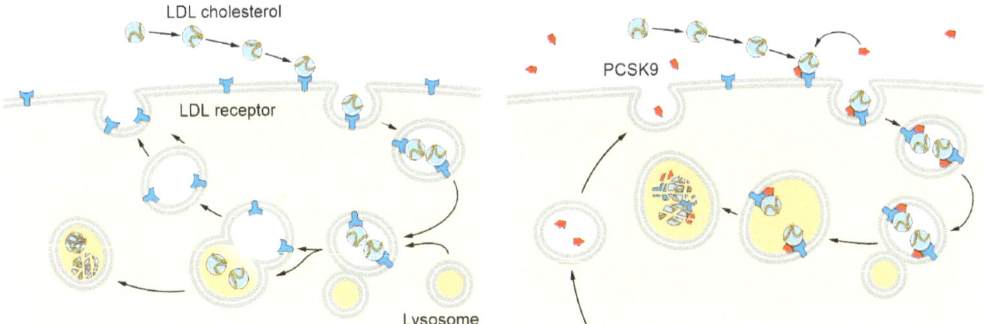

Figure 1. Mechanism of extracellular PCSK9 regulation to downregulate cell surface receptor levels. Left panel: Members of the low-density lipoprotein (LDL) receptor family bind to ligands (e.g., LDL receptors to LDL cholesterol) in the extracellular space. Clathrin-mediated endocytosis translocates the receptor/ligand complex into endosomes, where acidification causes the receptor and ligand to dissociate. The free receptor can be recycled back to the cell surface while the ligand remains in the endosomal/lysosomal degradation pathway. Right panel: Extracellular proprotein convertase subtilisin/kexin type 9 (PCSK9) is capable of binding to several members of the LDL receptor family, including its eponym, at the cell surface. After internalization, the binding strength of PCSK9 to the receptor increases at a slightly acidic pH, trapping the receptor in the intracellular compartment and leading to its lysosomal degradation and reduced cell surface receptor levels.

Although intracellular PCSK9 binding might occur, directing the bound LDL receptor directly from the trans-Golgi network to the lysosomes for degradation [121,126,127], a significant percentage of mature LDL receptor and PCSK9 proteins have to be integrated into the plasma membrane and secreted into the extracellular space, respectively, due to the principle of a delayed negative feedback loop. Subsequently, the majority of extracellular PCSK9 binds to target receptors in the immediate environment and is degraded, with only a small fraction entering the vascular system [126,128]. Plasma PCSK9, which is supposed to derive exclusively from the liver [128], is comparatively low in concentration (~0.05 µg/mL–~0.6 µg/mL) [105], which might indicate that circulating PCSK9, due to its self-destructive mode of action, primarily impacts cells and tissues with moderate to low target receptor levels and exerts a broader spectrum of regulatory functions than just cholesterol homeostasis.

3. Endocrine Regulation by PCSK9 and Its Inhibition

PCSK9 regulation might occur intracellularly, but especially extracellularly in an autocrine, paracrine, and endocrine fashion. In addition to cellular self-regulation of mature receptor concentrations at the cell surface, surrounding cells are capable of influencing the cell surface levels of PCSK9-regulated LDL receptor family members. There is evidence that neurons and human vascular smooth muscle cells secrete PCSK9, which might affect the abluminal side of the BBB [104,108]. Cerebral PCSK9 levels are considered to be independent of peripheral levels because PCSK9 does not cross an intact BBB [129]. Unlike for peripheral PCSK9, elevated levels of PCSK9 in brain fluids correlate with neurodegenerative disorders, including AD [130,131]. However, targeting cerebral PCSK9 levels might be as difficult as treating neurological symptoms and pathological mechanisms in the brain.

A more accessible PCSK9 reservoir is represented by liver-derived PCSK9 in the vascular system, which regulates cell surfaces in an endocrine manner. In a previous study, we identified circulating PCSK9 as a regulator of LRP1-mediated amyloid-β (Aβ) transport across the BBB (Figure 2) [132]. The development of the progressive neurodegenerative disorder AD is typically characterized by the increased levels of Aβ peptides [133,134], which accumulate extracellularly within the brain due to impaired brain clearance [135,136]. A central clearance pathway for brain Aβ is the continuous brain-to-blood transport across the BBB, which is highly dependent on cell surface LRP1 activity [90,95,137]. Microvascular LRP1 is supposed to be primarily located at the abluminal side and to a lesser extent at the luminal side of endothelial cells [90–92]. Using an established BBB in vitro model, we observed significantly reduced LRP1-mediated Aβ transport across a murine brain endothelial monolayer upon incubation with recombinant PCSK9 [132]. Consistently, we reversed this outcome by repetitively injecting FDA-approved monoclonal anti-PCSK9 antibodies into the peritoneum of an AD mouse model, resulting in substantially decreased cerebral Aβ concentrations and improved hippocampus-dependent learning behaviour compared to control-treated mice [132]. As peripheral PCSK9 inhibition was unable to reproduce these effects in brain endothelium-specific LRP1$^{-/-}$ AD mice, our data may reveal the potential to modify BBB receptor quantities for therapeutic contexts [132]. Since LRP1 is lowly concentrated in the brain microvasculature and PCSK9 expression has not been detected in brain microvascular endothelial cells, externally derived PCSK9 in a paracrine and endocrine fashion might play an important role in the tight control of LRP1 at the BBB. This observation is not exclusive to LRP1 but includes all PCSK9-regulated receptors.

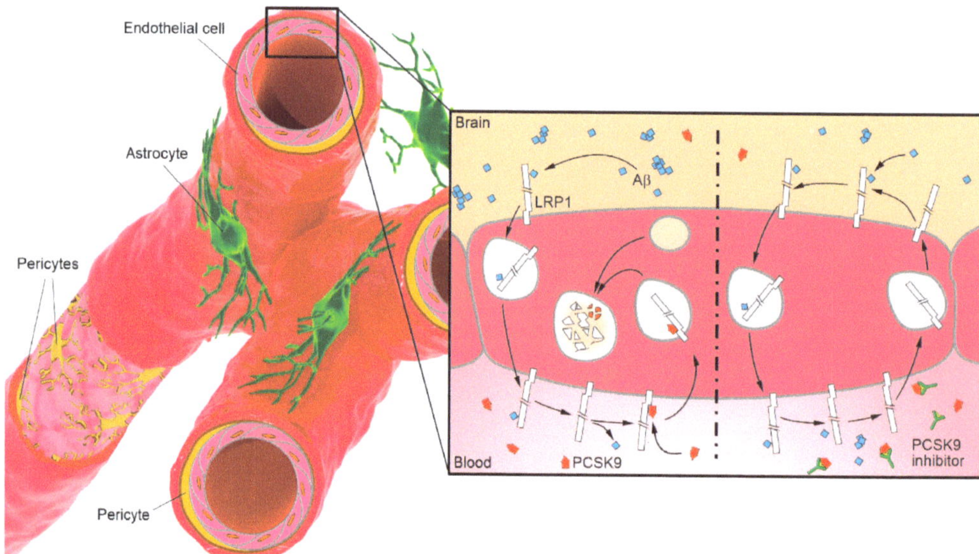

Figure 2. Simplified model of the PCSK9-regulated Aβ transcytosis across the BBB, mediated by LRP1, and its inhibition via monoclonal antibodies. Left panel: Low-density lipoprotein receptor-related protein 1 (LRP1) binds to cerebral amyloid-β (Aβ) and initiates its translocation into the vascular system across the blood–brain barrier (BBB) along its concentration gradient. Secreted PCSK9 in the brain and periphery binds to LRP1 at the endothelial cell surface and targets the receptor for lysosomal degradation, reducing the amount of available Aβ clearance receptors in the microvasculature. Right panel: The peripheral application of monoclonal antibodies targeting circulating PCSK9 diminishes the number of PCSK9 molecules capable of binding to LRP1, which increases the amount of LRP1 recycled back to the cell surface and consequently the rate of Aβ transcytosis from brain to blood.

The use of the receptor-binding epitopes of ApoB and E allows endocytosis by members of the LDL receptor family and subsequent transcellular transport from the luminal to the abluminal membrane of brain endothelial cells. Subsequently, Apo-like structures are exocytosed into the CNS and internalized by proximal neurons or astrocytes [138]. Studies fusing the binding epitope of ApoB or E to recombinant proteins [138], enzymes [138–140], or nanoparticles [141–143], displayed promising brain penetration efficiencies. Unfortunately, and in line with similar studies targeting other metabolic receptors for ligands such as transferrin or insulin, non-cerebral tissues, particularly the liver, also showed elevated concentrations of the Apo-like structures due to ubiquitous receptor expression. Therefore, strategies to increase brain penetration and minimize off-target delivery to peripheral tissues are required to effectively treat CNS disorders.

Our results demonstrate that receptor-mediated Aβ clearance across the BBB by LRP1 can be significantly enhanced due to peripheral PCSK9 inhibition. Besides the use of high-affinity monoclonal antibodies at substantial concentrations in therapeutic contexts to silence circulating PCSK9, PCSK9 receptor affinity is physiologically reduced by furin cleavage or the binding to Apo-containing lipoproteins as demonstrated for LDL particles [144]. In addition, total peripheral target receptor concentration directly influences the plasma PCSK9 concentration because PCSK9 is degraded at the end of the regulatory pathway. It is therefore conceivable that some of these mechanisms might affect circulating PCSK9 more effectively than cellular, or just secreted, PCSK9, suggesting the possibility that brain microvascular receptor levels could be increased not only generally but also selectively without favouring receptors of off-target peripheral tissues.

4. Limitations, Potential Risks, and Future Directions

Illuminating the complex crosstalk between peripheral tissues, which strongly influence the plasma composition, and the CNS at the BBB as a dynamic endocrine interface, might reveal potential approaches to increase the efficiency of drug penetration via endogenous pathways.

PCSK9 is a potent post-transcriptional regulator of members of the LDL receptor family. Circulating, liver-derived PCSK9 reduces cell surface receptor concentration within the vascular system, including the brain microvascular endothelium, whereas inhibition of peripheral PCSK9 has the opposite effect. Therapeutic FDA-approved inhibition of circulating PCSK9 by peripheral application of monoclonal antibodies to raise cell surface LDL receptor levels is a more recently developed but well-established treatment for patients with high plasma LDL cholesterol [145]. PCSK9 inhibitors are typically used in patients at very high risk of atherosclerotic cardiovascular disease events suffering from severe hypercholesterolemia, diabetes mellitus, or advanced chronic kidney disease, and despite these circumstances, exhibit an overall well-tolerated safety profile [146]. However, the consequences of PCSK9 inhibition over long-term treatment periods and in different pathophysiological contexts are still elusive. Beyond cholesterol metabolism, PCSK9 has been implicated in cancer cell immunity and the promotion of vascular inflammation [147–150], illustrating that the explicit role of PCSK9 in physiological and pathophysiological processes remains to be further explored. In addition to the observation that endocrine regulatory mechanisms via the vascular system could be relatively poorly understood, therapeutic interventions targeting circulating PCSK9 might inadvertently affect multiple physiological and pathophysiological processes throughout the body beyond the microvasculature. It is therefore all the more important to gain a deeper understanding of how a systemic regulatory mechanism might act site-specifically. One possibility could be the cell type-specific combination of co-receptors or adaptor molecules, that specify the interaction pattern, with extracellular ligands of a ubiquitously expressed cell surface receptor. Thus, a cell could be sensitized to a systemic regulatory mechanism, and vice versa, by the sufficient expression of a set of proteins. Further investigation is needed to clarify the potential implications of PCSK9 and its inhibition and to elucidate whether endocrine-regulating PCSK9 can be specifically targeted to increase receptor-mediated transcytosis across the BBB without favouring off-target internalization by PCSK9-producing peripheral organs. These considerations are not limited to PCSK9 and the LDL receptor family as metabolic receptors, including insulin and transferrin receptors, at the brain microvascular endothelium might also be sensitive to circulating regulatory proteins.

The mere knowledge that receptors at the BBB might be sensitive to, and their number regulated by, vascular circulating molecules could be beneficial in several ways. Transient upregulation could improve the translocation of peripherally administered therapeutic agents, such as nanoparticles, into the brain. In contrast, receptor downregulation could decrease off-target brain penetration and subsequent side effects of highly concentrated circulating drugs (e.g., t-Pa). As shown for enhanced LRP1-mediated Aβ clearance due to peripheral PCSK9 inhibition, altering the cell surface receptor concentration itself could be a therapeutic option by increasing or decreasing receptor-mediated transcytosis to or from the CNS. This would avoid the need to introduce a therapeutic agent into the CNS, and potential cross-reactions. BBB permeability changes dynamically due to aging or pathophysiological conditions, such as AD, which could be addressed by exploiting regulatory mechanisms as a preventive measure or therapeutic option after disease onset. Due to the relative simplicity of access and monitoring, endocrine regulatory pathways as described for circulating PCSK9 would be predestined. Thus, a deeper understanding of how BBB transport systems are organized and regulated might be crucial to efficiently exploit the BBB transcytosis machinery for drug delivery to the brain, and to modify the compromised BBB during pathophysiological states as a therapeutic option.

Author Contributions: Conceptualization, A.D.M.; writing—original draft preparation, A.D.M.; writing—review & editing, C.U.P.; visualization, A.D.M.; supervision, C.U.P.; project administration, C.U.P.; funding acquisition, C.U.P. All authors have read and agreed to the published version of the manuscript.

Funding: This research was funded by the Deutsche Forschungsgemeinschaft (Grant: PI 379/8-3), the Collaborative Research Center (Grant: CRC 877 A15), the Innovative Medicines Initiative 2 Joint Undertaking which receives support from the European Union's Horizon 2020 research and innovation program, the European Federation of Pharmaceutical Industries and Associations (EFPIA) (Grant: 807015-Im2Pact), and the Alzheimer Forschung Initiative e.V. (Grant: 20022).

Acknowledgments: We would like to thank Michael Plenikowski for his excellent technical support and expertise in visualizing the content.

Conflicts of Interest: The authors declare no conflict of interest.

Abbreviations

aa	amino acid
Aβ	amyloid-β
AD	Alzheimer's disease
Apo	apolipoprotein
BBB	blood–brain barrier
BCSFB	blood–cerebrospinal fluid barrier
CHRD	Cys/His-rich domain
CNS	central nervous system
CSF	cerebrospinal fluid
EGF	epidermal growth factor
ER	endoplasmic reticulum
LDL	low-density lipoprotein
LRP1	low-density lipoprotein receptor-related protein 1
PCSK9	proprotein convertase subtilisin/kexin type 9
SREBP	sterol-regulatory element binding protein

References

1. Pardridge, W.M. The blood-brain barrier: Bottleneck in brain drug development. *NeuroRx* **2005**, *2*, 3–14. [CrossRef]
2. Redzic, Z. Molecular biology of the blood-brain and the blood-cerebrospinal fluid barriers: Similarities and differences. *Fluids Barriers CNS* **2011**, *8*, 3. [CrossRef]

3. Daneman, R.; Prat, A. The blood-brain barrier. *Cold Spring Harb. Perspect. Biol.* **2015**, *7*, a020412. [CrossRef]

4. Johanson, C.E.; Stopa, E.G.; McMillan, P.N. The blood-cerebrospinal fluid barrier: Structure and functional significance. *Methods Mol. Biol.* **2011**, *686*, 101–131. [CrossRef]

5. Brightman, M.W.; Reese, T.S. Junctions between intimately apposed cell membranes in the vertebrate brain. *J. Cell Biol.* **1969**, *40*, 648–677. [CrossRef] [PubMed]

6. Coomber, B.L.; Stewart, P.A. Morphometric analysis of CNS microvascular endothelium. *Microvasc. Res.* **1985**, *30*, 99–115. [CrossRef]

7. Dallas, S.; Miller, D.S.; Bendayan, R. Multidrug resistance-associated proteins: Expression and function in the central nervous system. *Pharmacol. Rev.* **2006**, *58*, 140–161. [CrossRef]

8. Betz, A.L.; Goldstein, G.W. Polarity of the blood-brain barrier: Neutral amino acid transport into isolated brain capillaries. *Science* **1978**, *202*, 225–227. [CrossRef]

9. Sims, D.E. The pericyte–A review. *Tissue Cell* **1986**, *18*, 153–174. [CrossRef] [PubMed]

10. Shepro, D.; Morel, N.M. Pericyte physiology. *FASEB J.* **1993**, *7*, 1031–1038. [CrossRef] [PubMed]

11. Simard, M.; Arcuino, G.; Takano, T.; Liu, Q.S.; Nedergaard, M. Signaling at the gliovascular interface. *J. Neurosci.* **2003**, *23*, 9254–9262. [CrossRef] [PubMed]

12. Solár, P.; Zamani, A.; Lakatosová, K.; Joukal, M. The blood-brain barrier and the neurovascular unit in subarachnoid hemorrhage: Molecular events and potential treatments. *Fluids Barriers CNS* **2022**, *19*, 29. [CrossRef] [PubMed]

13. Abbott, N.J. Dynamics of CNS barriers: Evolution, differentiation, and modulation. *Cell. Mol. Neurobiol.* **2005**, *25*, 5–23. [CrossRef]

14. Dhuria, S.V.; Hanson, L.R.; Frey, W.H. Intranasal delivery to the central nervous system: Mechanisms and experimental considerations. *J. Pharm. Sci.* **2010**, *99*, 1654–1673. [CrossRef] [PubMed]

15. Born, J.; Lange, T.; Kern, W.; McGregor, G.P.; Bickel, U.; Fehm, H.L. Sniffing neuropeptides: A transnasal approach to the human brain. *Nat. Neurosci.* **2002**, *5*, 514–516. [CrossRef]

16. Mischley, L.K.; Conley, K.E.; Shankland, E.G.; Kavanagh, T.J.; Rosenfeld, M.E.; Duda, J.E.; White, C.C.; Wilbur, T.K.; de La Torre, P.U.; Padowski, J.M. Central nervous system uptake of intranasal glutathione in Parkinson's disease. *NPJ Park. Dis.* **2016**, *2*, 16002. [CrossRef]

17. Craft, S.; Raman, R.; Chow, T.W.; Rafii, M.S.; Sun, C.-K.; Rissman, R.A.; Donohue, M.C.; Brewer, J.B.; Jenkins, C.; Harless, K.; et al. Safety, Efficacy, and Feasibility of Intranasal Insulin for the Treatment of Mild Cognitive Impairment and Alzheimer Disease Dementia: A Randomized Clinical Trial. *JAMA Neurol.* **2020**, *77*, 1099–1109. [CrossRef]

18. Sikich, L.; Kolevzon, A.; King, B.H.; McDougle, C.J.; Sanders, K.B.; Kim, S.-J.; Spanos, M.; Chandrasekhar, T.; Trelles, M.D.P.; Rockhill, C.M.; et al. Intranasal Oxytocin in Children and Adolescents with Autism Spectrum Disorder. *N. Engl. J. Med.* **2021**, *385*, 1462–1473. [CrossRef]

19. Crowe, T.P.; Hsu, W.H. Evaluation of Recent Intranasal Drug Delivery Systems to the Central Nervous System. *Pharmaceutics* **2022**, *14*, 629. [CrossRef]

20. Fortuna, A.; Schindowski, K.; Sonvico, F. Editorial: Intranasal Drug Delivery: Challenges and Opportunities. *Front. Pharmacol.* **2022**, *13*, 868986. [CrossRef]

21. Frank, R.T.; Aboody, K.S.; Najbauer, J. Strategies for enhancing antibody delivery to the brain. *Biochim. Biophys. Acta* **2011**, *1816*, 191–198. [CrossRef] [PubMed]

22. Bellavance, M.-A.; Blanchette, M.; Fortin, D. Recent advances in blood-brain barrier disruption as a CNS delivery strategy. *AAPS J.* **2008**, *10*, 166–177. [CrossRef] [PubMed]

23. Meairs, S. Facilitation of Drug Transport across the Blood-Brain Barrier with Ultrasound and Microbubbles. *Pharmaceutics* **2015**, *7*, 275–293. [CrossRef]

24. Sweeney, M.D.; Sagare, A.P.; Zlokovic, B.V. Blood-brain barrier breakdown in Alzheimer disease and other neurodegenerative disorders. *Nat. Rev. Neurol.* **2018**, *14*, 133–150. [CrossRef]

25. McMahon, D.; Poon, C.; Hynynen, K. Evaluating the safety profile of focused ultrasound and microbubble-mediated treatments to increase blood-brain barrier permeability. *Expert Opin. Drug Deliv.* **2019**, *16*, 129–142. [CrossRef] [PubMed]

26. Karmur, B.S.; Philteos, J.; Abbasian, A.; Zacharia, B.E.; Lipsman, N.; Levin, V.; Grossman, S.; Mansouri, A. Blood-Brain Barrier Disruption in Neuro-Oncology: Strategies, Failures, and Challenges to Overcome. *Front. Oncol.* **2020**, *10*, 563840. [CrossRef] [PubMed]

27. Stewart, P.A. Endothelial vesicles in the blood-brain barrier: Are they related to permeability? *Cell. Mol. Neurobiol.* **2000**, *20*, 149–163. [CrossRef]

28. Sweeney, M.D.; Zhao, Z.; Montagne, A.; Nelson, A.R.; Zlokovic, B.V. Blood-Brain Barrier: From Physiology to Disease and Back. *Physiol. Rev.* **2019**, *99*, 21–78. [CrossRef]

29. Muller, W.A.; Gimbrone, M.A. Plasmalemmal proteins of cultured vascular endothelial cells exhibit apical-basal polarity: Analysis by surface-selective iodination. *J. Cell Biol.* **1986**, *103*, 2389–2402. [CrossRef]

30. Simionescu, M.; Gafencu, A.; Antohe, F. Transcytosis of plasma macromolecules in endothelial cells: A cell biological survey. *Microsc. Res. Tech.* **2002**, *57*, 269–288. [CrossRef]

31. Worzfeld, T.; Schwaninger, M. Apicobasal polarity of brain endothelial cells. *J. Cereb. Blood Flow Metab.* **2016**, *36*, 340–362. [CrossRef] [PubMed]

32. Zlokovic, B.V. The blood-brain barrier in health and chronic neurodegenerative disorders. *Neuron* **2008**, *57*, 178–201. [CrossRef] [PubMed]
33. Kumagai, A.K.; Eisenberg, J.B.; Pardridge, W.M. Absorptive-mediated endocytosis of cationized albumin and a beta-endorphin-cationized albumin chimeric peptide by isolated brain capillaries. Model system of blood-brain barrier transport. *J. Biol. Chem.* **1987**, *262*, 15214–15219. [CrossRef] [PubMed]
34. Hervé, F.; Ghinea, N.; Scherrmann, J.-M. CNS delivery via adsorptive transcytosis. *AAPS J.* **2008**, *10*, 455–472. [CrossRef]
35. Schlageter, K.E.; Molnar, P.; Lapin, G.D.; Groothuis, D.R. Microvessel organization and structure in experimental brain tumors: Microvessel populations with distinctive structural and functional properties. *Microvasc. Res.* **1999**, *58*, 312–328. [CrossRef]
36. Abbott, N.J.; Rönnbäck, L.; Hansson, E. Astrocyte-endothelial interactions at the blood-brain barrier. *Nat. Rev. Neurosci.* **2006**, *7*, 41–53. [CrossRef]
37. Pardridge, W.M. Drug transport across the blood-brain barrier. *J. Cereb. Blood Flow Metab.* **2012**, *32*, 1959–1972. [CrossRef]
38. Pardridge, W.M.; Boado, R.J. Reengineering biopharmaceuticals for targeted delivery across the blood-brain barrier. *Methods Enzymol.* **2012**, *503*, 269–292. [CrossRef]
39. Furtado, D.; Björnmalm, M.; Ayton, S.; Bush, A.I.; Kempe, K.; Caruso, F. Overcoming the Blood-Brain Barrier: The Role of Nanomaterials in Treating Neurological Diseases. *Adv. Mater.* **2018**, *30*, e1801362. [CrossRef]
40. Zhou, Y.; Peng, Z.; Seven, E.S.; Leblanc, R.M. Crossing the blood-brain barrier with nanoparticles. *J. Control. Release* **2018**, *270*, 290–303. [CrossRef]
41. Mitchell, M.J.; Billingsley, M.M.; Haley, R.M.; Wechsler, M.E.; Peppas, N.A.; Langer, R. Engineering precision nanoparticles for drug delivery. *Nat. Rev. Drug Discov.* **2021**, *20*, 101–124. [CrossRef]
42. Simonis, B.; Vignone, D.; Gonzalez Paz, O.; Donati, E.; Falchetti, M.L.; Bombelli, C.; Cellucci, A.; Auciello, G.; Fini, I.; Galantini, L.; et al. Transport of cationic liposomes in a human blood brain barrier model: Role of the stereochemistry of the gemini amphiphile on liposome biological features. *J. Colloid Interface Sci.* **2022**, *627*, 283–298. [CrossRef] [PubMed]
43. Mészáros, M.; Porkoláb, G.; Kiss, L.; Pilbat, A.-M.; Kóta, Z.; Kupihár, Z.; Kéri, A.; Galbács, G.; Siklós, L.; Tóth, A.; et al. Niosomes decorated with dual ligands targeting brain endothelial transporters increase cargo penetration across the blood-brain barrier. *Eur. J. Pharm. Sci.* **2018**, *123*, 228–240. [CrossRef] [PubMed]
44. Pardridge, W.M. Blood-brain barrier delivery of protein and non-viral gene therapeutics with molecular Trojan horses. *J. Control. Release* **2007**, *122*, 345–348. [CrossRef] [PubMed]
45. Pardridge, W.M.; Eisenberg, J.; Yang, J. Human blood-brain barrier insulin receptor. *J. Neurochem.* **1985**, *44*, 1771–1778. [CrossRef] [PubMed]
46. Pardridge, W.M.; Eisenberg, J.; Yang, J. Human blood-brain barrier transferrin receptor. *Metabolism* **1987**, *36*, 892–895. [CrossRef]
47. Duffy, K.R.; Pardridge, W.M.; Rosenfeld, R.G. Human blood-brain barrier insulin-like growth factor receptor. *Metabolism* **1988**, *37*, 136–140. [CrossRef]
48. Golden, P.L.; Maccagnan, T.J.; Pardridge, W.M. Human blood-brain barrier leptin receptor. Binding and endocytosis in isolated human brain microvessels. *J. Clin. Investig.* **1997**, *99*, 14–18. [CrossRef]
49. Dehouck, B.; Fenart, L.; Dehouck, M.P.; Pierce, A.; Torpier, G.; Cecchelli, R. A new function for the LDL receptor: Transcytosis of LDL across the blood-brain barrier. *J. Cell Biol.* **1997**, *138*, 877–889. [CrossRef]
50. Lu, Q.; Cai, X.; Zhang, X.; Li, S.; Song, Y.; Du, D.; Dutta, P.; Lin, Y. Synthetic Polymer Nanoparticles Functionalized with Different Ligands for Receptor-mediated Transcytosis across Blood-Brain Barrier. *ACS Appl. Bio Mater.* **2018**, *1*, 1687–1694. [CrossRef]
51. Lakkadwala, S.; Singh, J. Co-delivery of doxorubicin and erlotinib through liposomal nanoparticles for glioblastoma tumor regression using an in vitro brain tumor model. *Colloids Surf. B Biointerfaces* **2019**, *173*, 27–35. [CrossRef] [PubMed]
52. Larsen, J.M.; Martin, D.R.; Byrne, M.E. Recent advances in delivery through the blood-brain barrier. *Curr. Top. Med. Chem.* **2014**, *14*, 1148–1160. [CrossRef]
53. de Bock, M.; van Haver, V.; Vandenbroucke, R.E.; Decrock, E.; Wang, N.; Leybaert, L. Into rather unexplored terrain-transcellular transport across the blood-brain barrier. *Glia* **2016**, *64*, 1097–1123. [CrossRef] [PubMed]
54. Andreone, B.J.; Chow, B.W.; Tata, A.; Lacoste, B.; Ben-Zvi, A.; Bullock, K.; Deik, A.A.; Ginty, D.D.; Clish, C.B.; Gu, C. Blood-Brain Barrier Permeability Is Regulated by Lipid Transport-Dependent Suppression of Caveolae-Mediated Transcytosis. *Neuron* **2017**, *94*, 581–594.e5. [CrossRef] [PubMed]
55. Villaseñor, R.; Schilling, M.; Sundaresan, J.; Lutz, Y.; Collin, L. Sorting Tubules Regulate Blood-Brain Barrier Transcytosis. *Cell Rep.* **2017**, *21*, 3256–3270. [CrossRef]
56. Armulik, A.; Genové, G.; Mäe, M.; Nisancioglu, M.H.; Wallgard, E.; Niaudet, C.; He, L.; Norlin, J.; Lindblom, P.; Strittmatter, K.; et al. Pericytes regulate the blood-brain barrier. *Nature* **2010**, *468*, 557–561. [CrossRef]
57. Daneman, R.; Zhou, L.; Kebede, A.A.; Barres, B.A. Pericytes are required for blood-brain barrier integrity during embryogenesis. *Nature* **2010**, *468*, 562–566. [CrossRef]
58. Ben-Zvi, A.; Lacoste, B.; Kur, E.; Andreone, B.J.; Mayshar, Y.; Yan, H.; Gu, C. Mfsd2a is critical for the formation and function of the blood-brain barrier. *Nature* **2014**, *509*, 507–511. [CrossRef]
59. Yang, A.C.; Stevens, M.Y.; Chen, M.B.; Lee, D.P.; Stähli, D.; Gate, D.; Contrepois, K.; Chen, W.; Iram, T.; Zhang, L.; et al. Physiological blood-brain transport is impaired with age by a shift in transcytosis. *Nature* **2020**, *583*, 425–430. [CrossRef]
60. Nykjaer, A.; Willnow, T.E. The low-density lipoprotein receptor gene family: A cellular Swiss army knife? *Trends Cell Biol.* **2002**, *12*, 273–280. [CrossRef]

61. Lillis, A.P.; van Duyn, L.B.; Murphy-Ullrich, J.E.; Strickland, D.K. LDL receptor-related protein 1: Unique tissue-specific functions revealed by selective gene knockout studies. *Physiol. Rev.* **2008**, *88*, 887–918. [CrossRef]

62. Go, G.; Mani, A. Low-Density Lipoprotein Receptor (LDLR) Family Orchestrates Cholesterol Homeostasis. *Yale J. Biol. Med.* **2012**, *85*, 19–28.

63. Russell, D.W.; Brown, M.S.; Goldstein, J.L. Different combinations of cysteine-rich repeats mediate binding of low density lipoprotein receptor to two different proteins. *J. Biol. Chem.* **1989**, *264*, 21682–21688. [CrossRef] [PubMed]

64. Fagerberg, L.; Hallström, B.M.; Oksvold, P.; Kampf, C.; Djureinovic, D.; Odeberg, J.; Habuka, M.; Tahmasebpoor, S.; Danielsson, A.; Edlund, K.; et al. Analysis of the human tissue-specific expression by genome-wide integration of transcriptomics and antibody-based proteomics. *Mol. Cell. Proteom.* **2014**, *13*, 397–406. [CrossRef] [PubMed]

65. Brown, M.S.; Goldstein, J.L. A receptor-mediated pathway for cholesterol homeostasis. *Science* **1986**, *232*, 34–47. [CrossRef]

66. Beisiegel, U.; Weber, W.; Ihrke, G.; Herz, J.; Stanley, K.K. The LDL-receptor-related protein, LRP, is an apolipoprotein E-binding protein. *Nature* **1989**, *341*, 162–164. [CrossRef]

67. Rohlmann, A.; Gotthardt, M.; Hammer, R.E.; Herz, J. Inducible inactivation of hepatic LRP gene by cre-mediated recombination confirms role of LRP in clearance of chylomicron remnants. *J. Clin. Investig.* **1998**, *101*, 689–695. [CrossRef] [PubMed]

68. Bu, G.; Williams, S.; Strickland, D.K.; Schwartz, A.L. Low density lipoprotein receptor-related protein/alpha 2-macroglobulin receptor is an hepatic receptor for tissue-type plasminogen activator. *Proc. Natl. Acad. Sci. USA* **1992**, *89*, 7427–7431. [CrossRef]

69. Lenting, P.J.; Neels, J.G.; van den Berg, B.M.; Clijsters, P.P.; Meijerman, D.W.; Pannekoek, H.; van Mourik, J.A.; Mertens, K.; van Zonneveld, A.J. The light chain of factor VIII comprises a binding site for low density lipoprotein receptor-related protein. *J. Biol. Chem.* **1999**, *274*, 23734–23739. [CrossRef]

70. Rastegarlari, G.; Pegon, J.N.; Casari, C.; Odouard, S.; Navarrete, A.-M.; Saint-Lu, N.; van Vlijmen, B.J.; Legendre, P.; Christophe, O.D.; Denis, C.V.; et al. Macrophage LRP1 contributes to the clearance of von Willebrand factor. *Blood* **2012**, *119*, 2126–2134. [CrossRef]

71. Ashcom, J.D.; Tiller, S.E.; Dickerson, K.; Cravens, J.L.; Argraves, W.S.; Strickland, D.K. The human alpha 2-macroglobulin receptor: Identification of a 420-kD cell surface glycoprotein specific for the activated conformation of alpha 2-macroglobulin. *J. Cell Biol.* **1990**, *110*, 1041–1048. [CrossRef] [PubMed]

72. Kasza, A.; Petersen, H.H.; Heegaard, C.W.; Oka, K.; Christensen, A.; Dubin, A.; Chan, L.; Andreasen, P.A. Specificity of serine proteinase/serpin complex binding to very-low-density lipoprotein receptor and alpha2-macroglobulin receptor/low-density-lipoprotein-receptor-related protein. *Eur. J. Biochem.* **1997**, *248*, 270–281. [CrossRef] [PubMed]

73. Storm, D.; Herz, J.; Trinder, P.; Loos, M. C1 inhibitor-C1s complexes are internalized and degraded by the low density lipoprotein receptor-related protein. *J. Biol. Chem.* **1997**, *272*, 31043–31050. [CrossRef] [PubMed]

74. Meilinger, M.; Gschwentner, C.; Burger, I.; Haumer, M.; Wahrmann, M.; Szollar, L.; Nimpf, J.; Huettinger, M. Metabolism of activated complement component C3 is mediated by the low density lipoprotein receptor-related protein/alpha(2)-macroglobulin receptor. *J. Biol. Chem.* **1999**, *274*, 38091–38096. [CrossRef] [PubMed]

75. Gaultier, A.; Arandjelovic, S.; Niessen, S.; Overton, C.D.; Linton, M.F.; Fazio, S.; Campana, W.M.; Cravatt, B.F.; Gonias, S.L. Regulation of tumor necrosis factor receptor-1 and the IKK-NF-kappaB pathway by LDL receptor-related protein explains the antiinflammatory activity of this receptor. *Blood* **2008**, *111*, 5316–5325. [CrossRef]

76. Spijkers, P.P.; Denis, C.V.; Blom, A.M.; Lenting, P.J. Cellular uptake of C4b-binding protein is mediated by heparan sulfate proteoglycans and CD91/LDL receptor-related protein. *Eur. J. Immunol.* **2008**, *38*, 809–817. [CrossRef] [PubMed]

77. Zurhove, K.; Nakajima, C.; Herz, J.; Bock, H.H.; May, P. Gamma-secretase limits the inflammatory response through the processing of LRP1. *Sci. Signal.* **2008**, *1*, ra15. [CrossRef] [PubMed]

78. Gorovoy, M.; Gaultier, A.; Campana, W.M.; Firestein, G.S.; Gonias, S.L. Inflammatory mediators promote production of shed LRP1/CD91, which regulates cell signaling and cytokine expression by macrophages. *J. Leukoc. Biol.* **2010**, *88*, 769–778. [CrossRef] [PubMed]

79. Shinohara, M.; Tachibana, M.; Kanekiyo, T.; Bu, G. Role of LRP1 in the pathogenesis of Alzheimer's disease: Evidence from clinical and preclinical studies. *J. Lipid Res.* **2017**, *58*, 1267–1281. [CrossRef]

80. Rauch, J.N.; Luna, G.; Guzman, E.; Audouard, M.; Challis, C.; Sibih, Y.E.; Leshuk, C.; Hernandez, I.; Wegmann, S.; Hyman, B.T.; et al. LRP1 is a master regulator of tau uptake and spread. *Nature* **2020**, *580*, 381–385. [CrossRef]

81. Cooper, J.M.; Lathuiliere, A.; Migliorini, M.; Arai, A.L.; Wani, M.M.; Dujardin, S.; Muratoglu, S.C.; Hyman, B.T.; Strickland, D.K. Regulation of tau internalization, degradation, and seeding by LRP1 reveals multiple pathways for tau catabolism. *J. Biol. Chem.* **2021**, *296*, 100715. [CrossRef]

82. Espirito Santo, S.M.S.; Pires, N.M.M.; Boesten, L.S.M.; Gerritsen, G.; Bovenschen, N.; van Dijk, K.W.; Jukema, J.W.; Princen, H.M.G.; Bensadoun, A.; Li, W.-P.; et al. Hepatic low-density lipoprotein receptor-related protein deficiency in mice increases atherosclerosis independent of plasma cholesterol. *Blood* **2004**, *103*, 3777–3782. [CrossRef] [PubMed]

83. Hu, L.; Boesten, L.S.M.; May, P.; Herz, J.; Bovenschen, N.; Huisman, M.V.; Berbée, J.F.P.; Havekes, L.M.; van Vlijmen, B.J.M.; Tamsma, J.T. Macrophage low-density lipoprotein receptor-related protein deficiency enhances atherosclerosis in ApoE/LDLR double knockout mice. *Arterioscler. Thromb. Vasc. Biol.* **2006**, *26*, 2710–2715. [CrossRef] [PubMed]

84. Mueller, P.A.; Zhu, L.; Tavori, H.; Huynh, K.; Giunzioni, I.; Stafford, J.M.; Linton, M.F.; Fazio, S. Deletion of Macrophage Low-Density Lipoprotein Receptor-Related Protein 1 (LRP1) Accelerates Atherosclerosis Regression and Increases C-C Chemokine Receptor Type 7 (CCR7) Expression in Plaque Macrophages. *Circulation* **2018**, *138*, 1850–1863. [CrossRef] [PubMed]

85. Song, H.; Li, Y.; Lee, J.; Schwartz, A.L.; Bu, G. Low-density lipoprotein receptor-related protein 1 promotes cancer cell migration and invasion by inducing the expression of matrix metalloproteinases 2 and 9. *Cancer Res.* **2009**, *69*, 879–886. [CrossRef]

86. Langlois, B.; Perrot, G.; Schneider, C.; Henriet, P.; Emonard, H.; Martiny, L.; Dedieu, S. LRP-1 promotes cancer cell invasion by supporting ERK and inhibiting JNK signaling pathways. *PLoS ONE* **2010**, *5*, e11584. [CrossRef] [PubMed]

87. Gopal, U.; Bohonowych, J.E.; Lema-Tome, C.; Liu, A.; Garrett-Mayer, E.; Wang, B.; Isaacs, J.S. A novel extracellular Hsp90 mediated co-receptor function for LRP1 regulates EphA2 dependent glioblastoma cell invasion. *PLoS ONE* **2011**, *6*, e17649. [CrossRef] [PubMed]

88. Chen, J.; Su, Y.; Pi, S.; Hu, B.; Mao, L. The Dual Role of Low-Density Lipoprotein Receptor-Related Protein 1 in Atherosclerosis. *Front. Cardiovasc. Med.* **2021**, *8*, 682389. [CrossRef]

89. Moestrup, S.K.; Gliemann, J.; Pallesen, G. Distribution of the alpha 2-macroglobulin receptor/low density lipoprotein receptor-related protein in human tissues. *Cell Tissue Res.* **1992**, *269*, 375–382. [CrossRef]

90. Zlokovic, B.V.; Deane, R.; Sagare, A.P.; Bell, R.D.; Winkler, E.A. Low-density lipoprotein receptor-related protein-1: A serial clearance homeostatic mechanism controlling Alzheimer's amyloid β-peptide elimination from the brain. *J. Neurochem.* **2010**, *115*, 1077–1089. [CrossRef]

91. Pflanzner, T.; Janko, M.C.; André-Dohmen, B.; Reuss, S.; Weggen, S.; Roebroek, A.J.M.; Kuhlmann, C.R.W.; Pietrzik, C.U. LRP1 mediates bidirectional transcytosis of amyloid-β across the blood-brain barrier. *Neurobiol. Aging* **2011**, *32*, 2323.e1–2323.e11. [CrossRef]

92. Zhao, Z.; Sagare, A.P.; Ma, Q.; Halliday, M.R.; Kong, P.; Kisler, K.; Winkler, E.A.; Ramanathan, A.; Kanekiyo, T.; Bu, G.; et al. Central role for PICALM in amyloid-β blood-brain barrier transcytosis and clearance. *Nat. Neurosci.* **2015**, *18*, 978–987. [CrossRef]

93. Benchenane, K.; Berezowski, V.; Ali, C.; Fernández-Monreal, M.; López-Atalaya, J.P.; Brillault, J.; Chuquet, J.; Nouvelot, A.; MacKenzie, E.T.; Bu, G.; et al. Tissue-type plasminogen activator crosses the intact blood-brain barrier by low-density lipoprotein receptor-related protein-mediated transcytosis. *Circulation* **2005**, *111*, 2241–2249. [CrossRef] [PubMed]

94. Demeule, M.; Currie, J.-C.; Bertrand, Y.; Ché, C.; Nguyen, T.; Régina, A.; Gabathuler, R.; Castaigne, J.-P.; Béliveau, R. Involvement of the low-density lipoprotein receptor-related protein in the transcytosis of the brain delivery vector angiopep-2. *J. Neurochem.* **2008**, *106*, 1534–1544. [CrossRef]

95. Storck, S.E.; Meister, S.; Nahrath, J.; Meißner, J.N.; Schubert, N.; Di Spiezio, A.; Baches, S.; Vandenbroucke, R.E.; Bouter, Y.; Prikulis, I.; et al. Endothelial LRP1 transports amyloid-β(1-42) across the blood-brain barrier. *J. Clin. Investig.* **2016**, *126*, 123–136. [CrossRef] [PubMed]

96. Sakamoto, K.; Shinohara, T.; Adachi, Y.; Asami, T.; Ohtaki, T. A novel LRP1-binding peptide L57 that crosses the blood brain barrier. *Biochem. Biophys. Rep.* **2017**, *12*, 135–139. [CrossRef]

97. Kang, D.E.; Pietrzik, C.U.; Baum, L.; Chevallier, N.; Merriam, D.E.; Kounnas, M.Z.; Wagner, S.L.; Troncoso, J.C.; Kawas, C.H.; Katzman, R.; et al. Modulation of amyloid beta-protein clearance and Alzheimer's disease susceptibility by the LDL receptor-related protein pathway. *J. Clin. Investig.* **2000**, *106*, 1159–1166. [CrossRef] [PubMed]

98. Shibata, M.; Yamada, S.; Kumar, S.R.; Calero, M.; Bading, J.; Frangione, B.; Holtzman, D.M.; Miller, C.A.; Strickland, D.K.; Ghiso, J.; et al. Clearance of Alzheimer's amyloid-ss(1-40) peptide from brain by LDL receptor-related protein-1 at the blood-brain barrier. *J. Clin. Investig.* **2000**, *106*, 1489–1499. [CrossRef]

99. Silverberg, G.D.; Messier, A.A.; Miller, M.C.; Machan, J.T.; Majmudar, S.S.; Stopa, E.G.; Donahue, J.E.; Johanson, C.E. Amyloid efflux transporter expression at the blood-brain barrier declines in normal aging. *J. Neuropathol. Exp. Neurol.* **2010**, *69*, 1034–1043. [CrossRef]

100. Brown, M.S.; Goldstein, J.L. The SREBP pathway: Regulation of cholesterol metabolism by proteolysis of a membrane-bound transcription factor. *Cell* **1997**, *89*, 331–340. [CrossRef]

101. Llorente-Cortés, V.; Costales, P.; Bernués, J.; Camino-Lopez, S.; Badimon, L. Sterol regulatory element-binding protein-2 negatively regulates low density lipoprotein receptor-related protein transcription. *J. Mol. Biol.* **2006**, *359*, 950–960. [CrossRef] [PubMed]

102. von Arnim, C.A.F.; Kinoshita, A.; Peltan, I.D.; Tangredi, M.M.; Herl, L.; Lee, B.M.; Spoelgen, R.; Hshieh, T.T.; Ranganathan, S.; Battey, F.D.; et al. The low density lipoprotein receptor-related protein (LRP) is a novel beta-secretase (BACE1) substrate. *J. Biol. Chem.* **2005**, *280*, 17777–17785. [CrossRef]

103. Liu, Q.; Zhang, J.; Tran, H.; Verbeek, M.M.; Reiss, K.; Estus, S.; Bu, G. LRP1 shedding in human brain: Roles of ADAM10 and ADAM17. *Mol. Neurodegener.* **2009**, *4*, 17. [CrossRef] [PubMed]

104. Seidah, N.G.; Benjannet, S.; Wickham, L.; Marcinkiewicz, J.; Jasmin, S.B.; Stifani, S.; Basak, A.; Prat, A.; Chretien, M. The secretory proprotein convertase neural apoptosis-regulated convertase 1 (NARC-1): Liver regeneration and neuronal differentiation. *Proc. Natl. Acad. Sci. USA* **2003**, *100*, 928–933. [CrossRef]

105. Lagace, T.A.; Curtis, D.E.; Garuti, R.; McNutt, M.C.; Park, S.W.; Prather, H.B.; Anderson, N.N.; Ho, Y.K.; Hammer, R.E.; Horton, J.D. Secreted PCSK9 decreases the number of LDL receptors in hepatocytes and in livers of parabiotic mice. *J. Clin. Investig.* **2006**, *116*, 2995–3005. [CrossRef] [PubMed]

106. Poirier, S.; Mayer, G.; Benjannet, S.; Bergeron, E.; Marcinkiewicz, J.; Nassoury, N.; Mayer, H.; Nimpf, J.; Prat, A.; Seidah, N.G. The proprotein convertase PCSK9 induces the degradation of low density lipoprotein receptor (LDLR) and its closest family members VLDLR and ApoER2. *J. Biol. Chem.* **2008**, *283*, 2363–2372. [CrossRef] [PubMed]

107. Canuel, M.; Sun, X.; Asselin, M.-C.; Paramithiotis, E.; Prat, A.; Seidah, N.G. Proprotein convertase subtilisin/kexin type 9 (PCSK9) can mediate degradation of the low density lipoprotein receptor-related protein 1 (LRP-1). *PLoS ONE* **2013**, *8*, e64145. [CrossRef]

108. Ferri, N.; Tibolla, G.; Pirillo, A.; Cipollone, F.; Mezzetti, A.; Pacia, S.; Corsini, A.; Catapano, A.L. Proprotein convertase subtilisin kexin type 9 (PCSK9) secreted by cultured smooth muscle cells reduces macrophages LDLR levels. *Atherosclerosis* **2012**, *220*, 381–386. [CrossRef]

109. Cunningham, D.; Danley, D.E.; Geoghegan, K.F.; Griffor, M.C.; Hawkins, J.L.; Subashi, T.A.; Varghese, A.H.; Ammirati, M.J.; Culp, J.S.; Hoth, L.R.; et al. Structural and biophysical studies of PCSK9 and its mutants linked to familial hypercholesterolemia. *Nat. Struct. Mol. Biol.* **2007**, *14*, 413–419. [CrossRef]

110. Hampton, E.N.; Knuth, M.W.; Li, J.; Harris, J.L.; Lesley, S.A.; Spraggon, G. The self-inhibited structure of full-length PCSK9 at 1.9 A reveals structural homology with resistin within the C-terminal domain. *Proc. Natl. Acad. Sci. USA* **2007**, *104*, 14604–14609. [CrossRef]

111. Piper, D.E.; Jackson, S.; Liu, Q.; Romanow, W.G.; Shetterly, S.; Thibault, S.T.; Shan, B.; Walker, N.P.C. The crystal structure of PCSK9: A regulator of plasma LDL-cholesterol. *Structure* **2007**, *15*, 545–552. [CrossRef]

112. Seidah, N.G.; Prat, A. The Multifaceted Biology of PCSK9. *Endocr. Rev.* **2022**, *43*, 558–582. [CrossRef]

113. Naureckiene, S.; Ma, L.; Sreekumar, K.; Purandare, U.; Lo, C.F.; Huang, Y.; Chiang, L.W.; Grenier, J.M.; Ozenberger, B.A.; Jacobsen, J.S.; et al. Functional characterization of Narc 1, a novel proteinase related to proteinase K. *Arch. Biochem. Biophys.* **2003**, *420*, 55–67. [CrossRef]

114. Benjannet, S.; Rhainds, D.; Essalmani, R.; Mayne, J.; Wickham, L.; Jin, W.; Asselin, M.-C.; Hamelin, J.; Varret, M.; Allard, D.; et al. NARC-1/PCSK9 and its natural mutants: Zymogen cleavage and effects on the low density lipoprotein (LDL) receptor and LDL cholesterol. *J. Biol. Chem.* **2004**, *279*, 48865–48875. [CrossRef]

115. Davis, C.G.; Goldstein, J.L.; Südhof, T.C.; Anderson, R.G.; Russell, D.W.; Brown, M.S. Acid-dependent ligand dissociation and recycling of LDL receptor mediated by growth factor homology region. *Nature* **1987**, *326*, 760–765. [CrossRef]

116. Rudenko, G.; Henry, L.; Henderson, K.; Ichtchenko, K.; Brown, M.S.; Goldstein, J.L.; Deisenhofer, J. Structure of the LDL receptor extracellular domain at endosomal pH. *Science* **2002**, *298*, 2353–2358. [CrossRef]

117. Zhao, Z.; Michaely, P. The epidermal growth factor homology domain of the LDL receptor drives lipoprotein release through an allosteric mechanism involving H190, H562 and H586. *J. Biol. Chem.* **2008**, *283*, 26528–26537. [CrossRef]

118. Zhang, D.-W.; Lagace, T.A.; Garuti, R.; Zhao, Z.; McDonald, M.; Horton, J.D.; Cohen, J.C.; Hobbs, H.H. Binding of proprotein convertase subtilisin/kexin type 9 to epidermal growth factor-like repeat A of low density lipoprotein receptor decreases receptor recycling and increases degradation. *J. Biol. Chem.* **2007**, *282*, 18602–18612. [CrossRef] [PubMed]

119. Kwon, H.J.; Lagace, T.A.; McNutt, M.C.; Horton, J.D.; Deisenhofer, J. Molecular basis for LDL receptor recognition by PCSK9. *Proc. Natl. Acad. Sci. USA* **2008**, *105*, 1820–1825. [CrossRef] [PubMed]

120. Zhang, D.-W.; Garuti, R.; Tang, W.-J.; Cohen, J.C.; Hobbs, H.H. Structural requirements for PCSK9-mediated degradation of the low-density lipoprotein receptor. *Proc. Natl. Acad. Sci. USA* **2008**, *105*, 13045–13050. [CrossRef] [PubMed]

121. Saavedra, Y.G.L.; Day, R.; Seidah, N.G. The M2 module of the Cys-His-rich domain (CHRD) of PCSK9 protein is needed for the extracellular low-density lipoprotein receptor (LDLR) degradation pathway. *J. Biol. Chem.* **2012**, *287*, 43492–43501. [CrossRef] [PubMed]

122. Dubuc, G.; Chamberland, A.; Wassef, H.; Davignon, J.; Seidah, N.G.; Bernier, L.; Prat, A. Statins upregulate PCSK9, the gene encoding the proprotein convertase neural apoptosis-regulated convertase-1 implicated in familial hypercholesterolemia. *Arterioscler. Thromb. Vasc. Biol.* **2004**, *24*, 1454–1459. [CrossRef]

123. Jeong, H.J.; Lee, H.-S.; Kim, K.-S.; Kim, Y.-K.; Yoon, D.; Park, S.W. Sterol-dependent regulation of proprotein convertase subtilisin/kexin type 9 expression by sterol-regulatory element binding protein-2. *J. Lipid Res.* **2008**, *49*, 399–409. [CrossRef] [PubMed]

124. Strøm, T.B.; Tveten, K.; Leren, T.P. PCSK9 acts as a chaperone for the LDL receptor in the endoplasmic reticulum. *Biochem. J.* **2014**, *457*, 99–105. [CrossRef] [PubMed]

125. Nassoury, N.; Blasiole, D.A.; Tebon Oler, A.; Benjannet, S.; Hamelin, J.; Poupon, V.; McPherson, P.S.; Attie, A.D.; Prat, A.; Seidah, N.G. The cellular trafficking of the secretory proprotein convertase PCSK9 and its dependence on the LDLR. *Traffic* **2007**, *8*, 718–732. [CrossRef]

126. Poirier, S.; Mayer, G.; Poupon, V.; McPherson, P.S.; Desjardins, R.; Ly, K.; Asselin, M.-C.; Day, R.; Duclos, F.J.; Witmer, M.; et al. Dissection of the endogenous cellular pathways of PCSK9-induced low density lipoprotein receptor degradation: Evidence for an intracellular route. *J. Biol. Chem.* **2009**, *284*, 28856–28864. [CrossRef] [PubMed]

127. Susan-Resiga, D.; Girard, E.; Kiss, R.S.; Essalmani, R.; Hamelin, J.; Asselin, M.-C.; Awan, Z.; Butkinaree, C.; Fleury, A.; Soldera, A.; et al. The Proprotein Convertase Subtilisin/Kexin Type 9-resistant R410S Low Density Lipoprotein Receptor Mutation: A Novel Mechanism Causing Familial Hypercholesterolemia. *J. Biol. Chem.* **2017**, *292*, 1573–1590. [CrossRef]

128. Zaid, A.; Roubtsova, A.; Essalmani, R.; Marcinkiewicz, J.; Chamberland, A.; Hamelin, J.; Tremblay, M.; Jacques, H.; Jin, W.; Davignon, J.; et al. Proprotein convertase subtilisin/kexin type 9 (PCSK9): Hepatocyte-specific low-density lipoprotein receptor degradation and critical role in mouse liver regeneration. *Hepatology* **2008**, *48*, 646–654. [CrossRef]

129. Rousselet, E.; Marcinkiewicz, J.; Kriz, J.; Zhou, A.; Hatten, M.E.; Prat, A.; Seidah, N.G. PCSK9 reduces the protein levels of the LDL receptor in mouse brain during development and after ischemic stroke. *J. Lipid Res.* **2011**, *52*, 1383–1391. [CrossRef]

130. Zimetti, F.; Caffarra, P.; Ronda, N.; Favari, E.; Adorni, M.P.; Zanotti, I.; Bernini, F.; Barocco, F.; Spallazzi, M.; Galimberti, D.; et al. Increased PCSK9 Cerebrospinal Fluid Concentrations in Alzheimer's Disease. *J. Alzheimers Dis.* **2017**, *55*, 315–320. [CrossRef]

131. Courtemanche, H.; Bigot, E.; Pichelin, M.; Guyomarch, B.; Boutoleau-Bretonnière, C.; Le May, C.; Derkinderen, P.; Cariou, B. PCSK9 Concentrations in Cerebrospinal Fluid Are Not Specifically Increased in Alzheimer's Disease. *J. Alzheimers Dis.* **2018**, *62*, 1519–1525. [CrossRef] [PubMed]

132. Mazura, A.D.; Ohler, A.; Storck, S.E.; Kurtyka, M.; Scharfenberg, F.; Weggen, S.; Becker-Pauly, C.; Pietrzik, C.U. PCSK9 acts as a key regulator of Aβ clearance across the blood-brain barrier. *Cell. Mol. Life Sci.* **2022**, *79*, 212. [CrossRef]

133. Murphy, M.P.; LeVine, H. Alzheimer's disease and the amyloid-beta peptide. *J. Alzheimers Dis.* **2010**, *19*, 311–323. [CrossRef] [PubMed]

134. Kumar, A.; Singh, A.; Ekavali. A review on Alzheimer's disease pathophysiology and its management: An update. *Pharmacol. Rep.* **2015**, *67*, 195–203. [CrossRef]

135. Bateman, R.J.; Munsell, L.Y.; Morris, J.C.; Swarm, R.; Yarasheski, K.E.; Holtzman, D.M. Human amyloid-beta synthesis and clearance rates as measured in cerebrospinal fluid in vivo. *Nat. Med.* **2006**, *12*, 856–861. [CrossRef] [PubMed]

136. Mawuenyega, K.G.; Sigurdson, W.; Ovod, V.; Munsell, L.; Kasten, T.; Morris, J.C.; Yarasheski, K.E.; Bateman, R.J. Decreased clearance of CNS beta-amyloid in Alzheimer's disease. *Science* **2010**, *330*, 1774. [CrossRef]

137. Tarasoff-Conway, J.M.; Carare, R.O.; Osorio, R.S.; Glodzik, L.; Butler, T.; Fieremans, E.; Axel, L.; Rusinek, H.; Nicholson, C.; Zlokovic, B.V.; et al. Clearance systems in the brain-implications for Alzheimer disease. *Nat. Rev. Neurol.* **2015**, *11*, 457–470. [CrossRef]

138. Spencer, B.J.; Verma, I.M. Targeted delivery of proteins across the blood-brain barrier. *Proc. Natl. Acad. Sci. USA* **2007**, *104*, 7594–7599. [CrossRef]

139. Spencer, B.; Marr, R.A.; Gindi, R.; Potkar, R.; Michael, S.; Adame, A.; Rockenstein, E.; Verma, I.M.; Masliah, E. Peripheral delivery of a CNS targeted, metalo-protease reduces aβ toxicity in a mouse model of Alzheimer's disease. *PLoS ONE* **2011**, *6*, e16575. [CrossRef]

140. Sorrentino, N.C.; D'Orsi, L.; Sambri, I.; Nusco, E.; Monaco, C.; Spampanato, C.; Polishchuk, E.; Saccone, P.; de Leonibus, E.; Ballabio, A.; et al. A highly secreted sulphamidase engineered to cross the blood-brain barrier corrects brain lesions of mice with mucopolysaccharidoses type IIIA. *EMBO Mol. Med.* **2013**, *5*, 675–690. [CrossRef]

141. Wagner, S.; Zensi, A.; Wien, S.L.; Tschickardt, S.E.; Maier, W.; Vogel, T.; Worek, F.; Pietrzik, C.U.; Kreuter, J.; Briesen, H. von. Uptake mechanism of ApoE-modified nanoparticles on brain capillary endothelial cells as a blood-brain barrier model. *PLoS ONE* **2012**, *7*, e32568. [CrossRef] [PubMed]

142. Dal Magro, R.; Ornaghi, F.; Cambianica, I.; Beretta, S.; Re, F.; Musicanti, C.; Rigolio, R.; Donzelli, E.; Canta, A.; Ballarini, E.; et al. ApoE-modified solid lipid nanoparticles: A feasible strategy to cross the blood-brain barrier. *J. Control. Release* **2017**, *249*, 103–110. [CrossRef]

143. Kuo, Y.-C.; Chen, I.-Y.; Rajesh, R. Use of functionalized liposomes loaded with antioxidants to permeate the blood–brain barrier and inhibit β-amyloid-induced neurodegeneration in the brain. *J. Taiwan Inst. Chem. Eng.* **2018**, *87*, 1–14. [CrossRef]

144. Lagace, T.A. PCSK9 and LDLR degradation: Regulatory mechanisms in circulation and in cells. *Curr. Opin. Lipidol.* **2014**, *25*, 387–393. [CrossRef]

145. Chaudhary, R.; Garg, J.; Shah, N.; Sumner, A. PCSK9 inhibitors: A new era of lipid lowering therapy. *World J. Cardiol.* **2017**, *9*, 76–91. [CrossRef]

146. Jia, X.; Al Rifai, M.; Saeed, A.; Ballantyne, C.M.; Virani, S.S. PCSK9 Inhibitors in the Management of Cardiovascular Risk: A Practical Guidance. *Vasc. Health Risk Manag.* **2022**, *18*, 555–566. [CrossRef] [PubMed]

147. Mahboobnia, K.; Pirro, M.; Marini, E.; Grignani, F.; Bezsonov, E.E.; Jamialahmadi, T.; Sahebkar, A. PCSK9 and cancer: Rethinking the link. *Biomed. Pharmacother.* **2021**, *140*, 111758. [CrossRef] [PubMed]

148. Punch, E.; Klein, J.; Diaba-Nuhoho, P.; Morawietz, H.; Garelnabi, M. Effects of PCSK9 Targeting: Alleviating Oxidation, Inflammation, and Atherosclerosis. *J. Am. Heart Assoc.* **2022**, *11*, e023328. [CrossRef]

149. Wu, N.-Q.; Shi, H.-W.; Li, J.-J. Proprotein Convertase Subtilisin/Kexin Type 9 and Inflammation: An Updated Review. *Front. Cardiovasc. Med.* **2022**, *9*, 763516. [CrossRef]

150. Quagliariello, V.; Bisceglia, I.; Berretta, M.; Iovine, M.; Canale, M.L.; Maurea, C.; Giordano, V.; Paccone, A.; Inno, A.; Maurea, N. PCSK9 Inhibitors in Cancer Patients Treated with Immune-Checkpoint Inhibitors to Reduce Cardiovascular Events: New Frontiers in Cardioncology. *Cancers* **2023**, *15*, 1397. [CrossRef]

Review

Distribution of Monocarboxylate Transporters in Brain and Choroid Plexus Epithelium

Masaki Ueno [1,*], Yoichi Chiba [1], Ryuta Murakami [1], Yumi Miyai [1], Koichi Matsumoto [1], Keiji Wakamatsu [1], Genta Takebayashi [2], Naoya Uemura [2] and Ken Yanase [2]

1 Department of Pathology and Host Defense, Faculty of Medicine, Kagawa University, Takamatsu 761-0793, Kagawa, Japan; chiba.yoichi@kagawa-u.ac.jp (Y.C.); murakami.ryuta@kagawa-u.ac.jp (R.M.); miyai.yumi@kagawa-u.ac.jp (Y.M.); matsumoto.koichi@kagawa-u.ac.jp (K.M.); s20d727@kagawa-u.ac.jp (K.W.)
2 Department of Anesthesiology, Faculty of Medicine, Kagawa University, Takamatsu 761-0793, Kagawa, Japan; takebayashi.genta@kagawa-u.ac.jp (G.T.); uemura.naoya@kagawa-u.ac.jp (N.U.); yanase.ken@kagawa-u.ac.jp (K.Y.)
* Correspondence: ueno.masaki@kagawa-u.ac.jp; Tel.: +81-878-912-115

Abstract: The choroid plexus (CP) plays central roles in regulating the microenvironment of the central nervous system by secreting the majority of cerebrospinal fluid (CSF) and controlling its composition. A monolayer of epithelial cells of CP plays a significant role in forming the blood–CSF barrier to restrict the movement of substances between the blood and ventricles. CP epithelial cells are equipped with transporters for glucose and lactate that are used as energy sources. There are many review papers on glucose transporters in CP epithelial cells. On the other hand, distribution of monocarboxylate transporters (MCTs) in CP epithelial cells has received less attention compared with glucose transporters. Some MCTs are known to transport lactate, pyruvate, and ketone bodies, whereas others transport thyroid hormones. Since CP epithelial cells have significant carrier functions as well as the barrier function, a decline in the expression and function of these transporters leads to a poor supply of thyroid hormones as well as lactate and can contribute to the process of age-associated brain impairment and pathophysiology of neurodegenerative diseases. In this review paper, recent findings regarding the distribution and significance of MCTs in the brain, especially in CP epithelial cells, are summarized.

Keywords: choroid plexus; epithelial cell; lactate; monocarboxylate transporter; thyroid hormone

Citation: Ueno, M.; Chiba, Y.; Murakami, R.; Miyai, Y.; Matsumoto, K.; Wakamatsu, K.; Takebayashi, G.; Uemura, N.; Yanase, K. Distribution of Monocarboxylate Transporters in Brain and Choroid Plexus Epithelium. *Pharmaceutics* **2023**, *15*, 2062. https://doi.org/10.3390/pharmaceutics15082062

Academic Editors: Xavier Declèves and Inge S. Zuhorn

Received: 31 May 2023
Revised: 26 July 2023
Accepted: 28 July 2023
Published: 31 July 2023

1. Introduction

1.1. Barrier Function in Choroid Plexus

The brain restricts the entrance of ions and solutes circulating in the blood by two cellular barriers: the blood–brain barrier (BBB) and blood–cerebrospinal fluid (CSF) barrier (BCSFB) [1–4]. The BBB is composed of endothelial cells interconnected by tight junctions, two basement membranes, pericytes, and end-feet of astrocytes [1,5–9]. BBB endothelial cells have few vesicles and no fenestrations in the cytoplasm, whereas the endothelial cells are equipped with various transporters to supply energy from the blood to brain cells, such as (a) carbohydrate transporters, (b) monocarboxylate transporters (MCTs), (c) amino acid transporters, (d) fatty acid transporters, (e) nucleotide transporters, (f) hormone transporters, (g) organic anion and cation transporters, and (h) other transporters, including those for amines and choline [2–4,9]. However, barrier and carrier functions of BBB are affected by various invasions in disease states [7]. Accordingly, BBB dysfunction leads to various brain dysfunctions, such as cognitive dysfunction. On the other hand, endothelial cells of capillaries in the choroid plexus (CP) are originally fenestrated, allowing the passage of intravascular low-molecular-weight substances as well as ions in CP with a vascularized stroma [8,10–12]. Every CP has a vascularized stroma and is covered by a monolayer of epithelial cells interconnected by tight and adherens junctions. The tight junction contains occludins and claudins, which bind to cytosolic zonula occludens protein-1 (ZO-1) [10,11]. On the other hand, the adherens junction is distributed beneath the tight junction and contains cadherins, which bind to catenins distributed along the lateral surface of the cytoplasm of CP epithelial cells (CPEs) [10,11]. Accordingly, neighboring CPEs serve as a barrier between the blood and CSF, referred to as BCSFB, and restrict the entry of intravascular substances into ventricles [8,9].

1.2. Characterization of Choroid Plexus Epithelial Cells

CSF is produced through several kinds of ion and water transporters located in CPEs [1,10,11,13]. CPEs are characterized by the presence of various epithelial cytokeratins, β-catenin, vimentin, S-100 protein, podoplanin, and transthyretin/prealbumin [10,11,13–16]. Intermediate filaments in CPEs have been identified as keratins 8, 18, and possibly 19 [10,17,18]. Miettinen et al. [18] revealed immunoreactivity for cytokeratin (CK) 19 in CP tissues of human brains by immunoblotting, whereas Kasper et al. [17] found no immunoreactivity for CK 19 in CPEs of human brains by immunohistochemistry. We could not confirm immunoreactivity for CK19 (Progen, Biotechnik GmbH, Heidelberg, Germany, 61010) in human CPEs, although immunoreactivities for CK8 and CK18 were confirmed to be present in CPEs (Figure 1a–c). Justice et al. reported that strong immunoreactivity for α-1-antichymotrypsin was present in apical granular organelles in the cytoplasm of adult CPEs [19]. We confirmed, as shown here by immunostaining, that β-catenin, vimentin, S-100, podoplanin, transthyretin, and α-1-antichymotrypsin are expressed in the cytoplasmic membrane or cytoplasm of human CPEs (Figure 1d–i). In addition, representative transporters related to CSF production, such as Na$^+$, -K$^+$, -ATPase, aquaporin 1 (AQP1), and anion exchange protein 2 (AE2), are present in apical or basolateral cytoplasmic membranes of CPEs (Figure 1j–l) [10–13].

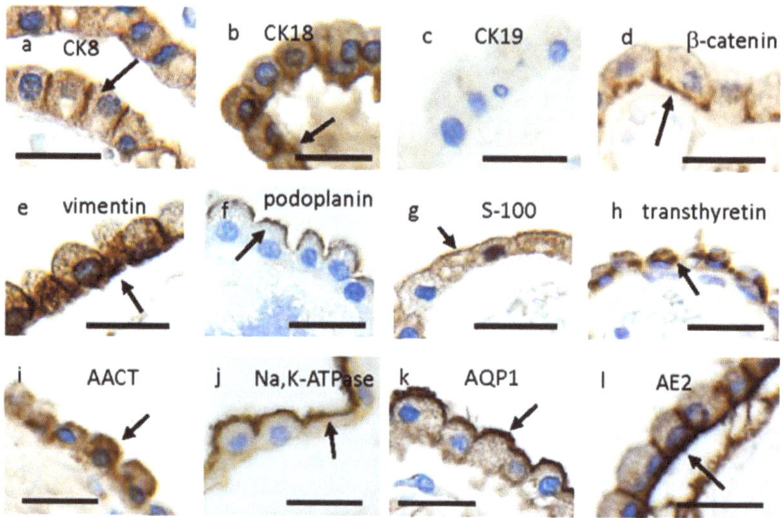

Figure 1. Distribution of immunoreactivities for CK8 (**a**), CK18 (**b**), CK19 (**c**), β-catenin (**d**), vimentin (**e**), podoplanin (**f**), S-100 (**g**), transthyretin/prealbumin (**h**), α1-antichymotrypsin (AACT) (**i**), Na$^+$, -K$^+$, -ATPase (**j**), AQP-1 (**k**), and AE2 (**l**), in epithelial cells of CP located in lateral ventricles of autopsied human brains. Arrows indicate immunoreactivity for these substances. Immunohistochemical findings from cases 2 (**i**), 3 (**d,g**), 6 (**a,k**), 7 (**f,h**), and 8 (**b,c,e,j,l**) are shown. Scale bars indicate 20 μm.

BBB and BCSFB not only have barrier functions but also carrier functions for intracerebral transport of essential nutrients such as glucose, lactate, and amino acids [1–4,9]. In addition, they play important roles in the removal of metabolic wastes and neurotoxic substances such as amyloid β (Aβ) [1,2,8,20,21]. It is well-known that one of the significant functions of CP is to produce and secrete CSF [11,20]. Accordingly, CPEs are equipped with transporters for ions and organic solutes that are different from those in BBB endothelial cells [1,2,8,21,22]. We previously reviewed the distribution of transporters for glucose, fructose, and urate [23]. Accordingly, in this review paper, we focused on recent findings on the distribution of MCTs in the brain, especially in CPEs. In the brain, MCTs 1, 2, 3, and 4 transport lactate, pyruvate, and ketone bodies, whereas MCT8 transports thyroid hormones. In this manuscript, previously reported results on the distribution of these MCTs in CPEs were summarized and confirmed by immunohistochemical staining. Then, their physiological function and the pathophysiological importance of their expression at BCSFB in neurological diseases were discussed. Table 1 shows a summary of clinicopathological profiles of autopsied human brains that were removed at Kagawa University Hospital, as introduced in our previously published papers [24,25]. Table 2 shows a summary of antibodies used in this manuscript. Before incubation with some antibodies, antigen retrieval was performed by heating sections in 10 mM sodium citrate buffer (pH 6) or 1 mM tris(hydroxymethyl)aminomethane (Tris)-ethylenediaminetetraacetic acid (EDTA) buffer (pH 9) for 20 min. These studies using autopsied human brains were approved by the institutional Ethics Committee of the Faculty of Medicine, Kagawa University, in accordance with the Declaration of Helsinki [24,25]. Then, these brain samples were used to confirm the distribution of substances, including MCTs, in this review manuscript.

Table 1. Summary of clinicopathological profiles.

(No.)	Age/Sex	Main Diagnosis
1	60/M	Dissecting aneurysm
2	64/M	Multiple system atrophy, Pneumonia
3	70/M	Myocardial infarction
4	71/M	Cerebellar tuberculosis
5	72/F	Pneumonia
6	74/M	Lung cancer
7	75/M	Gastric cancer
8	84/M	Myocardial infarction, Cerebral infarction

Table 2. Summary of antibodies used.

Antibody	Cat. No. (Clone Name)	Host Species and Usage
CK8	Progen, 61038	mouse, 1:50 (¶2)
CK18	ProteinTech, 66187-1-lg	mouse, 1:600 (¶2)
CK19	Progen, 61010	mouse, 1:10 (¶1)
β-catenin	SantaCruz, sc-7199	rabbit, 1:50 (-)
vimentin	DAKO, M0725 (V9)	mouse, 1:50 (¶1)
D2/40(podoplanin)	DAKO, M3619	mouse, 1:50 (¶1)
s-100	Nichirei, 422091	rabbit, diluted (-)
transthyretin	ProteinTech, 11891-1-AP	rabbit, 1:100 (-)
AACT	ProteinTech, 66078-1-lg	mouse, 1:500 (¶2)

Table 2. *Cont.*

Antibody	Cat. No. (Clone Name)	Host Species and Usage
Na$^+$,-K$^+$,-ATPase	SantaCruz, sc-48345	rabbit, 1:100 (¶1)
Aquaporin-1	ProteinTech, 20333-1-AP	rabbit, 1:250 (¶2)
AE2	sc-376632 (D-3)	mouse, 1:100 (¶1)
MCT1	Abcam, ab90582	mouse, 1:100 (¶1)
MCT2	Abcam, ab198272	rabbit, 1:100 (-)
MCT3	Abcam, ab60333	rabbit, 1:200 (-)
MCT4	Abcam, ab244385	rabbit, 1:50 (¶1)
MCT5	Abcam, ab191008	rabbit, 1:500 (¶1)
MCT8	Novus, NBP2-57308	rabbit, 1:200 (¶1)
HCA1-R	Novus, NLS-2095	rabbit, 1:200 (¶1)

(¶1, ¶2): Antigen retrieval with citrate buffer (pH 6) or Tris-EDTA buffer (pH 9) is needed prior to the application of the primary antibody. AACT: α1-antichymotrypsin.

2. Monocarboxylate Transporters in the Brain

MCTs catalyze the proton-linked influx or efflux of monocarboxylates such as L-lactate, pyruvate, and ketone bodies in various cells of several organs [26,27]. Consequently, MCTs enable the 1:1 exchange of monocarboxylate and protons across the cellular membrane [26,27]. The direction of transport depends on their intracellular and extracellular concentrations [27]. It is well-known that there are four isoforms, MCTs 1, 2, 3, and 4, in the brain. They belong to the SLC16 family of solute carriers, which has 14 members in total. The family includes MCTs to transport thyroid hormones. MCT8, which is a high-affinity transporter for 3,5,3′-triiodothyronine (T3), MCT10, which is the most homologous to MCT8, and eight orphan members also constitute the family [28]. MCTs 1–4 catalyze proton-coupled lactate transport, whereas MCT8 and MCT10 catalyze the sodium- and proton-independent transport of thyroid hormones [28,29]. In the brain, it is known that MCT1, MCT2, and MCT4 are widely expressed in several kinds of cells [30,31]. MCT1 is mainly expressed on endothelial cells with the barrier function both in humans and rodents [32,33]. MCT2 and MCT4 are mainly expressed in neurons and astrocytes, respectively. On the other hand, MCT3 is expressed in the retinal pigment epithelium and CPEs of mice [34,35]. Protein expression levels in plasma membrane fractions of isolated CP of humans and rats were measured using quantitative targeted absolute proteomics [36]. The study identified low-level expression of MCT1 and MCT3 in rats, low-level expression of MCT1 in humans, and very-low-level expression of MCT4 and MCT5 in humans [36]. It is likely that the expression level of transporters for lactate, pyruvate, and ketone bodies in CPEs affects their CSF concentrations. Their values in CSF have been reported in various diseased brains. Glucose transporters belong to either SLC2A/GLUT or SLC5A/SGLT families and are summarized in some papers [23,37,38]. In this review manuscript, the distributions of MCT1, MCT2, MCT3, MCT4, MCT5, and MCT8, which were previously reported to be present in CPEs, were first reviewed. In addition, detailed localization of MCTs in the brain, especially in cerebral microvessels and CPEs, was confirmed by immunohistochemical staining, as the microvessels and CPEs are important routes to transport intravascular substances from the blood into the brain.

2.1. MCT1 (SLC16A1) Distribution in the Brain and CPEs

MCT1 is well-known to be expressed ubiquitously in the brain. MCT1 expression was reported in endothelial cells, astrocytes [39–41], and oligodendrocytes [42]. Lactate is released from astrocytes via MCT4 and may be carried into oligodendrocytes via MCT1 [42]. It was reported that MCT1 was immunohistochemically expressed in microglia in healthy human brains [40] and on the apical side of the cytoplasm of CPEs in autopsied diseased human 2 brains [43]. Immunoreactivity for MCT1 (Abcam, ab90582) is confirmed to be expressed in endothelial cells, reactive astrocytes, and on the apical side of the cytoplasm of CPEs (Figure 2a–c).

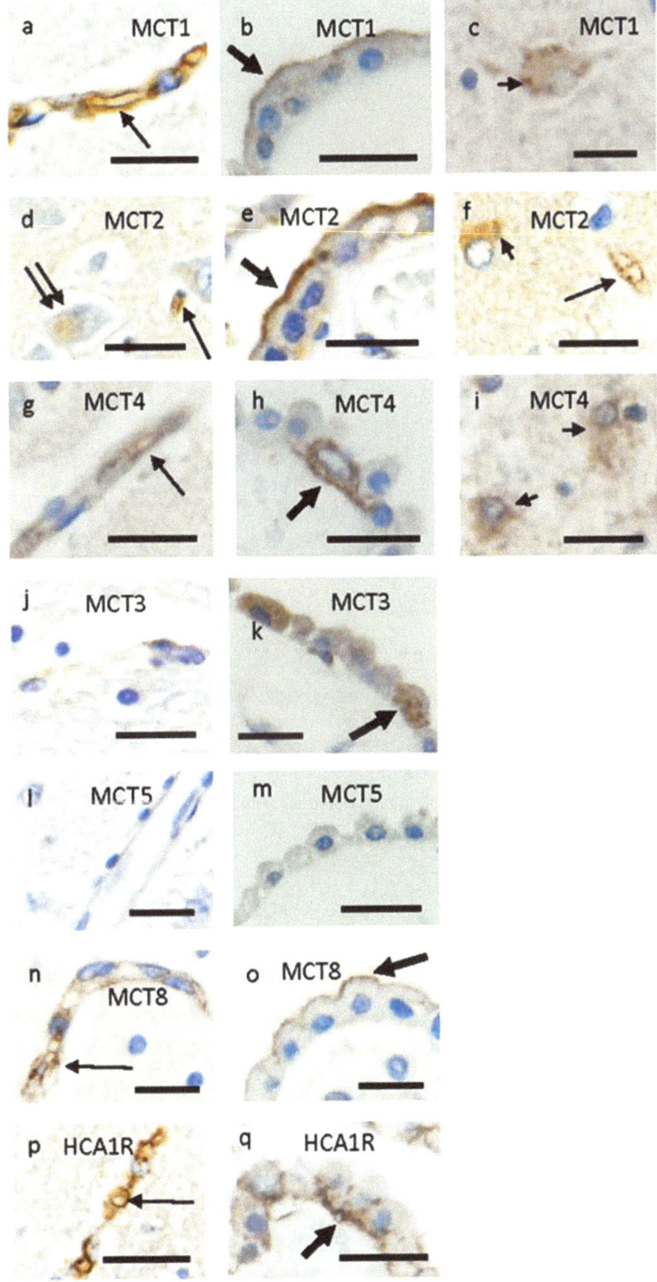

Figure 2. Distribution of immunoreactivities for MCT1 (**a–c**), MCT2 (**d–f**), MCT4 (**g–i**), MCT3 (**j,k**), MCT5 (**l,m**), MCT8 (**n,o**), and HCA1-R (**p,q**), in autopsied human brains. Immunohistochemical findings

from cases 1 (**g**), 3 (**a,h,j,k,m,p**), 4, (**c,f,i**), 5 (**b,d,l**), 6 (**e,q**), and 8 (**n,o**) are shown. Immunohistochemical images in microvessels indicated by thin arrows in hippocampal samples are shown in (**a,d,f,g,n,p**), whereas images in epithelial cells of CP located in lateral ventricles indicated by thick arrows are shown in (**b,e,h,k,o,q**). Immunohistochemical images in reactive astrocytes in hippocampal samples indicated by short arrows are shown in (**c,f,i**), whereas MCT2 immunostaining in neurons in hippocampal samples indicated by double arrows is shown in (**d**). No MCT3 immunostaining in microvessels is shown in (**j**), whereas no MCT5 immunostaining in microvessels and CPEs is shown in (**l,m**). Scale bars indicate 20 μm.

2.2. MCT2 (SLC16A7) Distribution in the Brain and CPEs

MCT2 catalyzes the proton-coupled transport of many monocarboxylates, including lactate, pyruvate, and ketone bodies, across the plasma membrane. MCT2 shows the highest affinity for lactate [29] and is also a high-affinity pyruvate transporter. The MCT2 gene is known to be transcribed with high-sensitivity in response to hypoxia, intracellular pH, and lactate [44]. MCT2 is expressed mainly in neurons and also in astrocytes [45]. Pierre et al. [46], using immunohistochemical techniques, reported that MCT1 was strongly expressed in astrocytes of mice, whereas MCT2 was expressed in a small subset of neurons of mice. These findings are consistent with the concept that lactate is released by astrocytes via MCT1 and is taken up into neurons via MCT2. They also reported [46] that CPEs of mice were heavily immunostained for MCT2 as well as MCT1. On MCT2 expression in human brains, immunoreactivity for MCT2 was present in neuronal axons, microglia, and endothelial cells in healthy human brains and additionally in astrocytes in brains of multiple sclerosis patients [41]. Immunoreactivity for MCT2 (Abcam, ab198272) is confirmed to be present in neuronal cytoplasm, endothelial cells, and reactive astrocytes, and on the apical side of the cytoplasm of human CPEs, as shown in Figure 2d–f.

2.3. MCT4 (SLC16A3) Distribution in the Brain and CPEs

MCT4 is a low-affinity high-capacity transporter and is expressed mainly in astrocytes [47]. MCT4 is known to facilitate the excretion of lactate in cells, in which glycolysis is highly active [48]. Immunoreactivity for MCT4 was present in microglia and endothelial cells as well as astrocytes of healthy human brains [41]. Murakami et al. [43] reported that MCT4 immunoreactivity was present in endothelial cells and reactive astrocytes, and on the basolateral side of the cytoplasm of CPEs in diseased human brains. Immunoreactivity for MCT4 (Abcam, ab244385) can be noted in the same location as reported previously (Figure 2g–i).

2.4. MCT3 (SLC16A8) Distribution in CPEs

MCT3 protein and mRNA of mice were detected in the retinal pigment epithelium and CPEs by immunohistochemistry and Western and Northern blot analyses [34]. Immunoreactivity for MCT3 was noted in the basolateral membrane of CPEs of mice. MCT3 expression was also reported to be detected in rat CP but not in human samples by quantitative targeted absolute proteomics [36]. MCT3 immunoreactivity (Abcam, ab60333) is seen in the cytoplasm of CPEs but not in microvessels (Figure 2j,k).

2.5. MCT5 (SLC16A4) Distribution in the Brain

The expression of MCT5 in the brain was reported in a paper by Halestrap et al. [26] and also shown in the isolated human CP by quantitative targeted absolute proteomics [36]. Beckner et al. [49] reported that MCT5 expression is increased in some glioblastoma cells. Immunoreactivity for MCT5 (Abcam, ab191008) is not seen in the diseased human brain, including CPEs (Figure 2l,m).

2.6. MCT8 (SLC16A2) Distribution in the Brain and CPEs

MCT8, also known as SLC16A2, was first cloned in 1994 and called XPCT because it was encoded by the XPCT gene in Xq13.2. [50]. After MCT10 was identified to be originally an aromatic amino acid transporter, MCT8 was established as an active transporter to transport iodothyronines, including the thyroid hormones T3 and T4 [51]. It is now considered that MCT8, MCT10, and organic anion transporting polypeptide 1C1 (OATP1C1) are the best-characterized specific thyroid hormone transporters [28,52]. MCT8 is widely expressed in most tissues, including the liver, kidney, heart, skeletal muscle, brain, pituitary, and thyroid [53]. It plays a major role in thyroid hormone uptake across the BBB [54]. In brains of humans and mice, MCT8 protein was reported to be expressed in the cortex, hippocampus, cerebellum, hypothalamus, tanycytes, cerebral vessels, and CP [55–57]. Roberts et al. [55] reported that MCT8 was immunohistochemically expressed in endothelial cells and was also visible on the apical and basal surfaces of human CPEs, whereas immunoreactivity for OATP-14, known as OATP1C1, was present on both apical and basolateral surfaces of CPEs. On the other hand, Roberts et al. stated [55] that MCT8 is expressed on the apical surface of CPEs and OATP14 is present primarily on the basolateral surface of CPEs in human and rodent brains. Wilpert et al. [58] reported that MCT8 protein in human brains was expressed in endothelial cells of BBB, CPEs, and tanycytes, whereas neuronal MCT8 protein was expressed in large quantities in specific brain regions. Alkemade et al. [59] reported that MCT8 was immunohistochemically expressed in neurons and glial cells of the human hypothalamus. In contrast, in mouse brains, MCT8 mRNA was expressed predominantly in neurons and also in CP [60]. Alkemade et al. [61] reported that MCT10 immunocytochemical staining was noted in neurons of hypothalamic nuclei. Mutations of MCT8 cause a severe neurodevelopmental disorder, Allan–Herndon–Dudley syndrome [62], which is an X-linked inherited disorder of brain development with hypomyelinating leukodystrophy. Patients developed several kinds of symptoms, such as hypotonia, primitive reflexes, scoliosis, muscular hypoplasia, and dystonia [62,63]. These indicate that thyroid hormone and MCT8 are essential for nervous system development. Immunoreactivity for MCT8 (Novus, NBP2-57308) is present in endothelial cells and on the apical side of the cytoplasm of CPEs (Figure 2n,o). Confirmatory immunohistochemical images in this manuscript are shown in Figures 1 and 2, whereas previously reported findings on regional distribution and cellular localization of MCTs in many papers are summarized in Table 3.

Table 3. Regional distribution and cellular localization of MCTs in the brain.

Isoform (Gene)	Predominat Substrates	Regional Distribution	Cellular Localization	References
MCT1 (*SLC16A1*)	Lactate, pyruvate,	Widespread	Endothelial cells, astrocytes	[24,27,30,33,39–45]
	ketone bodies	Cortex, hippocampus,	ependymocytes, microglia,	
		cerebellum,	oligodensrocyte,	
		choroid plexus	choroid plexus epithelium,	
			some neurons (Rt) [#1]	
MCT2 (*SLC16A7*)	Pyruvate, lactate,	Widespread	Neurons/axon, microglia,	[26,27,41,45]
	ketone bodies	Cortex, hippocampus,	endothelial cells,	
	(high affinity)	cerebellum,	choroid plexus epithelium	
		choroid plexus	*astrocytes (Rt, MS)* [#2]	
MCT3 (*SLC16A8*)	Lactate	Localized	*Retinal pigment epithelium (Ms)* [#3],	[26,34,35]
			choroid plexus epithelium (Ms) [#3]	
MCT4 (*SLC16A3*)	Lactate, pyruvate,	Widespread	Astrocytes, microglia	[26,27,41,43,46]
	ketone bodies	Cortex, hippocampus,	endothelial cells,	
	(low affinity)	cerebellum,	choroid plexus epithelium	
	(high capacity)	choroid plexus		

Table 3. *Cont.*

Isoform (Gene)	Predominat Substrates	Regional Distribution	Cellular Localization	References
MCT5 (*SLC16A4*)	Orphan	Localized	Isolated choroid plexus	[26,36]
MCT8 (*SLC16A2*)	Thyroid hormone	Widespread	Neurons, astrocytes,	[26,53–57]
	(high affinity)	Cortex, hippocampus,	endothelial cells	
		hypothalamus,	choroid plexus epithelium	
		choroid plexus		

Italics indicate findings in rodents or humans with neurodegenerative diseases. #1: MCT1 expression is detected in some neurons of rats (Rt) [39]. #2: MCT2 expression is detected in end-feet of astrocytes in rats (Rt) [45] and astrocytes in brains in the presence of multiple sclerosis (MS) [41]. #3: MCT3 expression is detected in retinal pigment and choroid plexus epithelia of mice (Ms) [34].

2.7. Lactate Transport in the Brain through Cerebral Microvessels and CPEs

MCT1 and MCT4 are involved in lactate release by astrocytes and contribute to energy supply by glycolysis in the cells [64–66]. MCT4 is a transporter with low affinity and high capacity for lactate and contributes to transport lactate from astrocytes to neurons. In contrast, MCT2 is mainly present in neurons, and lactate is taken up into neurons via MCT2 as an efficient oxidative energy substrate. MCT2 is known to be a transporter with a higher affinity for most monocarboxylates than MCT1 [26]. These MCTs contribute in concert to the astrocyte–neuron lactate shuttling [67–69]. Accordingly, the distribution of MCTs in the plasma membrane of neurons and astrocytes suggests a significant role of these transporters in the shuttling of energy metabolites between neurons and astrocytes [67–70].

It was reported that lactate values in CSF of 7614 individuals increased with aging [71]. Results from a CSF-based study indicated that lactate levels in CSF of Parkinson's disease (PD) patients increased compared with controls and were correlated with clinical disease progression [72]. On the other hand, lower lactate levels in CSF were reported in patients with dementia, including Alzheimer's disease (AD) and frontotemporal dementia, compared with non-demented individuals [73]. Accordingly, lactate levels in CSF may be useful for understanding the degree of aging and progression of neurodegenerative diseases. The hydroxy-carboxylic acid 1 receptor (HCA1 receptor), a receptor for lactate, is highly expressed in principal neurons, whereas the receptor is also expressed in astrocytes and endothelial cells [35,74]. In addition, it was recently reported that the HCA1 receptor was expressed in CPEs [43]. Immunoreactivity for the HCA1 receptor (Novus, NLS-2095) is shown to be present in cerebral endothelial cells and on the basolateral membrane of CPEs (Figure 2p,q).

3. Discussion

In this manuscript, we first reviewed the localization and significance of MCTs in cerebral microvessels and CP, and subsequently confirmed the detailed localization of cytoplasmic and membranous molecules reported previously to be expressed in CPEs. As lactate is transported with protons through MCTs in CPEs, representative transporters for water and electrolytes are also described in this review paper. Although MCT5 was not immunohistochemically confirmed to be localized in CPEs, MCT1, MCT2, and MCT4, which are the main transporters for lactate in the brain, were shown to be immunohistochemically expressed in CPEs as well as astrocytes and endothelial cells. MCT2 was also expressed in the cytoplasm of some neurons. MCT3 was shown to be immunohistochemically expressed in the cytoplasm of CPEs. In addition, the HCA1 receptor, which is a receptor for lactate and mediates a decrease in cellular cAMP levels, was immunohistochemically expressed in cerebral microvessels and CPEs. Polarized distributions of MCT1, MCT2, MCT3, MCT4, and HCA1-R and putative directions of lactate through CPEs are indicated by dashed arrows in Figure 3. Considering the movement of lactate between astrocytes and neurons via MCTs, it can be suggested that lactate is transported from CSF into the cytoplasm of CPEs via MCT2 and is released to the CP stroma via MCT4, whereas MCT1 may facilitate the transport of lactate from the cytoplasm of CPEs into CSF (Figure 3). At present, however, the directions cannot be determined. It remains to be clarified whether MCT3 is involved in the transport of lactate in CPEs.

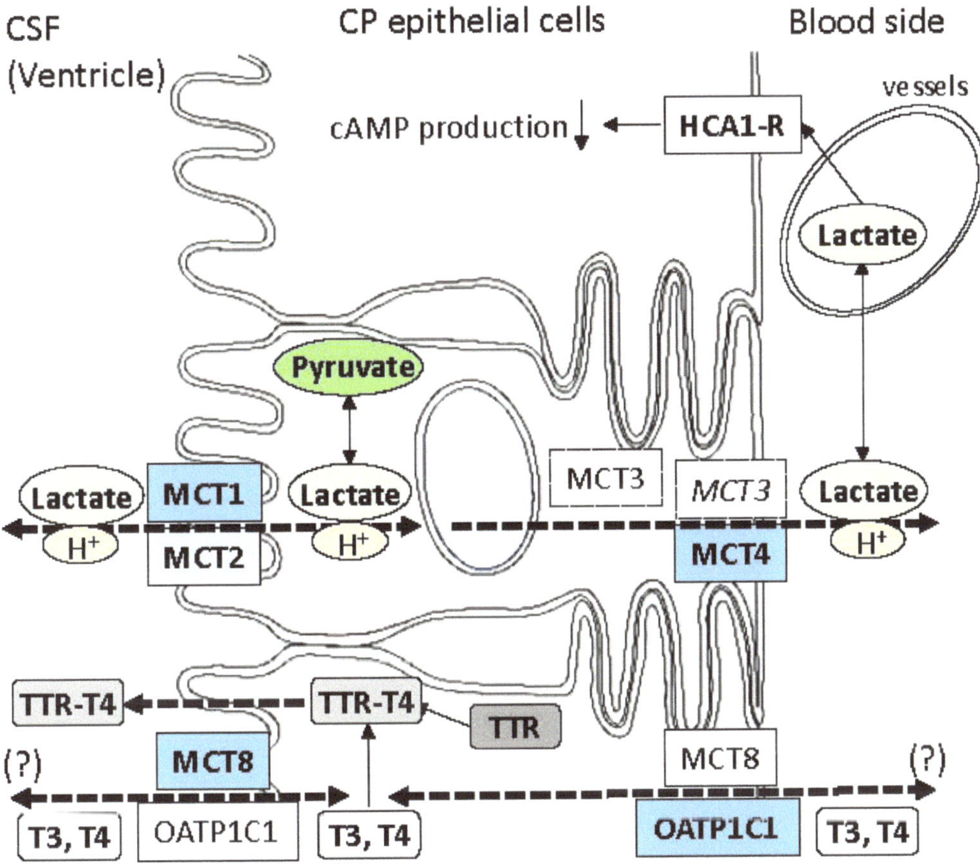

Figure 3. Polarized distribution of a receptor for lactate and transporters for lactate and thyroid hormones in CPEs. HCA1-R is expressed in the basolateral membrane of the cytoplasm of CPEs and induces decreased cyclic AMP production in the CPE cytoplasm. MCT1 and MCT2 are present on the apical (CSF-facing) side of CPEs, whereas MCT4 is present on the basal (CP stroma-facing) side of CPEs. MCT3 expression on the basolateral side of CPEs has been reported only in mice [34] but has not been confirmed in human brains, including CPEs [36]. Accordingly, MCT3 is written in italics and surrounded by dotted square lines. Putative directions of lactate through CPEs are indicated by dashed arrows. MCT8 and OATP1C1, which are known to be transporters for thyroid hormones, are considered to be distributed on the apical and basolateral sides of CPEs [55]. Thyroid hormones are considered to move between CSF and the CPE cytoplasm via these transporters. According to the paper reported by Roberts et al. [55], MCT8 is distributed on the apical surface of CPEs, whereas OATP1C1 is present primarily on the basolateral side of CPEs. Putative directions of thyroid hormones through CPEs are indicated by dashed arrows. However, the directions cannot be confirmed. As transthyretin (TTR) is synthesized in CPEs, T4 bound to TTR (TTR-T4) is considered to move from the cytoplasm of CPEs into ventricles.

Lactate values in CSF are known to increase with aging [71]. Lactate levels in PD patients increased compared with those of controls [72], whereas lower CSF lactate levels were reported in patients suffering from dementia, including AD, compared with non-demented individuals [73]. It is likely that excess exposure to lactate in brain tissues causes acidic tissue injury. On the other hand, lower CSF lactate levels may suggest the impaired function of lactate transport through MCT1, MCT2, MCT3, and/or MCT4 in CPEs in patients with dementia. The specific mechanism of energy supply to the brain in patients with neurodegenerative diseases has not yet been elucidated.

Thyroid hormones are considered to be involved not only in in neurogenesis and neurodifferentiation, but also in cognitive functions. It is interesting that hippocampal neurons are considered to be affected by thyroid hormone levels [75]. It has been pointed out that hypothyroidism is frequently associated with cognitive impairment and/or depressive-like behavior [76]. At present, however, specific foci responsible for symptoms of brain disorders, such as cognitive impairment, in patients with hypothyroidism, remain to be clarified. In addition, it is unclear why hippocampal neurons are affected in hypothyroidism. A large-scale study [77] showed that patients with hypothyroidism had a higher risk of memory impairment and also had a more than three-fold increase in the dementia risk. Several studies have reported an association between thyroid disorders and AD. However, there remains no consensus regarding the precise role of thyroid dysfunction in AD. A meta-analysis using clinical subject headings and keywords from databases [78] showed no significant association of hypothyroidism and the risk of cognitive dysfunction without adjustment for vascular comorbidities. On the other hand, another meta-analysis using some databases [79] showed that hypothyroidism was significantly more prevalent in patients with AD than in controls. It is likely that these discrepancies in findings on hypothyroidism in patients with dementia are due to the multiple causes of cognitive impairment.

MCT8, an active transporter that transports thyroid hormones [51,52], was reported to be expressed in endothelial cells but also in CPEs [55,58]. Also in this paper, MCT8 immunoreactivity was shown in CPEs as well as endothelial cells. These findings suggest that thyroid hormones can be transported from the blood into the hippocampus through these cells. There are two hypothesized mechanisms for thyroid hormones to move out of CP into CSF: one is the secretion of thyroid hormones bound to CP-derived transthyretin, and the other is the efflux of thyroid hormones via thyroid hormone transporters in CPEs such as MCT8 [80], as shown in Figure 3. Although putative directions of thyroid hormones through CPEs are also indicated by dashed arrows in Figure 3, these cannot be confirmed.

4. Conclusions and Future Direction

This paper reviewed the distribution of MCTs as transporters for lactate and thyroid hormones in the brain, especially in endothelial cells and CPEs. Lactate and thyroid hormones are important for the maintenance of several kinds of brain function. It is likely that their insufficiency or excess exposure to them due to CP damage causes brain dysfunction. Recent developments in brain imaging have increased the capacity to diagnose brain diseases. Functional ^{1}H magnetic resonance spectroscopy (fMRS) is becoming a powerful diagnostic tool for brain diseases [81]. It was reported that lactate evaluated with fMRS has the potential to be a new diagnostic and prognostic marker for AD [81]. As the mitochondrial glycolysis pathway is considered to be disrupted in many brain disorders, the activation of anaerobic glycolysis with increased lactate production must occur in the presence of various brain disorders. Along with the development of brain imaging, the importance of lactate measurement will likely increase.

Author Contributions: M.U., Y.C. and K.M. wrote the manuscript. R.M., Y.M., K.W., G.T., N.U. and K.Y. reviewed the draft. M.U., Y.C., R.M., K.M. and K.W. made photos and figures. Y.C., R.M., Y.M., K.M., K.W., G.T., N.U. and K.Y. prepared samples and performed staining. All authors have read and agreed to the published version of the manuscript.

Funding: This paper was supported by grants from JSPS KAKENHI, 19K07508 (YC), 20K16193 (RM), and 20K16550 (NU) of Japan.

Conflicts of Interest: The authors declare no conflict of interest.

References

1. Redzic, Z. Molecular biology of the blood-brain and the blood-cerebrospinal fluid barriers: Similarities and differences. *Fluids Barriers CNS* **2011**, *8*, 3. [CrossRef] [PubMed]
2. Sweeney, M.D.; Zhao, Z.; Montagne, A.; Nelson, A.R.; Zlokovic, B.V. Blood-Brain Barrier: From Physiology to Disease and Back. *Physiol. Rev.* **2019**, *99*, 21–78. [CrossRef] [PubMed]
3. Uchida, Y.; Goto, R.; Takeuchi, H.; Luczak, M.; Usui, T.; Tachikawa, M.; Terasaki, T. Abundant Expression of OCT2, MATE1, OAT1, OAT3, PEPT2, BCRP, MDR1, and xCT Transporters in Blood-Arachnoid Barrier of Pig and Polarized Localizations at CSF- and Blood-Facing Plasma Membranes. *Drug Metab. Dispos.* **2020**, *48*, 135–145. [CrossRef] [PubMed]
4. Huttunen, K.M.; Terasaki, T.; Urtti, A.; Montaser, A.B.; Uchida, Y. Pharmacoproteomics of Brain Barrier Transporters and Substrate Design for the Brain Targeted Drug Delivery. *Pharm. Res.* **2022**, *39*, 1363–1392. [CrossRef]
5. Reese, T.S.; Karnovsky, M.J. Fine structural localization of a blood-brain barrier to exogenous peroxidase. *J. Cell Biol.* **1967**, *34*, 207–217. [CrossRef] [PubMed]
6. Brightman, M.W.; Reese, T.S. Junctions between intimately apposed cell membranes in the vertebrate brain. *J. Cell Biol.* **1969**, *40*, 648–677. [CrossRef]
7. Liebner, S.; Dijkhuizen, R.M.; Reiss, Y.; Plate, K.H.; Agalliu, D.; Constantin, G. Functional morphology of the blood–brain barrier in health and disease. *Acta Neuropathol.* **2018**, *135*, 311–336. [CrossRef] [PubMed]
8. Ueno, M.; Chiba, Y.; Murakami, R.; Matsumoto, K.; Fujihara, R.; Uemura, N.; Yanase, K.; Kamada, M. Disturbance of Intracerebral Fluid Clearance and Blood–Brain Barrier in Vascular Cognitive Impairment. *Int. J. Mol. Sci.* **2019**, *20*, 2600. [CrossRef]
9. Morris, M.E.; Rodriguez-Cruz, V.; Felmlee, M.A. SLC and ABC Transporters: Expression, Localization, and Species Differences at the Blood-Brain and the Blood-Cerebrospinal Fluid Barriers. *AAPS J.* **2017**, *19*, 1317–1331. [CrossRef]
10. Damkier, H.; Praetorius, J. Structure of the mammalian choroid plexus. In *Role of the Choroid Plexus in Health and Disease*; Praetorius, J., Blazer-Yost, B., Damkier, H., Eds.; Springer: New York, NY, USA, 2020; pp. 1–33.
11. Praetorius, J.; Damkier, H.H. Transport across the choroid plexus epithelium. *Am. J. Physiol. Physiol.* **2017**, *312*, C673–C686. [CrossRef]
12. Pardridge, W.M. CSF, blood-brain barrier, and brain drug delivery. *Expert Opin. Drug Deliv.* **2016**, *13*, 963–975. [CrossRef] [PubMed]
13. Wakamatsu, K.; Chiba, Y.; Murakami, R.; Miyai, Y.; Matsumoto, K.; Kamada, M.; Nonaka, W.; Uemura, N.; Yanase, K.; Ueno, M. Metabolites and Biomarker Compounds of Neurodegenerative Diseases in Cerebrospinal Fluid. *Metabolites* **2022**, *12*, 343. [CrossRef] [PubMed]
14. Lach, B.; Scheithauer, B.W.; Gregor, A.; Wick, M.R. Colloid cyst of the third ventricle. A comparative immunohisto-chemical study of neuraxis cysts and choroid plexus epithelium. *J. Neurosurg.* **1993**, *78*, 101–111. [CrossRef] [PubMed]
15. Shibahara, J.; Kashima, T.; Kikuchi, Y.; Kunita, A.; Fukayama, M. Podoplanin is expressed in subsets of tumors of the central nervous system. *Virchows Arch.* **2006**, *448*, 493–499. [CrossRef] [PubMed]
16. Kirik, O.V.; Sufieva, D.A.; Nazarenkova, A.; Korzhevskiy, D.E. Cell contact protein beta-catenin in ependymal and epithelial cells of the choroid plexus of the cerebral lateral ventricles. *Morphology* **2016**, *149*, 33–37. [PubMed]
17. Kasper, M.; Karsten, U.; Stosiek, P. Detection of cytokeratin(s) in epithelium of human Plexus choroideus by monoclonal antibodies. *Acta Histochem.* **1986**, *78*, 101–103. [CrossRef]
18. Miettinen, M.; Clark, R.; Virtanen, I. Intermediate filament proteins in choroid plexus and ependyma and their tumors. *Am. J. Pathol.* **1986**, *123*, 231–240.
19. Justice, D.L.; Rhodes, R.H.; Tökés, Z.A. Immunohistochemical demonstration of proteinase inhibitor alpha-1-antichymotrypsin in normal human central nervous system. *J. Cell. Biochem.* **1987**, *34*, 227–238. [CrossRef]
20. Damkier, H.H.; Brown, P.D.; Praetorius, J. Cerebrospinal Fluid Secretion by the Choroid Plexus. *Physiol. Rev.* **2013**, *93*, 1847–1892. [CrossRef]
21. Spector, R.; Keep, R.F.; Snodgrass, S.R.; Smith, Q.R.; Johanson, C.E. A balanced view of choroid plexus structure and function: Focus on adult humans. *Exp. Neurol.* **2015**, *267*, 78–86. [CrossRef]
22. Johanson, C.E.; Keep, R.F. Blending Established and New Perspectives on Choroid Plexus-CSF Dynamics. In *Role of the Choroid Plexus in Health and Disease*; Springer: Berlin/Heidelberg, Germany, 2020; pp. 35–81. [CrossRef]
23. Chiba, Y.; Murakami, R.; Matsumoto, K.; Wakamatsu, K.; Nonaka, W.; Uemura, N.; Yanase, K.; Kamada, M.; Ueno, M. Glucose, Fructose, and Urate Transporters in the Choroid Plexus Epithelium. *Int. J. Mol. Sci.* **2020**, *21*, 7230. [CrossRef] [PubMed]
24. Matsumoto, K.; Chiba, Y.; Fujihara, R.; Kubo, H.; Sakamoto, H.; Ueno, M. Immunohistochemical analysis of transporters related to clearance of amyloid-β peptides through blood–cerebrospinal fluid barrier in human brain. *Histochem. Cell Biol.* **2015**, *144*, 597–611. [CrossRef]

25. Wakamatsu, K.; Chiba, Y.; Murakami, R.; Matsumoto, K.; Miyai, Y.; Kawauchi, M.; Yanase, K.; Uemura, N.; Ueno, M. Immuno-histochemical expression of osteopontin and collagens in choroid plexus of human brains. *Neuropathology* **2022**, *42*, 117–125. [CrossRef] [PubMed]

26. Halestrap, A.P. The SLC16 gene family—Structure, role and regulation in health and disease. *Mol. Asp. Med.* **2013**, *34*, 337–349. [CrossRef]

27. Iwanaga, T.; Kishimoto, A. Cellular distributions of monocarboxylate transporters: A review. *Biomed. Res.* **2015**, *36*, 279–301. [CrossRef]

28. Friesema, E.C.H.; Jansen, J.; Jachtenberg, J.-W.; Visser, W.E.; Kester, M.H.A.; Visser, T.J. Effective Cellular Uptake and Efflux of Thyroid Hormone by Human Monocarboxylate Transporter 10. *Mol. Endocrinol.* **2008**, *22*, 1357–1369. [CrossRef]

29. Halestrap, A.P. The monocarboxylate transporter family-Structure and functional characterization. *IUBMB Life* **2012**, *64*, 1–9. [CrossRef] [PubMed]

30. Pierre, K.; Pellerin, L. Monocarboxylate transporters in the central nervous system: Distribution, regulation and function. *J. Neurochem.* **2005**, *94*, 1–14. [CrossRef]

31. Vijay, N.; Morris, M.E. Role of Monocarboxylate Transporters in Drug Delivery to the Brain. *Curr. Pharm. Des.* **2014**, *20*, 1487–1498. [CrossRef]

32. Smith, J.P.; Drewes, L.R. Modulation of Monocarboxylic Acid Transporter-1 Kinetic Function by the cAMP Signaling Pathway in Rat Brain Endothelial Cells. *J. Biol. Chem.* **2006**, *281*, 2053–2060. [CrossRef]

33. Uhernik, A.L.; Li, L.; LaVoy, N.; Velasquez, M.J.; Smith, J.P. Regulation of Monocarboxylic Acid Transporter-1 by cAMP Dependent Vesicular Trafficking in Brain Microvascular Endothelial Cells. *PLoS ONE* **2014**, *9*, e85957. [CrossRef] [PubMed]

34. Philp, N.J.; Yoon, H.; Lombardi, L.; Daniele, L.L.; Sauer, B.; Gallagher, S.M.; Pugh, E.N.; Brauchi, S.; Rauch, M.C.; Alfaro, I.E.; et al. Mouse MCT3 gene is expressed preferentially in retinal pigment and choroid plexus epithelia. *Am. J. Physiol. Cell Physiol.* **2001**, *280*, C1319–C1326. [CrossRef] [PubMed]

35. Bergersen, L.H. Lactate transport and signaling in the brain: Potential therapeutic targets and roles in body-brain interaction. *J. Cereb. Blood Flow. Metab.* **2015**, *35*, 176–185. [CrossRef] [PubMed]

36. Uchida, Y.; Zhang, Z.; Tachikawa, M.; Terasaki, T. Quantitative targeted absolute proteomics of rat blood-cerebrospinal fluid barrier transporters: Comparison with a human specimen. *J. Neurochem.* **2015**, *134*, 1104–1115. [CrossRef] [PubMed]

37. Mueckler, M.; Thorens, B. The SLC2 (GLUT) family of membrane transporters. *Mol. Asp. Med.* **2013**, *34*, 121–138. [CrossRef]

38. Wright, E.M.; Loo, D.D.F.; Hirayama, B.A. Biology of Human Sodium Glucose Transporters. *Physiol. Rev.* **2011**, *91*, 733–794. [CrossRef]

39. Leino, R.L.; Gerhart, D.Z.; Drewes, L.R. Monocarboxylate transporter (MCT1) abundance in brains of sucking and adult rats: A quantitative electron microscopic immunogold study. *Brain Res.* **1999**, *113*, 47–54. [CrossRef]

40. Smith, J.P.; Uhernik, A.L.; Li, L.; Liu, Z.; Drewes, L.R. Regulation of Mct1 by cAMP-dependent internalization in rat brain endothelial cells. *Brain Res.* **2012**, *1480*, 1–11. [CrossRef]

41. Nijland, P.G.; Michailidou, I.; Witte, M.E.; Mizee, M.R.; van der Pol, S.M.A.; van her Hof, B.; Reijerkerk, A.; Pellerin, L.; van der Valk, P.; de Vries, H.E.; et al. Cellular distribution of glucose and monocarboxylate transporters in human brain white matter and multiple sclerosis lesions. *Glia* **2014**, *62*, 1125–1141. [CrossRef]

42. Rinholm, J.E.; Hamilton, N.B.; Kessaris, N.; Richardson, W.D.; Bergersen, L.H.; Attwell, D. Regulation of Oligodendrocyte Development and Myelination by Glucose and Lactate. *J. Neurosci.* **2011**, *31*, 538–548. [CrossRef]

43. Murakami, R.; Chiba, Y.; Nishi, N.; Matsumoto, K.; Wakamatsu, K.; Yanase, K.; Uemura, N.; Nonaka, W.; Ueno, M. Immunoreac-tivity of receptor and transporters for lactate located in astrocytes and epithelial cells of choroid plexus of human brain. *Neurosci. Lett.* **2021**, *741*, 135479. [CrossRef]

44. Caruso, J.P.; Koch, B.J.; Benson, P.D.; Varughese, E.; Monterey, M.D.; Lee, A.E.; Dave, A.M.; Kiousis, S.; Sloan, A.E.; Mathupala, S.P. pH, Lactate, and Hypoxia: Reciprocity in Regulating High-Affinity Monocarboxylate Transporter Expression in Glioblastoma. *Neoplasia* **2017**, *19*, 121–134. [CrossRef] [PubMed]

45. Hanu, R.; McKenna, M.; O'Neill, A.; Resneck, W.G.; Bloch, R.J.; Takimoto, M.; Hamada, T.; Puchowicz, M.A.; Xu, K.; Sun, X.; et al. Monocarboxylic acid transporters, MCT1 and MCT2, in cortical astrocytes in vitro and in vivo. *Am. J. Physiol. Physiol.* **2000**, *278*, C921–C930. [CrossRef]

46. Pierre, K.; Pellerin, L.; Debernardi, R.; Riederer, B.; Magistretti, P. Cell-specific localization of monocarboxylate transporters, MCT1 and MCT2, in the adult mouse brain revealed by double immunohistochemical labeling and confocal microscopy. *Neuroscience* **2000**, *100*, 617–627. [CrossRef] [PubMed]

47. Lundquist, A.J.; Llewellyn, G.N.; Kishi, S.H.; Jakowec, N.A.; Cannon, P.M.; Petzinger, G.M.; Jakowec, M.W. Knockdown of Astrocytic Monocarboxylate Transporter 4 in the Motor Cortex Leads to Loss of Dendritic Spines and a Deficit in Motor Learning. *Mol. Neurobiol.* **2022**, *59*, 1002–1017. [CrossRef] [PubMed]

48. Dimmer, K.-S.; Friedrich, B.; Lang, F.; Deitmer, J.W.; Bröer, S. The low-affinity monocarboxylate transporter MCT4 is adapted to the export of lactate in highly glycolytic cells. *Biochem. J.* **2000**, *350*, 219–227. [CrossRef]

49. Beckner, M.E.; Pollack, I.F.; Nordberg, M.L.; Hamilton, R.L. Glioblastomas with copy number gains in EGFR and RNF139 show increased expressions of carbonic anhydrase genes transformed by ENO1. *BBA Clin.* **2016**, *5*, 1–15. [CrossRef]

50. Lafrenière, R.G.; Carrel, L.; Willard, H.F. A novel transmembrane transporter encoded by the *XPCT* gene in Xq13.2. *Hum. Mol. Genet.* **1994**, *3*, 1133–1139. [CrossRef]

51. Friesema, E.C.H.; Ganguly, S.; Abdalla, A.; Manning Fox, J.E.; Halestrap, A.P.; Visser, T.J. Identification of Monocarboxylate Transporter 8 as a Specific Thyroid Hormone Transporter. *J. Biol. Chem.* **2003**, *278*, 40128–40135. [CrossRef]

52. Visser, W.E.; Friesema, E.C.; Jansen, J.; Visser, T.J. Thyroid hormone transport in and out of cells. *Trends Endocrinol. Metab.* **2008**, *19*, 50–56. [CrossRef]

53. Visser, W.E.; Friesema, E.C.H.; Visser, T.J. Minireview: Thyroid Hormone Transporters: The Knowns and the Unknowns. *Mol. Endocrinol.* **2011**, *25*, 1–14. [CrossRef]

54. Wirth, E.K.; Schweizer, U.; Kohrle, J. Transport of Thyroid Hormone in Brain. *Front. Endocrinol.* **2014**, *5*, 98. [CrossRef]

55. Roberts, L.M.; Woodford, K.; Zhou, M.; Black, D.S.; Haggerty, J.E.; Tate, E.H.; Grindstaff, K.K.; Mengesha, W.; Raman, C.; Zerangue, N. Expression of the Thyroid Hormone Transporters Monocarboxylate Transporter-8 (SLC16A2) and Organic Ion Transporter-14 (SLCO1C1) at the Blood-Brain Barrier. *Endocrinology* **2008**, *149*, 6251–6261. [CrossRef] [PubMed]

56. Wirth, E.K.; Roth, S.; Blechschmidt, C.; Hölter, S.M.; Becker, L.; Racz, I.; Zimmer, A.; Klopstock, T.; Failus-Durner, V.; Fuchs, H.; et al. Neuronal 3′,3,5-triiodothyronine (T3) uptake and behavioral phenotype of mice deficient in Mct8, the neuronal T3 transporter mutated in Allan-Herndon-Dudley syndrome. *J. Neurosci.* **2009**, *30*, 9439–9449. [CrossRef] [PubMed]

57. Braun, D.; Kinne, A.; Bräuer, A.U.; Sapin, R.; Klein, M.O.; Köhrle, J.; Wirth, E.K.; Schweizer, U. Developmental and cell type-specific expression of thyroid hormone transporters in the mouse brain and in primary brain cells. *Glia* **2011**, *59*, 463–471. [CrossRef]

58. Wilpert, N.-M.; Krueger, M.; Opitz, R.; Sebinger, D.; Paisdzior, S.; Mages, B.; Schulz, A.; Spranger, J.; Wirth, E.K.; Stachelscheid, H.; et al. Spatiotemporal Changes of Cerebral Monocarboxylate Transporter 8 Expression. *Thyroid* **2020**, *30*, 1366–1383. [CrossRef]

59. Alkemade, A.; Friesema, E.C.; Unmehopa, U.A.; Fabriek, B.O.; Kuiper, G.G.; Leonard, J.L.; Wiersinga, W.M.; Swaab, D.F.; Visser, T.J.; Fliers, E. Neuroanatomical Pathways for Thyroid Hormone Feedback in the Human Hypothalamus. *J. Clin. Endocrinol. Metab.* **2005**, *90*, 4322–4334. [CrossRef] [PubMed]

60. Heuer, H.; Maier, M.K.; Iden, S.; Mittag, J.; Friesema, E.C.H.; Visser, T.J.; Bauer, K. The Monocarboxylate Transporter 8 Linked to Human Psychomotor Retardation Is Highly Expressed in Thyroid Hormone-Sensitive Neuron Populations. *Endocrinology* **2005**, *146*, 1701–1706. [CrossRef] [PubMed]

61. Alkemade, A.; Friesema, E.C.H.; Kalsbeek, A.; Swaab, D.F.; Visser, T.J.; Fliers, E. Expression of thyroid hormone trans-porters in the human hypothalamus. *J. Clin. Endocrinol. Metab.* **2011**, *96*, E967–E971. [CrossRef]

62. Schwartz, C.E.; May, M.M.; Carpenter, N.J.; Rogers, R.C.; Martin, J.; Bialer, M.G.; Ward, J.; Sanabria, J.; Marsa, S.; Lewis, J.A.; et al. Allan-Herndon-Dudley syndrome and the mono-carboxylate transporter 8 (MCT8) gene. *Am. J. Hum. Genet.* **2005**, *77*, 41–53.

63. Wolff, T.M.; Veil, C.; Dietrich, J.W.; Müller, M.A. Mathematical modeling and simulation of thyroid homeostasis: Implications for the Allan-Herndon-Dudley syndrome. *Front. Endocrinol.* **2022**, *13*, 882788. [CrossRef] [PubMed]

64. Bröer, S.; Rahman, B.; Pellegri, G.; Pellerin, L.; Martin, J.-L.; Verleysdonk, S.; Hamprecht, B.; Magistretti, P.J. Comparison of Lactate Transport in Astroglial Cells and Monocarboxylate Transporter 1 (MCT 1) Expressing Xenopus laevis Oocytes. Expression of two different monocarboxylate transporters in astroglial cells and neurons. *J. Biol. Chem.* **1997**, *272*, 30096–30102. [CrossRef] [PubMed]

65. Maekawa, E.; Minehira, K.; Kadomatsu, K.; Pellerin, L. Basal and stimulated lactate fluxes in primary cultures of astrocytes are differentially controlled by distinct proteins. *J. Neurochem.* **2008**, *107*, 789–798. [CrossRef] [PubMed]

66. Rootafio, K.; Pellerin, L. Oxygen tension controls the expression of the monocarboxylate transporter MCT4 in cultures mouse cortical astrocytes via a hypoxia-inducible factor-1alpha-mediated transcriptional regulation. *Glia* **2014**, *62*, 477–490. [CrossRef]

67. Pellerin, L.; Pellegri, G.; Bittar, P.G.; Charnay, Y.; Bouras, C.; Martin, J.-L.; Stella, N.; Magistretti, P.J. Evidence Supporting the Existence of an Activity-Dependent Astrocyte-Neuron Lactate Shuttle. *Dev. Neurosci.* **1998**, *20*, 291–299. [CrossRef]

68. Pellerin, L. Brain energetics (thought needs fond). *Curr. Opin. Clin. Nutr. Metab. Care* **2008**, *11*, 701–705. [CrossRef]

69. Pérez-Escuredo, J.; Van Hée, V.F.; Sboarina, M.; Falces, J.; Payen, V.L.; Pellerin, L.; Sonveaux, P. Monocarboxylate transporters in the brain and in cancer. *Biochim. Biophys. Acta (BBA)—Mol. Cell Res.* **2016**, *1863*, 2481–2497. [CrossRef]

70. Bittar, P.G.; Charnay, Y.; Pellerin, L.; Bouras, C.; Magistretti, P.J. Selective Distribution of Lactate Dehydrogenase Isoenzymes in Neurons and Astrocytes of Human Brain. *J. Cereb. Blood Flow Metab.* **1996**, *16*, 1079–1089. [CrossRef]

71. Leen, W.G.; Willemsen, M.A.; Wevers, R.A.; Verbeek, M.M. Cerebrospinal Fluid Glucose and Lactate: Age-Specific Reference Values and Implications for Clinical Practice. *PLoS ONE* **2012**, *7*, e42745. [CrossRef]

72. Liguori, C.; Stefani, A.; Fernandes, M.; Cerroni, R.; Mercuri, N.B.; Pierantozzi, M. Biomarkers of Cerebral Glucose Metabolism and Neurodegeneration in Parkinson's Disease: A Cerebrospinal Fluid-Based Study. *J. Park. Dis.* **2022**, *12*, 537–544. [CrossRef]

73. Bonomi, C.G.; De Lucia, V.; Mascolo, A.P.; Assogna, M.; Motta, C.; Scaricamazza, E.; Sallustio, F.; Mercuri, N.B.; Koch, G.; Martorana, A. Brain energy metabolism and neurodegeneration: Hints from CSF lactate levels in dementias. *Neurobiol. Aging* **2021**, *105*, 333–339. [CrossRef] [PubMed]

74. Lauritzen, K.H.; Morland, C.; Puchades, M.; Holm-Hansen, S.; Hagelin, E.M.; Lauritzen, F.; Attramadal, H.; Storm-Mathisen, J.; Gjedde, A.; Bergersen, L.H. Lactate Receptor Sites Link Neurotransmission, Neurovascular Coupling, and Brain Energy Metabolism. *Cereb. Cortex* **2014**, *24*, 2784–2795. [CrossRef] [PubMed]

75. Madeira, M.D.; Sousa, N.; Lima-Andrade, M.T.; Calheiros, F.; Cadete-Leite, A.; Paula-Barbosa, M.M. Selective vulnerability of the hippocampal pyramidal neurons to hypothyroidism in male and female rats. *J. Comp. Neurol.* **1992**, *322*, 501–518. [CrossRef] [PubMed]

76. Bakalov, D.; Iliev, P.; Sabit, Z.; Tafradjiiska-Hadjiolova, R.; Bocheva, G. Attenuation of Hypothyroidism-Induced Cognitive Impairment by Modulating Serotonin Mediation. *Vet. Sci.* **2023**, *10*, 122. [CrossRef]

77. Stern, M.; Finch, A.; Haskard-Zolnierek, K.B.; Howard, K.; Deason, R.G. Cognitive decline in mid-life: Changes in memory and cognition related to hypothyroidism. *J. Health Psychol.* **2022**, *28*, 388–401. [CrossRef]

78. Ye, Y.; Wang, Y.; Li, S.; Guo, J.; Ding, L.; Liu, M. Association of Hypothyroidism and the Risk of Cognitive Dysfunction: A Meta-Analysis. *J. Clin. Med.* **2022**, *11*, 6726. [CrossRef]

79. Salehipour, A.; Dolatshahi, M.; Haghshomar, M.; Amin, J. The Role of Thyroid Dysfunction in Alzheimer's Disease: A Systematic Review and Meta-Analysis. *J. Prev. Alzheimer's Dis.* **2023**, *10*, 276–286. [CrossRef]

80. Richardson, S.J.; Wijayagunaratne, R.C.; D'Souza, D.G.; Darras, V.M.; Van Herck, S.L.J. Transport of thyroid hormones via the choroid plexus into the brain: The roles of transthyretin and thyroid hormone transmembrane transporters. *Front. Neurosci.* **2015**, *9*, 66. [CrossRef]

81. Shirbandi, K.B.; Rikhtegar, R.; Khalafi, M.; Attari, M.M.A.; Rahmani, F.; Javanmardi, P.B.; Iraji, S.M.; Aghdam, Z.B.; Rashnoudi, A.M.R. Functional Magnetic Resonance Spectroscopy of Lactate in Alzheimer Disease: A Comprehensive Review of Alzheimer Disease Pathology and the Role of Lactate. *Top. Magn. Reson. Imaging* **2023**, *32*, 15–26. [CrossRef]

MDPI AG

Grosspeteranlage 5

4052 Basel

Switzerland

Tel.: +41 61 683 77 34

Pharmaceutics Editorial Office

E-mail: pharmaceutics@mdpi.com

www.mdpi.com/journal/pharmaceutics

www.ingramcontent.com/pod-product-compliance
Lightning Source LLC
LaVergne TN
LVHW071935160726
843515LV00011B/2618